Dawsons
532.517.11 GIR

24 0374089 2

AF598307

RECEIVED
2 MAY 1999
DATE DUE
20 OCT 1999
10 Feb 00
30 JUN 2000
02 MAY 2001
25.9.02
16 MAY 2005
11 Sept 08
Return Material Promptly

INTRODUCTION TO TURBULENCE

Combustion: An International Series
Norman Chigier, *Editor*

Bayvel and *Orzechowski,* Liquid Atomization
Chigier, Combustion Measurements
Kuznetsov and *Sabel'nikov,* Turbulence and Combustion
Lefebvre, Atomization and Sprays
Li, Applied Thermodynamics: Availability Method and Energy Conversion
Libby, Introduction to Turbulence

Forthcoming Titles

Chen and *Jaw,* Fundamentals of Turbulence Modeling

INTRODUCTION TO TURBULENCE

Paul A. Libby

University of California, San Diego
Department of Applied Mechanics and Engineering Sciences
La Jolla, California

USA	Publishing Office:	Taylor & Francis 1101 Vermont Avenue, N.W., Suite 200 Washington, D.C. 20005-3521 Tel: (202) 289-2174 Fax: (202) 289-3665
	Distribution Center:	Taylor & Francis 1900 Frost Road, Suite 101 Bristol, PA 19007-1598 Tel: (215) 785-5800 Fax: (215) 785-5515
UK		Taylor & Francis Ltd. 1 Gunpowder Square London EC4A 3DE Tel: 171 583 0490 Fax: 171 583 0581

INTRODUCTION TO TURBULENCE

Copyright © 1996 Taylor & Francis. All rights reserved. Printed in the United States of America. Except as permitted under the United States Copyright Act of 1976, no part of this publication may be reproduced or distributed in any form or by any means, or stored in a database or retrieval system, without the prior written permission of the publisher.

1 2 3 4 5 6 7 8 9 0 BRBR 9 8 7 6

This book was set in Times Roman by Pro Image, Inc. The editors were Lynne Lachenbach and Holly Seltzer. Cover design by Michelle Fleitz. Prepress supervisor was Miriam Gonzalez. Printing and binding by Braun-Brumfield, Inc.

A CIP catalog record for this book is available from the British Library.
♾ The paper in this publication meets the requirements of the ANSI Standard Z39.48-1984 (Permanence of Paper)

Library of Congress Cataloging-in-Publication Data

Libby, Paul A.
Introduction to turbulence/Paul A. Libby.
p. cm.—(Combustion: an international series)
Includes bibliographical references.

1. Turbulence. I. Title. II. Series: Combustion (New York, N.Y.: 1989)
TA357.5.T87L53 1996
620.1′064—dc20 96-14285
CIP

ISBN 1-56032-100-8
ISSN 1040-2756

CONTENTS

PREFACE

Turbulence is the most common state of fluid flow in a wide range of technologies and natural conditions. As a consequence of its practical importance and fundamental difficulties, there is a considerable and growing research literature and a number of textbooks devoted to turbulent flows. The orientation of the present volume is different in several respects from that of available texts. "Introduction" in the title is meant to imply inclusion of a wide range of topics relevant to the study of turbulence; in brief, we do not limit our attention to "modeling" or to the "physics" or to the calculation of a particular class of turbulent flows. Moreover, the text incorporates current perspectives on turbulence and provides a large number of references to the literature on many topics without encumbering the exposition unduly; it thereby provides background for further study of the available monographs and current research literature if the reader is so inclined. For example, the book discusses intermittency and the resultant conditional sampling and averaging which arises in current experimental, theoretical, and computational studies of many turbulent flows. In addition, it covers the role of large-scale computation of the fundamental equations of fluid mechanics in providing information on variables which are difficult if not impossible to measure experimentally. Furthermore, asymptotic methods are used frequently to expose important features of turbulent flows. In this author's view, it is important for students in the engineering sciences to appreciate and be able to use these methods, since they play an essential role in the theory of fluid flows in particular and in other engineering fields as well. Although the approach is essentially theoretical, the text emphasizes complementary experimental techniques and results so as to reinforce theoretical concepts. Throughout, we discuss the turbulence of both velocity and scalar fields such as temperature, since the turbulent transport of thermal energy and passive scalars—e.g., contaminants—as well as the possible effects of buoyancy when the two fields are dynamically coupled, arise in many of the most interesting and important flows.

This book should be useful to students in various engineering fields and to practicing engineers who encounter turbulent phenomena. With respect to classroom use, it can be assigned for a one-year course on turbulence for students with a sound, senior or first-year-graduate course in fluid dynamics as background. Indeed, the point of view adopted for the presentation of various topics will enhance the fluid mechanical capabilites of the students. Selected chapters can be used for a semester or quarter course in turbulence. For example, it is possible to consider in the classroom only the highlights of Chapters 1–7, and to focus on Chapters 9–10, which are devoted to applications.

Chapter 1 provides a description of turbulence, its various manifestations, and a brief history of its study. Chapter 2 takes as a point of departure the conservation and transport equations for a Newtonian fluid with constant fluid properties. Although many interesting and important applications—for example, those arising in turbulent combustion—are thereby excluded from consideration, there are sound pedagogical reasons for this restriction. In particular, most experimental, theoretical, and computational research is concerned with turbulence in this simple fluid. Furthermore, a firm understanding of the characteristics of such turbulence is needed before advancing to more complex fluids. Chapter 2 concludes with a discussion of the direct numerical simulation and large eddy simulation of turbulence and the need for a statistical description of many turbulent flows of applied interest.

Chapter 3 discusses the techniques necessary to deal with such a description and to understand the various statistical tools used in current experimental, theoretical, and computational research on turbulence. Although some of the topics in this chapter are not used explicitly in subsequent considerations, they involve concepts that are important for understanding the statistical nature of turbulence and lead naturally in Chapter 4 to the averaging of the conservation and transport equations of fluid flow. We thus encounter the Reynolds stresses and fluxes, the closure problem, and the need for approximation methods for the description of turbulence.

Chapter 5 is devoted to the scales of turbulence; in particular, we establish that there exists a range of length and time scales of turbulent flows and the various contributions to the averaged conservation and transport equations are dominated by processes with different scales. Central to the discussion of this chapter are the sections on isotropic, homogeneous turbulence; here for the first time we use asymptotic methods based in this instance on two length scales, the large scales which relate to the energy-containing fluctuations and the small scales at which viscosity plays a dominant role.

Further perspective on the fundamental aspects of turbulence is provided in Chapter 6, which considers various simple flows to expose the competition among the processes of convection, production, diffusion, redistribution, and dissipation. In some of these flows two or more of these processes are dominant, so that a sense of the competition among them is developed.

Chapter 7 returns to the full conservation and transport equations, to the closure problem, and to classical closure models based on gradient transport. The conceptual and practical shortcomings of these models, and the special forms they assume for various flows, are discussed.

The foregoing is prologue to Chapters 8 and 9: applications of the ideas developed earlier to the two fundamental classes of flows, to free shear flows such as wakes and jets, and to wall-bounded flows arising in channels, pipes, and boundary layers. In treating the wall-bounded flows we again use asymptotic methods which expose their two-layer structure: the flow remote from the wall, which is dominated by turbulent transport; and the flow near the wall, where turbulent transport gives way to transport due to viscosity and thermal conductivity. In discussing boundary layers we restrict attention to equilibrium flows which are analogous to the similarity flows of laminar boundary-layer theory. The reader will thereby see once again, in this case within the context of turbulent flows, the essential simplification and physical insight which result from similarity considerations.

Chapter 10 takes up methods based on second-moment closure in order to indicate the extensions of earlier perspectives and the additional physical insights they afford. Free shear and wall-bounded flows are reexamined by one of the current second-moment methods. In the case of wall-bounded flows, asymptotic methods are again employed. Although no examples are worked out in detail, the full second-moment methods represented by Reynolds stress and Reynolds flux closure are described and then specialized to yield the algebraic stress theory of turbulent transport. At several points in earlier chapters the important influence of buoyancy on turbulent transport is noted; indeed, in Chapter 6, one of the simple flows discussed involves coupling between velocity and temperature fluctuations. However, a special perspective on Reynolds stress closure permits a more complete discussion of this influence. Chapter 10 concludes with reference to other methods of analyzing turbulent flows, methods whose detailed treatment is beyond the scope of the present work.

I am greatly indebted to many friends and colleagues who read and made helpful suggestions relative to various earlier versions of this book. I particularly want to acknowledge in this regard my faculty colleagues in the Department of Applied Mechanics and Engineering Sciences at the University of California San Diego, namely, Profs. C. H. Gibson, S. Sarkar, F. A. Williams, and C. W. van Atta. Special thanks are due to Prof. J. D. Goddard, who carefully read and made invaluable suggestions for some of the chapters, and to Prof. Peter Bradshaw for helpful comments on various versions of the manuscript as it evolved.

Finally, I am greatly indebted to Dr. S. C. Li, who was kind enough to draw the figures on his computer. Despite this invaluable and diverse assistance, sole responsibility for remaining flaws resides with the author.

Paul A. Libby

CHAPTER
ONE
INTRODUCTION

This chapter gives a brief history of turbulence research and discusses the various manifestations of turbulence.[†]

1.1 EARLY DEVELOPMENTS

The sketches of whirlpools of Leonardo da Vinci (1452–1519) suggest that the tortuous, disordered fluid motion which we characterize as turbulence has long been known to exist. Indeed, Leonardo was apparently fascinated by the motion of water; his sketch pads contain many realistic drawings of vortical flows which testify to his considerable qualitative understanding of fluid mechanics (cf., e.g., Lugt, 1983). Nevertheless, the first quantitative investigation of this state of fluid motion is generally attributed to Osborne Reynolds (1883, 1894), who used flow visualization with injected dye to study the conditions under which water flowing in a glass tube is either laminar or turbulent. The parameter he established for distinguishing these two states, the Reynolds number, is forever associated with his name.

Reynolds also first suggested the decomposition of the flow variables into mean or average quantities and their residuals, the fluctuations. When this

[†]For a more complete history, see Monin and Yaglom (1971).

decomposition is applied to the conservation and transport equations of fluid mechanics, there result additional terms due to the fluctuations, terms which have the character of stresses. These Reynolds stresses introduce an essential complication into the analysis of turbulent flows but at the same time describe important processes determining the behavior of such flows. These observations make it easy to appreciate why the study of turbulence is associated with Osborne Reynolds, and why his name is indelibly stamped on the subject.

1.2 DEVELOPMENTS IN THE FIRST HALF OF THE TWENTIETH CENTURY

Important advances in the theory of turbulence occurred during the period between the two World Wars and are associated with the names of von Karman (1937), Prandtl (1925), and Taylor (1921, 1935a, 1935b), scientists who made many significant contributions to the aeronautical sciences. Two different threads of turbulence research during this period can be identified. One relates to the semiempirical theory which provides information of direct interest in applications—for example, the skin friction in pipes, on airfoils, and on ships. A second thread concerns the theory of turbulence under highly idealized conditions—that is, isotropic, homogeneous turbulence. Advances in this second thread are due to the scientists cited earlier and to Kolmogorov (1941); this whole subject is described in detail by Batchelor (1967). The important perspective which evolves from this theory is that fluctuations in turbulent flows involve a wide range of scales, with the large scales having a relative permanence and the small scales having relatively short lifetimes. Indeed, turbulence is distinguished from unsteady laminar flow in terms of the continuous spectrum of scales of the fluctuations involved. There also arises from this theory a description of the transfer of energy from the large-scale fluctuations into the smaller scales which are directly influenced by molecularity. This cascade process is associated with a famous "ditty" due to Richardson (1922):

> Big whorls have little whorls,
> Which feed on their velocity;
> And Little whorls have lesser whorls,
> And so on to viscosity
> (in the molecular sense).

In recent years it has been found that in some flows—e.g., in the two-dimensional mixing layer (cf. Brown and Roshko, 1974, and Section 5.8) and in the highly idealized situation described by two-dimensional turbulence (cf. Section 6.9), large-scale vortices periodically pair so that larger fluid structures are created, a situation not foreseen by Richardson. However, no suitable poem to highlight such structural growth has yet been proposed. Central aspects of the cascade process theory of small-scale fluctuations are discussed in Sections 5.2–5.4.

The principal semiempirical theories of turbulence developed prior to roughly 1960 are based on certain representations of the Reynolds stresses and their scalar counterparts, namely, the Reynolds fluxes, by means of an analogy with molecular stress and flux but with the exchange coefficients dependent on flow characteristics rather than on the viscosity and thermal conductivity of the fluid. The various theories differ from one another in the details of the analogy, i.e., in the description of the exchange coefficients, and each is generally restricted to a particular class of flows. They have as their common purpose the description of turbulent jets, wakes, and mixing layers and of turbulent pipe, channel, and boundary-layer flows. The analyses of these flows according to these early theories are taken up in Chapters 8 and 9.

Although this discussion emphasizes theoretical developments, complementary experimental research in the laboratory and in the field play important roles in both the fundamental and applied aspects of turbulence, and such research has been carried out continuously since the work of Reynolds. Thus, even in monographs devoted largely to theory, e.g., Batchelor (1967) and Townsend (1956), frequent comparison of theory and experiment is made. Moreover, Bradshaw (1971) provides a useful short introduction to turbulence based on the relevant measuring techniques. Finally, we note that monographs are available on the theory and practice of the most important instrument in experimental turbulence, the hot-wire anemometer and its variants (cf., e.g., Comte-Bellot, 1976, & Bruun, 1976). In recent times the laser Doppler anemometer has supplemented and in some applications supplanted the hot-wire anemometer (cf. Durst et al., 1976).

1.3 RECENT DEVELOPMENTS

A significant change in perspective relative to the semiempirical theories of turbulence occurred in the 1960s, when the computation of large systems of partial differential equations became feasible with the widespread availability of high-speed digital computers. The early work of Keller and Fridman (1924) showed that a hierarchy of equations can be systematically developed from the conservation equations of fluid flow by taking successively higher moments, i.e., by weighted averaging. The original notion of Reynolds corresponds in this sense to taking a simple time average, and the early semiempirical theories—those emphasized in Chapters 7–9—close the resultant equations by approximating the second-moment terms which arise from such averaging, i.e., the Reynolds stresses and Reynolds fluxes. However, the second level in the hierarchy provides a set of second-moment equations, and the third moment and other terms which appear—e.g., those involving pressure fluctuations—can in turn be approximated so that the resultant system of equations is closed. Such a system contains more of the physics of turbulence and can therefore be applied to more complex flow but requires numerical solution by high-speed computers (cf., e.g., Rodi, 1980; Launder, 1989a). These developments are discussed in Chapter 10. Much current research related to semiempirical theories of turbulence is concerned with the exploitation, assessment, and improvement of these second-moment formulations.

Another development evolving from the availability of large-scale computers relates to exact numerical solutions of the equations describing fluid motion for flows with limited, relatively low Reynolds numbers and simple geometries. Such solutions provide information on fluid mechanical quantities which are difficult if not impossible to measure experimentally and are thus invaluable for the assessment and improvement of turbulent theories of applied interest. These developments and their limitations are discussed in Section 2.7 (cf. Rogallo and Moin, 1984; Reynolds, 1990).

Turbulence provides an intellectual challenge for a significant number of research workers, with the consequence that the journal literature contains many articles involving advanced theoretical concepts. It is the inherent complexity and difficulty of treating turbulent flows with their essential nonlinearity and inherent richness which will continue to pose this challenge for the foreseeable future. However, it must be observed that the articles in question have little applied significance.

1.4 MANIFESTATIONS OF TURBULENCE

Turbulent flows occur in a wide variety of circumstances.† Some examples occur in everyday life, e.g., the billowing of clouds in the sky. Indeed, the length and velocity scales of the atmospheric boundary layer surrounding the earth are such that turbulent processes play a dominant role in determining both global and local weather. Examples occur within the household; the meandering of smoke from a cigarette (see Fig. 1.4.1) can be recorded by simple photographic means. Figure 1.4.2 shows the flow of water from a vertical pipe at two different rates. For low flow rates the stream of water has the smooth exterior we associate with laminar flow; while at higher rates the exterior is irregular, suggesting fluctuations of the velocity within the jet and thus turbulent flow. The discharge from smoke stacks (see Fig. 1.4.3) almost always displays similar irregularities on its boundary.

†Several examples of turbulent flows in this book are taken from Corrsin (1961), who provides an excellent, general discussion of turbulence.

Figure 1.4.1 Free convection from a cigarette. (From Corrsin, 1961, with permission.)

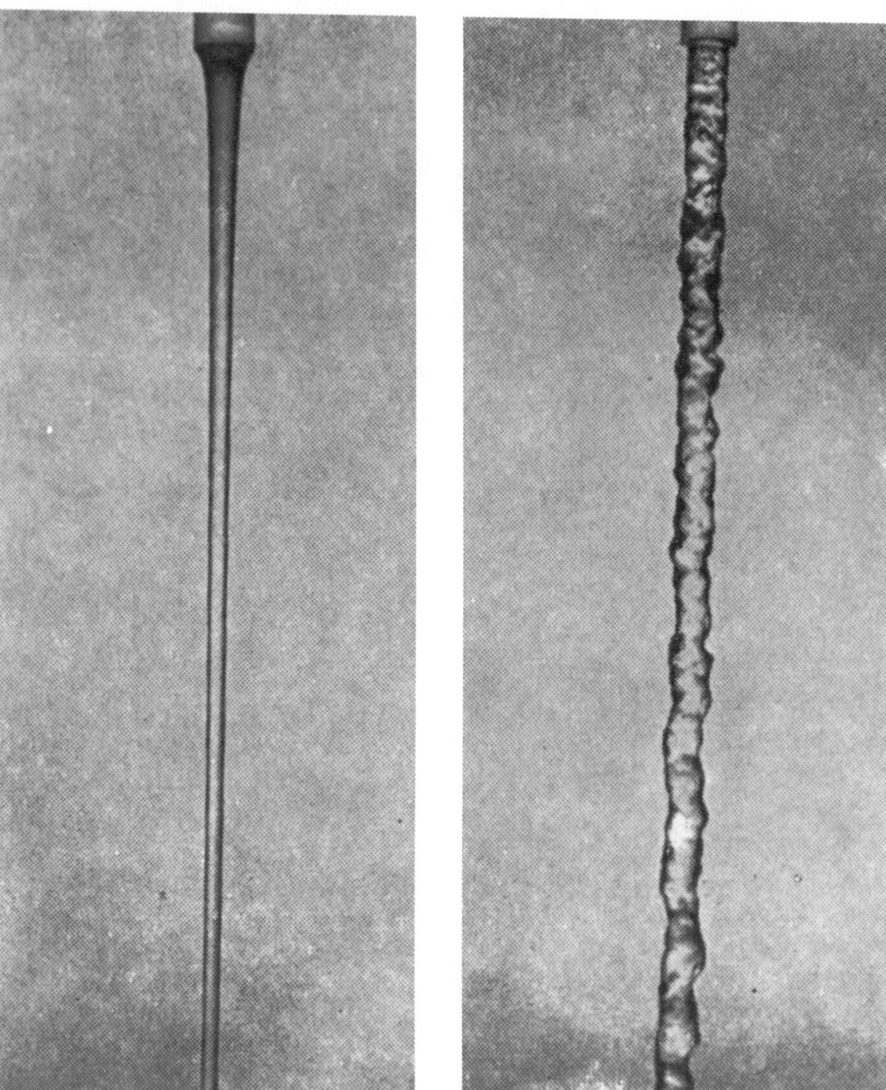

Figure 1.4.2 Water flow from a vertical jet at two different flow rates. (From Corrsin, 1961, with permission.)

The observation of other examples of turbulent flows requires more elaborate techniques. Figure 1.4.4 is a shadowgraph of a supersonic projectile showing the configuration of the shock waves determining the global features of the flow and, more important for our discussion, the turbulent wake downstream of the body. Astronomical observations of various heavenly bodies and of the surface

Figure 1.4.3 Dispersion from a smoke stack. (From Corrsin, 1961, with permission.)

of the sun suggest the common existence of highly irregular gaseous structures which may be forms of turbulence, possibly involving magnetohydrodynamic effects—i.e., heavenly turbulence. Finally, note that the performance and efficiency of most devices and systems of technological interest, such as power plants, jet engines, pipelines, and chemical factories, are determined by turbulent processes. This is a consequence of the large length and velocity scales required for many practical operations.

The many manifestations of turbulent flow and their technological importance provide compelling reasons for studying and understanding this common fluid mechanical state.

1.5 ESTABLISHING TURBULENT FLOWS

In our discussions we assume that turbulence is established in one of several ways. After passing through a grid or baffle at a suitably high velocity, the downstream fluid is turbulent; indeed, grid turbulence is the subject of extensive research because it represents one of the fundamentally important and readily studied turbulent flows. Modifications to grid turbulence involving nonuniform mean velocities and mean temperatures are also of interest. We occassionally consider turbulence in the absence of mean motion, for example, the situation

Figure 1.4.4 A supersonic projectile with a turbulent wake. (From Corrsin, 1961, with permission.)

which arises from vigorous stirring of fluid in a container. This same situation arises when grid turbulence is viewed in a frame of reference that is moving with the mean motion. Finally, under suitable conditions of velocity and length scale, shear flows such as jets, wakes, and boundary layers which are initially laminar experience transition to turbulence and form the fully developed turbulent flows we consider. The fundamental quantity determining when transition occurs in such layers is their Reynolds number based on the velocity, the thickness of the layer, and the kinematic viscosity of the fluid, but many other factors also significantly influence transition. Some of these factors are free-stream turbulence, free-stream pressure gradient, surface roughness, surface curvature, and surface temperature. Because of its practical importance in determining drag and heat transfer and its complexity, transition is the subject of extensive research, although there does not appear to be available a current review of the subject (cf., e.g., Dryden, 1959).

1.6 SUMMARY

In this chapter we have sketched the long history of turbulence research extending over a century from the seminal studies of Osborne Reynolds in the 1880s. The significant change around the 1960s in our ability to treat more fully and more accurately simple and complex turbulent flows as a consequence of the advent of powerful computers has been noted. We have also displayed several examples of turbulent flows so that their irregular motion and complexity can be appreciated. The means for generating turbulence in containers and wind tunnels and the transition from laminar to turbulent flow have been discussed.

CHAPTER

TWO

CONSERVATION AND TRANSPORT EQUATIONS

In this and subsequent chapters we use Cartesian tensor notation in order to achieve compactness. In doing so we reserve the indices k and l to imply summation, with the consequence that they appear only in repeated form. Other indices introduced are i, j, and m; they rarely appear repeated, and when we are forced to use them in that form, we warn of the violation of our reservation. For clarity, our discussion is restricted largely to consideration of Cartesian coordinates, but some flows—e.g., turbulent flow in a pipe—are more naturally and efficiently treated in cylindrical coordinates. Thus, at appropriate points in the discussion we present without details the relevant equations in those coordinates.

We consider a fluid flow described in terms of the three Cartesian coordinates x_i with related velocity components $\tilde{u}_i$, $i = 1, 2, 3$. It is sometimes convenient to introduce the vector equivalent of these coordinates and velocity components, $\mathbf{x}$ and $\tilde{\mathbf{u}}$, respectively. We consider the temperature $\tilde{\theta}$ to represent a general passive scalar, i.e., either temperature or a trace contaminant.[†] The adjective "passive" in this context implies that heated fluid elements are convected and diffused by molecular motion without significantly altering the velocity field in which they coexist. This contrasts with flows involving significant density

[†]The restriction to small concentrations of a foreign species arises from the assumption of constant density and constant viscosity coefficient. Thus the concentration of a species with negligible influence on the mixture density and viscosity coefficient need not be small.

changes; in such flows there is a dynamic coupling between the velocity and temperature fields. In the case of flows influenced by buoyancy, the temperature is not passive since such coupling occurs; but if the temperature changes are suitably small, the Boussinesq approximation applies and the influence of density changes can be simply treated.

We need to distinguish among instantaneous, mean, and fluctuating values of various quantities. To do so we follow the notation of Tennekes and Lumley (1972) and let lowercase letters with a superposed tilde denote the first, capital letters the second, and lowercase letters without tilde the third. Thus the velocity components are $\tilde{u}_i(\mathbf{x}, t) = U_i(\mathbf{x}) + u_i(\mathbf{x}, t)$, the pressure $\tilde{p}(\mathbf{x}, t) = P(\mathbf{x}) + p(\mathbf{x}, t)$, and the temperature $\tilde{\theta}(\mathbf{x}, t) = \Theta(\mathbf{x}) + \theta(\mathbf{x}, t)$. Here we write the mean quantities as time independent; this is the case in most of our discussion, although in Chapters 3 and 10 we deal with turbulent flows requiring more general mean values and in Chapters 5 and 6 we discuss the *time evolution* of various simple flows.

We give special attention to the functional dependencies of the variables under consideration with the notion that the physical implications of the discussion are thereby clarified. Thus, in the previous paragraph we display the space and time dependencies as appropriate. Elsewhere we introduce variables with altered notation to emphasize a particular perspective; for example, $\tilde{u}_i(t; \mathbf{x})$ indicates the variation *with time* of the quantity $\tilde{u}_i$ at a *fixed* spatial location $\mathbf{x}$. We also are careful with sub- and superscript notation, in hopes of providing clarity and emphasis.

The relevant physical properties of the fluid, taken to be constant, are density, ρ, viscosity coefficient, μ, and thermal conductivity, λ. The first two of these frequently appear in the combination $\nu = \mu/\rho$, the kinematic viscosity. The single thermodynamic property of the fluid which arises is the coefficient of specific heat at constant pressure, c_p.†

Some of these quantities appear combined in a dimensionless, important parameter, the Prandtl number, $N_\sigma \equiv \mu\, c_p/\lambda$, which in gases is close to unity but which in many liquids is large compared to unity; for example, in liquid water at room temperature $N_\sigma \approx 6$. Mercury and liquid metals are special fluids with $N_\sigma << 1$; thus the behavior of temperature fluctuations in mercury is the subject of research study (cf. Clay, 1973). The Prandtl number is a measure of the relative diffusivity of momentum and heat; if $N_\sigma >> 1$, the thermal conductivity is small relative to the viscosity coefficient, with the consequence that large temperature gradients can arise in regions of the flow involving small velocity gradients. In many flows the temperature is initially convected by the

†Water and air are the two fluids of greatest interest to us. Their fluid properties at standard pressure and temperature are as follows: For water, $\rho = 1$ g/cm^3, $\nu = 1.0 \times 10^{-2}$ cm^2/s, $\lambda = 6.0 \times 10^4$ dynes/s K, and $c_p = 4.2 \times 10^7$ cm^2/s^2 K; for air, $\rho = 1.3 \times 10^{-3}$ g/cm^3, $\nu = 1.9 \times 10^{-1}$ cm^2/s, $\lambda = 9.2 \times 10^2$ dynes/s K, and $c_p = 1.0 \times 10^7$ cm^2/s^2 K.

velocity so that the fluctuations of temperature and velocity are intimately connected. However, if in the fluid under consideration $N_\sigma >> 1$, then as the flow evolves, the velocity fluctuations decay more rapidly than those of the temperature, with the consequence that relatively large temperature gradients can occur in a relatively quiescent velocity field. If, on the contrary, the fluid is such that $N_\sigma << 1$, then the opposite situation, large velocity gradients in a relatively smooth temperature field, can arise.[†]

A uniform, generally accepted terminology for the fundamental equations of mechanics does not exist. Here we identify a conservation equation as one describing the invariance of a quantity connected with a fluid element as it passes through a flow field. For example, if the mass fraction of a species Y_i is constant along streamlines, then Y_i is *conserved* so that

$$\frac{DY_i}{Dt} = 0$$

where the derivative is associated with the local streamline. Conservation equations are sometimes termed *balance equations*. A transport equation describes the *variation* of a variable connected with a fluid element as it passes through a flow field. Thus, if chemical reaction changes the composition of a fluid element, the previous equation becomes

$$\frac{DY_i}{Dt} = \omega_i$$

where ω_i is the rate of change of the mass fraction of species i, a quantity which determines the *transport* of that species along a streamline. Thus we shall identify conservation and transport equations.

2.1 CONSERVATION OF MASS

In a fluid of constant density, the fundamental requirement that mass be conserved reduces to a kinematic condition on the velocity field:

$$\frac{\partial \tilde{u}_k}{\partial x_k} = 0 \tag{2.1.1}$$

This equation implies that the rate of strain experienced by a fluid element is

[†]The counterparts of the thermal conductivity and Prandtl number for flows with trace contaminants are the diffusion coefficient, D, and the Schmidt number, $N_{Sc} = \mu/\rho D$ respectively. Our comments regarding the importance of the size of the Prandtl number relative to unity apply to the Schmidt number as well. An exception arises for a gaseous contaminant with a molecular weight considerably different from that of the background gas. In this case the Schmidt number can differ significantly from unity. For example, trace amounts of helium in air involve a Schmidt number of roughly 0.25.

restrained so that its volume is unaltered. Although it derives from a statement of the conservation of mass, Eq. (2.1.1) is frequently termed the continuity equation because of the implied kinematic restraint it describes.

To emphasize the importance of Eq. (2.1.1) for our considerations, it is worth noting its counterpart in variable density flows:

$$\frac{\partial \tilde{u}_k}{\partial x_k} = -\frac{1}{\tilde{\rho}}\left(\frac{\partial \tilde{\rho}}{\partial t} + \tilde{u}_k \frac{\partial \tilde{\rho}}{\partial x_k}\right) \tag{2.1.2}$$

where $\tilde{\rho}$ is the instantaneous density. If, as is the case in some flows of interest (e.g., in turbulent combustion and high-speed turbulent boundary layers and wakes), the right side of Eq. (2.1.2) differs significantly from zero, the velocity components are clearly restrained in a more complex fashion. In fact, the consequence of this restraint is that the density plays the dominant role in coupling the various conservation equations in these flows.[†]

2.2 TRANSPORT OF MOMENTUM

The equations of motion are

$$\rho\left[\frac{\partial \tilde{u}_i}{\partial t} + \frac{\partial}{\partial x_k}(\tilde{u}_k\, \tilde{u}_i)\right] = -\frac{\partial \tilde{p}}{\partial x_i} + \tilde{f}_i + \frac{\partial \tilde{\tau}_{ik}}{\partial x_k} \tag{2.2.1}$$

where $\tilde{f}_i$ denotes a general body force per unit volume. For example, if the x_2 coordinate is taken to be vertical, positive upward, then the gravitational force, an example of a body force, is

$$f_i = -\rho g \delta_{i2} \tag{2.2.2}$$

here δ_{ij} is the Kronecker delta and g is the gravitational constant.[††,‡]

In Eqs. (2.2.1) $\tilde{\tau}_{ij}$ is the shear stress tensor, which is related to the rate of strain tensor $\tilde{s}_{ij}$ through the viscosity coefficient according to the constitutive relation

$$\tilde{\tau}_{ij} = 2\mu \tilde{s}_{ij} = \mu\left(\frac{\partial \tilde{u}_i}{\partial x_j} + \frac{\partial \tilde{u}_j}{\partial x_i}\right) \tag{2.2.3}$$

In Eqs. (2.2.1) the terms on the left side describe the convection of momentum, while the terms on the right side are the forces acting on a fluid element due to the pressure gradient, body force, and gradients of the shear stresses, respectively.

[†]We note that the right side of Eq. (2.1.2) involves an Eulerian derivative of the density $D\rho/Dt$. In a variable-density flow with negligible molecular diffusion, e.g., with stable stratification, $D\rho/Dt \approx 0$ and Eq. (2.1.1) applies although ρ is not constant everywhere.

[††]Nearly all except the most elementary texts in fluid dynamics give thorough derivations of Eqs. (2.2.1) and their counterparts in cylindrical coordinates (cf. White, 1974; Bird et al., 1960).

[‡]The value of g is 980 cm/s^2.

If Eq. (2.2.3) is substituted into Eq. (2.2.1), we obtain

$$\rho \left[\frac{\partial \tilde{u}_i}{\partial t} + \frac{\partial}{\partial x_k} (\tilde{u}_k \, \tilde{u}_i) \right] = -\frac{\partial \tilde{p}}{\partial x_i} + \tilde{f}_i + \mu \frac{\partial^2 \tilde{u}_i}{\partial x_k \, \partial x_k} \tag{2.2.4}$$

Note that in the conservation equations introduced so far, the velocity components appear only in differential form. This implies that these equations apply to any inertial coordinate system and thus are essentially unaltered by the addition of any constant velocity vector. In the more advanced calculation methods discussed in Chapter 10 it is desirable if the various approximations needed to complete the describing equations respect this property of Galilean invariance.

2.3 TRANSPORT OF THERMAL ENERGY

The principle of conservation of thermal energy in fluid flows can be expressed in various forms. For our purposes we require a version applicable to low-speed flows with constant fluid properties, chemical homogeneity, and no radiative transfer. Thus we take

$$\rho c_p \left[\frac{\partial \tilde{\theta}}{\partial t} + \frac{\partial}{\partial x_k} (\tilde{u}_k \, \tilde{\theta}) \right] = -\frac{\partial q_k}{\partial x_k} \tag{2.3.1}$$

where q_i is the ith component of the heat flux vector. Here the terms on the left side describe the convection of thermal energy due to fluid motion, while those on the right indicate the change in that energy as a consequence of molecular heat conduction. Equation (2.3.1) is completed by the Fourier law of heat conduction:

$$q_i = -\lambda \frac{\partial \tilde{\theta}}{\partial x_i} \tag{2.3.2}$$

where λ is the coefficient of thermal conductivity. Thus Eq. (2.3.1) can be written in the convenient form

$$\rho \left[\frac{\partial \tilde{\theta}}{\partial t} + \frac{\partial}{\partial x_k} (\tilde{u}_k \, \tilde{\theta}) \right] = \frac{\mu}{N_\sigma} \frac{\partial^2 \tilde{\theta}}{\partial x_k \, \partial x_k} \tag{2.3.3}$$

Note that there is a significant and fundamental difference in Eqs. (2.2.1) and (2.3.3). The former involves two effects contributing to the momentum balance, the force due to the pressure gradient and the body force, both explicitly absent in the energy balance but, of course, implicitly present via their influence on the velocity components $\tilde{u}_i(\mathbf{x}, t)$.

The derivation of the equations of conservation of energy begins with the first law of thermodynamics applied to a fluid element, including the work done by the pressure and the viscous stresses and the transfer of heat by conduction and radiation. Introduction of thermodynamic relations and the Fourier law of heat conduction and use of the momentum equations to account for the work done by the stresses leads to equations for the stagnation enthalpy or the static enthalpy or the static temperature, each having utility in special circumstances. The essential assumptions required to arrive at Eq. (2.3.1) relate to the kinetic energy per unit mass being negligible compared with the static enthalpy, i.e., to the Mach number being small compared with unity, and to the rate of work done by the stresses being negligible compared to the convection and conduction of heat.

2.4 PRESSURE

Although Eqs. (2.1.1), (2.2.1), and (2.3.1) are the fundamental equations we shall deal with, there are alternative forms derivable therefrom which provide different perspectives. One such form is obtained by differentiating Eq. (2.2.4) with respect to x_i and summing over i. As a consequence of continuity there results, after some rearrangement,

$$\frac{\partial^2 \tilde{p}}{\partial x_k \, \partial x_k} = -\rho \frac{\partial^2}{\partial x_l \, \partial x_k} (\tilde{u}_l \, \tilde{u}_k) + \frac{\partial \tilde{f}_k}{\partial x_k} \tag{2.4.1}$$

If we have an appropriate form for the body force $\tilde{f}_i$, this is a Poisson equation whose solution can be written in terms of an integral over all fluid space with the solid boundaries taken into account. An implication is that the pressure field at any instant of time is *determined* by the corresponding instantaneous velocity field, but the instantaneous pressure gradient at each point in space makes an important, in some cases dominant, contribution to the balance of momentum and thus to the right side of Eq. (2.4.1). A further implication from Eq. (2.4.1) is that pressure is a global variable, a feature which contributes to the central problem in turbulence, namely closure (cf. Section 4.8), and leads to an essential difficulty in advanced methods for analyzing turbulent flows (cf. Section 10.1). In brief, the influence of a variable which depends on *global features* of the flow is approximated by *local features of the turbulence*; from this perspective it is not surprising that descriptions of this influence are subject to large uncertainties and are the topic of current investigation. From Eq. (2.4.1) we also see that the pressure depends on body forces via the $\tilde{f}_i$ terms. Not as evident is the dependence of pressure on translational and rotational acceleration, but this may be understood if the **x** coordinate system used in developing Eq. (2.4.1), an inertial system, is transformed to a noninertial frame. The additional terms which result imply that the pressure depends on these accelerations. Furthermore, if we transform to a *curvilinear* coordinate system, other addition terms arise that reflect the influence of curvature on pressure. The implication is that the description of pressure in buoyant, rotating, and curved flows calls for special attention (cf. Sections 6.4, 6.6, and 9.14).

2.5 THE VORTICITY EQUATION

A second supplementary form of our equations arises from a consideration of vorticity, which is determined by the velocity field. We have as a definition

$$\tilde{\omega}_i = \varepsilon_{ilk} \frac{\partial \tilde{u}_k}{\partial x_l} \tag{2.5.1}$$

where ε_{ijm} is the unit alternating tensor with the value unity if the indices are in the order 1, 2, 3, 1, etc.; the value minus one if in the order 1, 3, 2, 1, etc.; and the value zero otherwise.[†] Thus in expanded form we have

$$\begin{aligned} \tilde{\omega}_1 &= \frac{\partial \tilde{u}_3}{\partial x_2} - \frac{\partial \tilde{u}_2}{\partial x_3} \\ \tilde{\omega}_2 &= \frac{\partial \tilde{u}_1}{\partial x_3} - \frac{\partial \tilde{u}_3}{\partial x_1} \\ \tilde{\omega}_3 &= \frac{\partial \tilde{u}_2}{\partial x_1} - \frac{\partial \tilde{u}_1}{\partial x_2} \end{aligned} \tag{2.5.2}$$

If Eqs. (2.2.1) are written for two different values of the index i, cross-differentiated, and subtracted, there results after some rearrangement the balance equation for the vorticity component corresponding to the third index. The general result is

$$\frac{\partial \tilde{\omega}_i}{\partial t} + \frac{\partial}{\partial x_k}(\tilde{u}_k \tilde{\omega}_i) = \tilde{\omega}_k \frac{\partial \tilde{u}_i}{\partial x_k} + \nu \frac{\partial^2 \tilde{\omega}_i}{\partial x_k \, \partial x_k} + \frac{1}{\rho}\left(\frac{\partial \tilde{f}_j}{\partial x_m} - \frac{\partial \tilde{f}_m}{\partial x_j}\right)\varepsilon_{ijm} \tag{2.5.3}$$

The left side of this equation describes the convection of the ith vorticity component due to fluid motion, while the second term on the right side describes the diffusion of that component due to molecularity. The third term on the right side is clearly a body force term which is absent for many forms of the body force. The most interesting term in this equation is the first one on the right side; it describes the influence on the vorticity component $\tilde{\omega}_i$ of the interaction between the various vorticity components and the rate of strain in the flow. These interactions consist of two tilting and one stretching contribution; for example, if $i = 1$, $\tilde{\omega}_2 \, \partial \tilde{u}_1/\partial x_2$ and $\tilde{\omega}_3 \, \partial \tilde{u}_1/\partial x_3$ describe the influence on $\tilde{\omega}_1$ of the tilting of the $\tilde{\omega}_2$ and $\tilde{\omega}_3$ components due to the gradients $\partial \tilde{u}_1/\partial x_2$ and $\partial \tilde{u}_1/\partial x_3$, respectively, while $\tilde{\omega}_1 \, \partial \tilde{u}_1/\partial x_1$ accounts for the effect of stretching due to the gradient $\partial \tilde{u}_1/\partial x_1$. These effects are shown schematically in Fig. 2.5.1, which indicates distributions of the $\tilde{u}_1$ velocity component in the x_2 and x_3 coordinate directions. Suggested is the tilting of the $\tilde{\omega}_2$ and the $\tilde{\omega}_3$ vorticity components and the altering thereby of the $\tilde{\omega}_1$ component. The stretching component is difficult to indicate graphically but can be appreciated intuitively.

[†]The sequences may start with any entry.

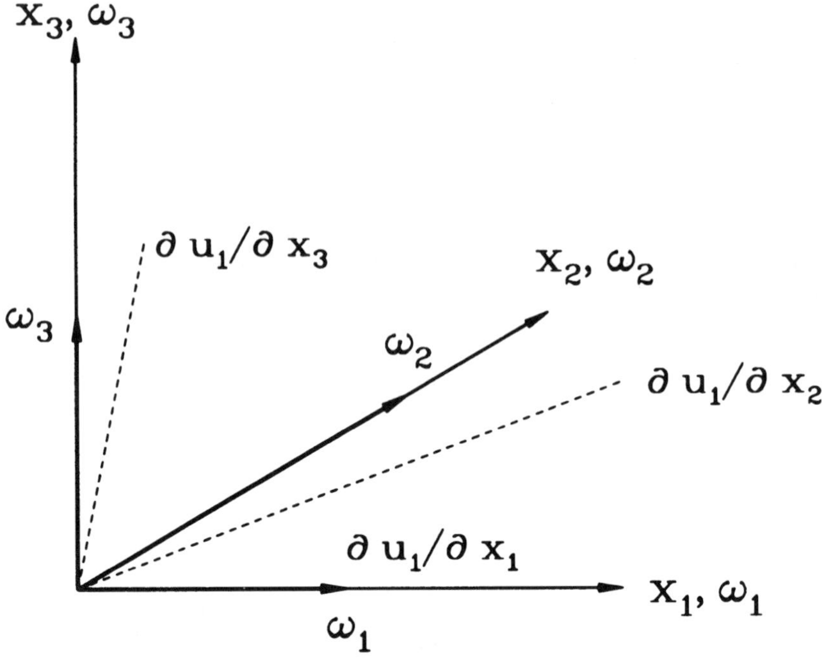

Figure 2.5.1 Dynamics of the vorticity components contributing to the ω_1 vorticity.

It is important to note that in a two-dimensional flow, e.g., one with $\tilde{u}_3 \equiv 0$ and $\partial/\partial x_3 \equiv 0$, only one component of vorticity is nonzero and Eq. (2.5.3) reduces to a convective-diffusive balance for that component, $\tilde{\omega}_3$.†

2.6 BOUNDARY AND INITIAL CONDITIONS

If the body force terms are inessential or independent of temperature, then Eqs. (2.1.1) and (2.2.1) constitute four equations for the four variables, $\tilde{u}_i$, $i = 1, 2, 3$ and $\tilde{p}$, while Eq. (2.3.1) determines the temperature $\tilde{\theta}$. In more general prob-

†Three-dimensionality and the resultant interaction between vorticity and the rate-of-strain field are essential features of turbulence. Nevertheless, two-dimensional and indeed one-dimensional turbulence are studied because of their simplicity and the possibly useful implications for three-dimensional turbulence (cf., e.g., Ashurst, 1979; Carnevale and Vallis, 1990). Batchelor (1967, pp. 186–187) gracefully states the concern called for in examining such implications as follows: "Motion in two-dimensions has the simple property that the vorticity of a fluid element is unchanged, except by molecular diffusion, as the element follows the motion (and precisely because of this special characteristic of two-dimensional motion we should beware of assuming too close a relation between two- and three-dimensional turbulence)." Bradshaw (1978, p. 10) is more forceful in stating: "Three dimensionality is essential to the genesis and maintenance of turbulence." In Section 6.6 we discuss two-dimensional turbulence briefly.

lems, such as those in which buoyancy plays a role, body forces couple the five equations and simultaneous solutions are required. In Section 10.13 we discuss turbulent flows influenced by buoyancy.

To apply these equations to the description of the velocity field of a particular flow, it is necessary to specify boundary conditions, on solid surfaces if they are present and at infinity if a uniform stream is involved. The conditions on a solid surface require that the fluid adjacent to the surface have the velocity of the surface which in the usual coordinate system is zero. It is also necessary to specify initial conditions on the velocity components throughout the flow field, i.e., to specify $\tilde{u}_i(\mathbf{x}, 0)$, satisfying continuity, Eq. (2.1.1). The related initial pressure and vorticity fields are not independent but are given by Eqs. (2.4.1) and (2.5.2).

Similar but somewhat simpler considerations apply to the determination of the temperature field in a particular flow. Boundary conditions on solid boundaries are generally specified in terms of either temperature or temperature gradient or a relation between the two. When uniform streams at infinity are involved, the temperature there is specified. The initial conditions simply call for specification of a temperature distribution $\tilde{\theta}(\mathbf{x}, 0)$.

It is worth noting that in some fundamental applications of these equations the domain for numerical solution is either a rectangle or a rectangular parallelopiped, depending on whether a two- or three-dimensional solution is sought with periodic boundary conditions on the flow variables on its boundary. Thus the solution domain is one of an infinite number of identical cells in either two or three coordinate directions. Variations in this perspective permit flows with simple rates of strain and simple bulk motions, e.g., rotation about an edge, to be treated. Although such geometries are highly idealized, as discussed later, they permit invaluable numerical solutions of the describing equations to be obtained.

2.7 DIRECT AND LARGE EDDY SIMULATIONS

The implications of these considerations relative to turbulence can be understood from the following perspective. Consider an isothermal fluid within a cubical container with a side of length L_c. Consider further that an initial velocity distribution is characterized by a velocity measure U_g (for example, the root mean square of the modulus of the velocity vector averaged over the entire volume), and by a length $L_g << L_c$ (for example, the distance beyond which the velocities at two points are essentially independent). Thus we can form a global Reynolds number $N_{Rg} = U_g L_g/\nu$ associated with each particular initial distribution.

We then consider a series of calculations each for a fixed value of N_{Rg}, a fixed value of the ratio L_g/L_c and extending over a fixed time from $t = 0$ to $t = t_f$. The resulting velocity distributions at $t = t_f$ are a realization. Thus the picture we have is fluid in a container subject to stirring which at $t = 0$ is terminated so that the velocity fluctuations decay with time under the influence of molecularity; that is, as t_f increases, the root mean square of the modulus of the velocity vector averaged over space decreases. If N_{Rg} is suitably small, then small changes in the initial distribution from one realization to another lead to comparably small changes in the flow at $t = t_f$, and thus with some experience the consequence of changing the initial velocity in some fashion can be anticipated. However, if N_{Rg} is suitably large, then inordinately large changes in the velocity distribution at $t = t_f$ result from even small changes in the initial velocity distribution and no amount of experience will allow detailed predictions of the consequences of such changes. Thus we find the history of the flow to be exquisitely sensitive to initial conditions.

This situation can be considered from an alternative point of view. Suppose we wish to compute the flow about a slender body of revolution with a characteristic length L aligned with a uniform stream of velocity U so that we can identify a Reynolds number UL/ν. Independent of the initial conditions imposed at the start of the calculation, if the Reynolds number is suitably small, the results of a direct numerical simulation (DNS) will in due course settle down to an essentially steady flow. On the contrary, at a suitably large Reynolds number the flow in the immediate neighborhood of the body, i.e., in its boundary layer, and in its wake will be unsteady. In this case, if we wish to characterize the flow in a relatively simple fashion, some averaging technique must be applied.

These situations are characteristic of turbulence and suggest clearly the need to resort in many situations to a statistical description of the flow, one which characterizes the evolution with time of the first flow considered in terms of statistical measures at as many spatial locations as desired at each time t_f. Similarly, the flow in the boundary layer and wake can usefully be characterized in a statistical fashion depending on the use to which the results are to be put. For example, the crudest description of the flow in a closed container might be the evolution in time of the rms velocity averaged over the entire volume, while a more detailed description would provide the evolution of the root mean square of each velocity component at many spatial locations within the container.

Although we have indicated the appropriateness of a statistical description of many turbulent flows, it is important to consider direct numerical simulations of the conservation equations. After all, if we have a well-defined set of partial differential equations subject to well-defined boundary and initial conditions, and if we have suitably large computers, it seems totally reasonable simply to solve the equations numerically. This is indeed possible for a limited set of conditions, but because of the limitations of this approach, we must resort to statistical methods for most flows of applied interest. However, DNS is extremely useful as a means of providing the data for statistical methods. This procedure corresponds to *averaging solutions* rather than pursuing the more usual procedure developed in later chapters of *solving averaged equations*. To facilitate our exposition we return to the flow within a cube of sides L_c, which we consider to be one of a triply infinite array of identical cells with periodic boundary conditions imposed on their boundaries. In the present discussion we imagine that the flow is nonisothermal so that the initial data involve specification of the temperature distribution as well as that of the velocity. Although we establish the specific requirements later, it is sufficient for present purposes to state that spatial resolution of the five flow variables at arbitrary points within and on the boundaries of the cell, i.e., of $\tilde{u}_i$, $i = 1, 2, 3$, $\tilde{p}$, and $\tilde{\theta}$, at each time t necessitates that each side of the cube be divided into a suitably large number N of segments which we take to be of uniform length. Thus for $N >> 1$ the flow at each time is determined by $5N^3$ numbers. The requirement on N is set by the necessity of resolving all fluctuations (cf. Section 5.3), with the consequence that

$$N > N_{Rg}^{3/4} \tag{2.7.1}$$

Since for computational reasons N is generally taken to be a power of 2, it is appropriate to consider the values shown in Table 2.7.1.

We see from this table that the scope of a calculation rapidly increases as the Reynolds number N_{Rg} increases, such that it is easy to anticipate limitations imposed by the available computing power. In the early 1990s the most advanced supercomputers require running times measured in tens of hours to carry out calculations of a single Reynolds number on a cube with 128 grid points on each side. But a Reynolds number N_{Rg} of 600 corresponds to a relatively benign turbulence compared to values of applied interest. Moreover, it can be appreciated that the behavior of turbulence in a cubic cell does not correspond to flows of technological concern; rather, engineers are interested in turbulent flows in pipes and channels and about wings, fuselages, and hulls. However, despite the highly idealized geometry and limited Reynolds numbers which can be treated, calculations of the sort suggested in this discussion provide information which is invaluable for the assessment and development of turbulence theories of more applied interest. Of particular value from DNS are variables and quantities which are difficult, if not impossible, to measure experimentally, such as fluctuating pressures, the rate of strain tensor, and the vorticity, and analyses identifying the structure of turbulence, such as concentrations of vorticity. Accordingly, there

Table 2.7.1 Parameters for direct numerical calculations of flow in a cell of a cubic array

N	$(N_{Rg})_{max}$	$5N^3$
32	102	1.64(5)
64	256	1.31(6)
128	645	1.05(7)
256	1,635	8.39(7)

are ongoing efforts by computational fluid dynamicists to increase the scope, duration, and complexity of high-Reynolds-number turbulent flows treated by DNS.

The first application of DNS was due to Orzag and Patterson (1972), who considered unforced, isotropic turbulence, which corresponds to a special case of decaying turbulence in a container as discussed earlier. Subsequently, Siggia and Patterson (1978) incorporated forcing at low frequencies so that the turbulence while isotropic and homogeneous is maintained. A conceptually important advance is due to Rogallo (1981), who included uniform shear in DNS so that the evolution of turbulence with a production mechanism could be examined (see also Rogers et al., 1987). Several flows have been studied by DNS: the mixing layer which evolves in time (cf. Metcalfe et al., 1987) or in space (cf. Lowery et al., 1987); fully developed channel flow (cf. Kim et al., 1987); and the turbulent boundary layer (cf. Spalart, 1992). Recent further applications of DNS involve both mean shear and buoyancy (cf. Gerz et al., 1989) and the response of particles in turbulent flows (cf. Elghobashi and Truesdell, 1992).

Some indication of the limitations, scope, and utilization of these DNS calculations is obtained by considering further the channel flow of Kim et al. (1987). The Reynolds number based on the centerline mean velocity and the channel half-height is 3,300, a value considerably less than those relevant to applications; for example, the corresponding number in pipe flows arising in industry is frequently 10^5 or more. The computational domain in the channel flow calculation is periodic in the downstream and spanwise directions and contains 2 million grid points, a number which suggests the complexity of the data handling called for in DNS. The results of these calculations are widely used for the study of large turbulent structures and for the assessment and improvement of turbulence theories of engineering interest (cf. Mansour et al., 1988; Reynolds, 1990).

Although growth in the computing speed and power actually experienced in past years can be expected to continue indefinitely, it does not appear that direct numerical solution of the conservation equations for flow configurations and Reynolds numbers of applied interest will be feasible in the foreseeable future. This prospect implies that the approaches used in the past to deal with turbulent flows of such interest, namely, those based on solutions of averaged equations, the moment methods we discuss, will continue to be useful.

We now take up an attractive alternative to DNS which provides a theory that is intermediate between DNS and moment theories. This alternative is large eddy simulation (LES), in which the limitation of Eq. (2.7.1) is abandoned and the flow variables at grid points are determined by large-scale calculations in an averaged sense, i.e., averaged over a volume in the neighborhood of each grid point (cf. Section 3.8). The equations solved in this approach are not exact, so there arise some uncertainities connected with descriptions of the influence of the unresolved small turbulence scales; this is especially the case in flows involving walls, where the turbulence is dominated by small turbulence scales. However, flows of more direct applied interest, e.g., with higher Reynolds numbers, but still with simple geometries can be studied with results intermediate in accuracy and completeness between those given by DNS and the results of moment methods discussed later. LES proves to be especially useful in calculations of the atmospheric boundary layer, the domain in which it was first proposed and exploited (cf. Wyngaard, 1981). Further references are Reynolds (1990), Piomelli et al. (1990), and Galperin and Orzag (1993). It should also be noted that when complex, transient turbulent flows are treated by advanced moment methods, e.g., by the Reynolds stress theory discussed in Section 10.10, the computational effort required for LES and such methods is comparable; in this case the former, with its less extensive approximations, may be preferable.

2.8 SUMMARY

In this chapter we have set forth the fundamental equations of fluid mechanics and supplementary equations which follow from them. Although it may be appealing to consider a straightforward application of well-developed numerical techniques and powerful computers to provide solutions to these equations subject to the applicable initial and boundary conditions, we have shown that for the foreseeable future only flows involving modest Reynolds numbers and simple geometries can be dealt with in this fashion. We discussed the usefulness for fundamental turbulence developments of these large-scale computations. However, the implication is that the treatment of turbulent flows of applied interest must be based on the *averaged* conservation and transport equations.

CHAPTER

THREE

STATISTICAL TOOLS

This chapter takes up the various statistical tools necessary for subsequent discussions by considering one or more arbitrary variables, for example, $\tilde{a}(\mathbf{x}, t)$, which can be thought of as a velocity component, the pressure, or the temperature and which has a mean value, generally $A(\mathbf{x})$, and an associated fluctuation $a(\mathbf{x}, t)$. Although our principal attention relates to systems with statistical values that depend only on $\mathbf{x}$, that is, statistically stationary systems in the usual sense, we also discuss the characterization of transitory and periodic systems. The former involve statistical values that depend on space and on time from the onset of the process, while periodic systems involve values that depend on space and a variable related to the time within each periodic interval—e.g., a phase angle. At the outset we note that our discussion of averaging techniques is more extensive than is called for solely by the requirements of the subsequent analysis, but these techniques are conceptually important and frequently applied in experimental turbulence research.

3.1 AVERAGING

Our starting point is the several means of averaging. In Fig. 3.1.1 we show the variation with time at a given spatial location of $\tilde{a}(t; \mathbf{x})$, a variation we use for

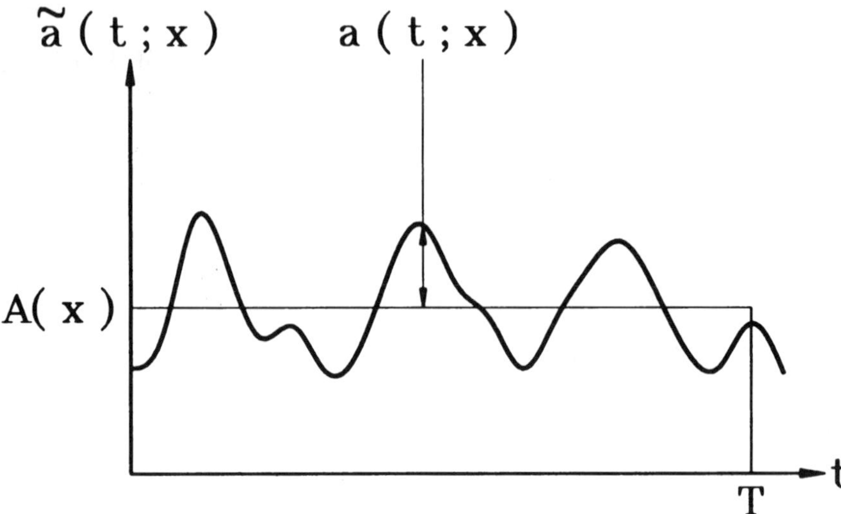

Figure 3.1.1 The time variation of a variable $\tilde{a}(t; \mathbf{x})$ and its mean value.

several considerations in this chapter. We first introduce time averaging defined by the equation

$$\bar{\tilde{a}}(\mathbf{x}) = A(\mathbf{x}) = \lim_{T \to \infty} \frac{1}{T} \int_0^T dt\, \tilde{a}(t; \mathbf{x}) \tag{3.1.1}$$

where the overbar indicates the averaging operation. The implication of this definition is that a random variable at a fixed spatial location is integrated over a suitably long time interval T, the averaging time. Under the assumption that the limiting process does indeed converge to a well-defined value, the variable $\tilde{a}(\mathbf{x}, t)$ is said to be statistically stationary. This mode of averaging is the most widely used in both turbulence theory and experimental turbulence, since the output from a probe held at a fixed spatial location in a flow can be readily averaged in the sense of Eq. (3.1.1). In practice, the appropriateness of a particular averaging time to establish stationarity is determined by repeatedly increasing T until $A(\mathbf{x})$ is unchanged within an acceptable tolerance.

A variant of time averaging, ensemble averaging, arises in statistically stationary systems when for one or more reasons a continuous history of the variable $\tilde{a}(\mathbf{x}, t)$ is not available.† Rather, we form the average from a suitably large

†Some complex optical systems used in turbulence research provide data at a low sampling rate, so a continuous record is not obtained. This same perspective is called for in the widespread practice of digitizing analog signals so as to provide time series suitable for treatment by digital computers.

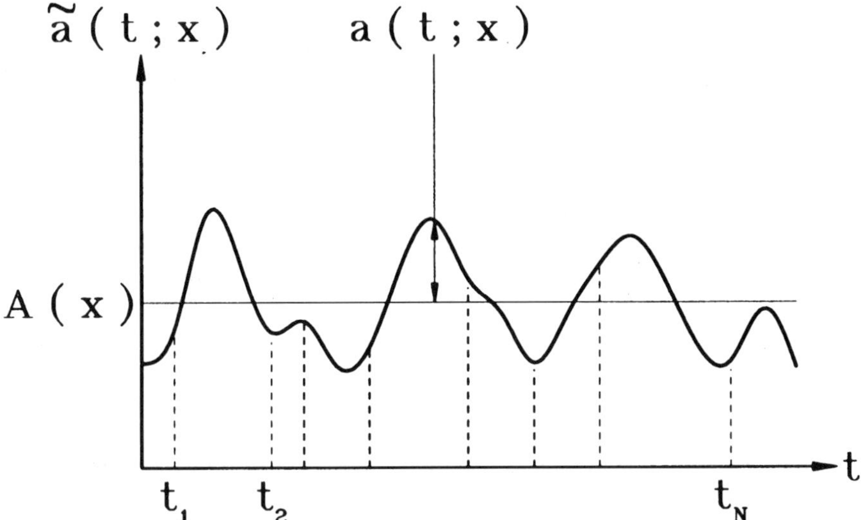

Figure 3.1.2 The sampling of a variable $\tilde{a}(t;\, \mathbf{x})$ at discrete times.

number of realizations of $\tilde{a}$, denoted in this case as $\tilde{a}(t_n;\, \mathbf{x})$, $n = 1, 2, \ldots, N$. Then we have

$$\langle \tilde{a}(\mathbf{x}, t) \rangle = A(\mathbf{x}) = \lim_{N \to \infty} \frac{1}{N} \sum_{n=1}^{N} \tilde{a}(t_n;\, \mathbf{x}) \tag{3.1.2}$$

where the symbol $\langle\ \rangle$ replaces the overbar to denote the averaging operation. In this case the suitability of a particular value of N is determined by recalculating $A(\mathbf{x})$ with increasing values of N until the mean is suitably constant.† In Fig. 3.1.2 we show the same arbitrary variable $\tilde{a}(t;\, \mathbf{x})$ but indicate its value at the discrete times $t_1, t_2, \ldots, t_N$ used to form $A(\mathbf{x})$ according to Eq. (3.1.2).

By an extension of the perspective suggested by Eq. (3.1.2), we can define suitable averages for flows which are transitory, i.e., not statistically stationary. Consider as an example the flow variables at a particular spatial location within a tube emptying a container of high-pressure air. In this case we consider a transitory flow which starts at a time $t = 0$ when the valve is opened and terminates at $t = t_f$ when the flow ceases. Again we focus on a general variable $\tilde{a}$, perhaps one of the velocity components or the pressure or the temperature,

†Statistical theory indicates that the error in $A(\mathbf{x})$ equals $(\overline{a^2}/N)^{1/2}/A$, so N must be increased if the relative intensity $(\overline{a^2})^{1/2}(\mathbf{x})/A(\mathbf{x})$ is large. Similar considerations apply to the averaging period T in Eq. (3.1.1). Thus if $N = n\,T$, where n is the number of samples recorded per unit time, we can obtain an estimate of the averaging time required to achieve a specified accuracy in the mean value $A(\mathbf{x})$.

at a fixed spatial location associated with $\mathbf{x}$ within the tube but consider a sequence of N realizations so that we have N records, denoted by $\tilde{a}_n(t; \mathbf{x})$, $n = 1, 2, \ldots, N$. At a fixed time $\hat{t}$, where $0 \leq \hat{t} \leq t_f$, we can determine a mean value according to

$$\langle \tilde{a}(\mathbf{x}, \hat{t}) \rangle = A(\mathbf{x}, \hat{t}) = \lim_{N \to \infty} \frac{1}{N} \sum_{n=1}^{N} \tilde{a}_n(\mathbf{x}, \hat{t}) \tag{3.1.3}$$

Again the implications of the indicated functional dependencies should be noted. Clearly, if we change the location $\mathbf{x}$ and/or the time $\hat{t}$ at which the samples are taken, we change the average value. For a fixed location $\mathbf{x}$, the N time records permit the mean value given by Eq. (3.1.3) to be determined for arbitrary values of $\hat{t}$ within the entire range $0 < \hat{t} < \hat{t}_f$. Although other examples of transitory systems involving random variables are of interest, e.g., the decay of turbulence in a container after a stirring mechanism is removed (cf. Sections 5.2–5.4), this emptying of a pressurized tank is used repeatedly to illustrate various additional averaging considerations.

The statistical description of periodic flows, such as the flow in piston chamber of an internal combustion engine, calls for a perspective close to that leading to Eq. (3.1.3) but with the time $\hat{t}$ replaced by a phase angle ϕ, for example, the crank angle in the case of an engine, so that Eq. (3.1.3) becomes

$$\langle \tilde{a}(\mathbf{x}, \phi) \rangle = \lim_{N \to \infty} \frac{1}{N} \sum_{n=1}^{N} \tilde{a}_n(\mathbf{x}, \phi)$$

Note that although the standard nomenclature "statistically nonstationary" appears to be applied to these cases, if as $N \to \infty$ a definite limit is achieved, the averaging technique yields valid statistical "stationary" quantities even if they depend on the sampling time $\hat{t}$ in the case of the emptying tank and on the angle ϕ in the case of periodic systems.

For each of the methods of averaging, it is easy to establish by straightforward operations on Eqs. (3.1.1)–(3.1.3) their operative algebra and calculus. We illustrate relative to time averaging of Eq. (3.1.1) by noting the following useful rules:

$$\begin{aligned} \overline{\tilde{a}(t; \mathbf{x}) + \tilde{b}(t; \mathbf{x})} &= A(\mathbf{x}) + B(\mathbf{x}) \\ \overline{\alpha \tilde{a}(t; \mathbf{x})} &= \alpha\, A(\mathbf{x}) \\ \overline{\frac{\partial \tilde{a}}{\partial x_i}(t; \mathbf{x})} &= \frac{\partial A(\mathbf{x})}{\partial x_i} \end{aligned} \tag{3.1.4}$$

where α is any constant.

We sometimes must confront the *time derivative* of a statistical quantity calculated according to Eq. (3.1.1). One class of such time derivatives relates to turbulent flows involving two or more distinct time scales. For example, consider the measurement of the wind velocity and direction at a particular location within the atmospheric boundary layer. These variables involve seasonal, diurnal, and even hourly variations, but an anemometer output for a 5- or 10-min period can yield a nearly "statistically stationary" velocity and direction. However, when the average over one such period is compared with a corresponding average over a longer period, for example 1 hour, there will in general be a significant difference and the flow therefore cannot be considered to be statistically stationary. Nevertheless, under these circumstances we can approximate the time derivatives of the mean velocity and mean flow direction from averages of these quantities taken over relatively short periods spaced considerably longer periods apart in time. Thus, if we accept the mean of the velocity taken over a 5-min interval to be suitably stationary, we can estimate the time derivative of that mean velocity if we difference two such values taken perhaps an hour apart.

Time derivatives of mean quantities are also relevant in transitory turbulent flows. For example, we discussed earlier the emptying of a tank through a tube and the determination of $A(\mathbf{x}, \hat{t})$. In this case we can clearly determine

$$\frac{\partial A}{\partial \hat{t}}(\mathbf{x}, \hat{t}) = \lim_{\delta_{\hat{t}} \to 0} \frac{A(\mathbf{x}, \hat{t} + \delta_{\hat{t}}) - A(\mathbf{x}, \hat{t})}{\delta_{\hat{t}}} \tag{3.1.5}$$

In selecting the time increment $\delta_{\hat{t}}$ in Eq. (3.1.5), the usual considerations calling for a judicious choice between too large and too small values must be made; if δ_t is too large, the derivative can be inaccurate because of the curvature of the $A(\mathbf{x}, t)$, the truncation error; if δ_t is too small, the numerator in Eq. (3.1.5) involves the small difference in relatively large numbers and the derivative is again inaccurate, the round-off error. Similarly, if we wish to describe the decay of turbulence in a container after the stirring mechanism is removed, we can again consider a suitably large number of realizations, determine mean values, for example, $A(\mathbf{x}, \hat{t})$ and $A(\mathbf{x}, \hat{t} + \delta_{\hat{t}})$, and apply Eq. (3.1.5) to determine the rate of change of those values with time.

3.2 FLUCTUATIONS AND THEIR CHARACTERIZATION

In each of the averaging methods we can identify the deviation either of an instantaneous value or of one realization from an associated mean value, the fluctuation, as a new random variable but with a zero mean, i.e., by definition $\overline{a(\mathbf{x}, t)} = \langle a(\mathbf{x}, t)\rangle = \langle a(\mathbf{x}; \hat{t})\rangle = 0$. The fluctuations of $\tilde{a}(t; \mathbf{x})$ are shown in Figs. 3.1.1 and 3.1.2. For clarity, in this section we restrict our attention to statistically stationary variables.

The mean value of various functions of the fluctuations identify their nature and characteristics. The mean square value, $\overline{a^2}(\mathbf{x})$, is the variance and is indicative of the extent of the variations about the mean value. The relative intensity, $\overline{(a^2)}^{1/2}(\mathbf{x})/A(\mathbf{x})$, is a more useful measure of that variation. Additional information concerning fluctuations is given by the higher moments; the skewness, $S \equiv \overline{a^3}(\mathbf{x})/[\overline{a^2}(\mathbf{x})]^{3/2}$, indicates the relative extent of negative and positive fluctuations about the mean. If $S > 0$, the positive fluctuations dominate; while negative fluctuations dominate if $S < 0$. Finally, the kurtosis, defined as $K \equiv \overline{a^4}/\overline{a^2}^2$, also called the flatness factor, determines the extent of symmetric but remote deviations from the mean.

We are frequently interested in the correlation of the fluctuations of two variables. If we have $\overline{\tilde{a}(\mathbf{x}, t)\tilde{b}(\mathbf{x}, t)} = A(\mathbf{x})B(\mathbf{x}) + \overline{a(\mathbf{x}, t)b(\mathbf{x}, t)}$, we are concerned with the last term, which can be normalized in various ways. However, the correlation coefficient,

$$R_{ab}(\mathbf{x}) = \frac{\overline{ab}(\mathbf{x})}{[\overline{a^2}(\mathbf{x})\overline{b^2}(\mathbf{x})]^{1/2}} \tag{3.2.1}$$

indicates in a meaningful way the extent of the interdependence of the two variables. A fundamental property of this coefficient is that it is bounded between $+1$ and -1. If $|R_{ab}| << 1$, the two variables are poorly correlated; while if $R_{ab} \approx 1, -1$, the two variables are nearly perfectly correlated or nearly perfectly anticorrelated, respectively. This latter property can be seen, for example, by assuming that $a = \alpha b$, where α is an arbitrary constant. If $\alpha > 0$, then $R_{ab}(\mathbf{x}) = 1$; while if $\alpha < 0$, then $R_{ab}(\mathbf{x}) = -1$.

The correlation coefficients of higher moments are sometimes of interest. Thus we could define, for example,

$$R_{a^2b}(\mathbf{x}) = \frac{\overline{a^2b}(\mathbf{x})}{[\overline{a^4}(\mathbf{x})\overline{b^2}(\mathbf{x})]^{1/2}} \tag{3.2.2}$$

Other coefficients are clearly suggested. If a and b are either perfectly correlated or perfectly anticorrelated, then $R_{a^2b}(\mathbf{x})$ is proportional to the skewness of either a or b.

The correlation coefficients defined by Eqs. (3.2.1) and (3.2.2) are readily extended to transitory systems in which the correlations between different variables depend on the spatial location $\mathbf{x}$ and the sampling time $\hat{t}$. Extension is also readily made to periodic systems in which that time is replaced by the phase angle ϕ. In these systems, ensemble rather than time averaging is applied.

3.3 PROBABILITY DENSITY FUNCTIONS

A different perspective on the characterization of the fluctuations of one or more variables is provided by the probability density functions (pdfs).† Consider a

†With the data recording and time series techniques in current use in experimental turbulence, the pdf assumes an important role in the statistical description of turbulence and in the presentation of data.

statistically stationary variable $\tilde{a}(\mathbf{x}, t)$ whose values are in principle unbounded and a function $P(\tilde{a}; \mathbf{x})$ with the following properties:

$$
\begin{aligned}
1 &= \int_{-\infty}^{\infty} d\,\tilde{a}P(\tilde{a}; \mathbf{x}) \\
A(\mathbf{x}) &= \int_{-\infty}^{\infty} d\,\tilde{a}\,\tilde{a}P(\tilde{a}; \mathbf{x}) \\
\overline{a^2}(\mathbf{x}) &= \int_{-\infty}^{\infty} d\tilde{a}\,(\tilde{a} - A)^2 P(\tilde{a}; \mathbf{x}) \\
\overline{a^3}(\mathbf{x}) &= \int_{-\infty}^{\infty} d\tilde{a}\,(\tilde{a} - A)^3 P(\tilde{a}; \mathbf{x})
\end{aligned}
\tag{3.3.1}
$$

Higher-order moments without limit can be obtained in an obvious way from $P(\tilde{a}; \mathbf{x}b)$

Experimentally, $P(\tilde{a}; \mathbf{x})$ can be determined as follows. Consider that for all practical purposes the variable $\tilde{a}(t; \mathbf{x})$ is bounded in the range from $\tilde{a}_{\min}$ to $\tilde{a}_{\max}$ at the particular spatial location in question; i.e., the likehood of $\tilde{a}$ being either less than $\tilde{a}_{\min}$ or greater than $\tilde{a}_{\max}$ is negligible. Divide that range into a large number $M - 1$ of equal subranges defined by the values $\tilde{a}_1, \tilde{a}_2, \ldots, \tilde{a}_M$, where $\tilde{a}_1 = \tilde{a}_{\min}$ and $\tilde{a}_M = \tilde{a}_{\max}$. Consider a "bin" to be the subrange between $\tilde{a}_n$ and $\tilde{a}_{n+1} = \tilde{a}_n + \Delta_{\tilde{a}}$. If $\tilde{a}(t; \mathbf{x})$ is continuous as shown in Fig. 3.3.1, the incremental times during which the variable is in this subrange are $\Delta_{t1}, \Delta_{t2}, \ldots$, and sum to $T_{\tilde{a}_n}$ within the entire interval T. As T increases, the fraction of time during

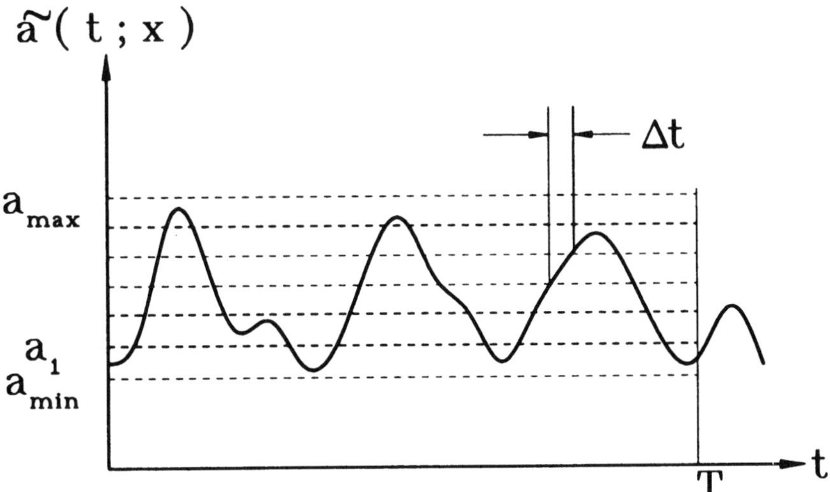

Figure 3.3.1 The distribution of $\tilde{a}(t; \mathbf{x})$ interpreted in terms of the related pdf.

which the variable is within the subrange approaches a constant, provided, as we assume, that the variable relates to a statistically stationary system. The result of letting $T \rightarrow \infty$ for each of the $M - 1$ intervals is the histogram of $a(t; \mathbf{x})$ as shown in Fig. 3.3.2. It is clear that if the entries in the histogram in the subranges near the edges of the range of $\tilde{a}(t; \mathbf{x})$ are to be statistically reliable, long record lengths, i.e., large values of T, are required.

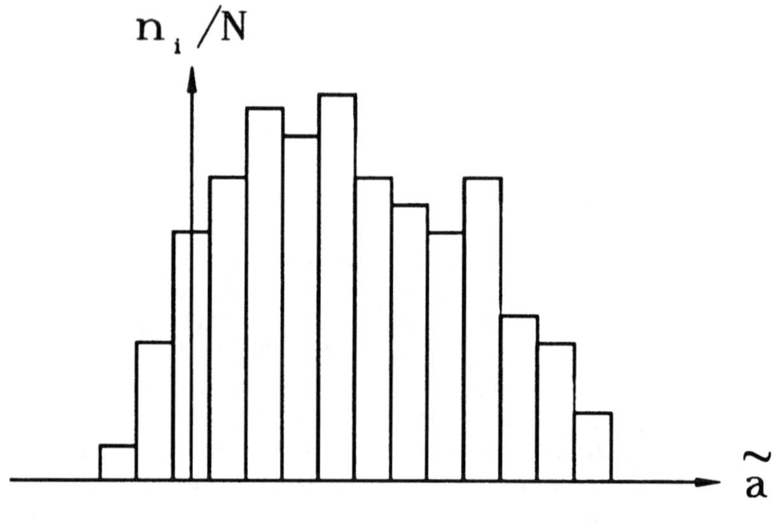

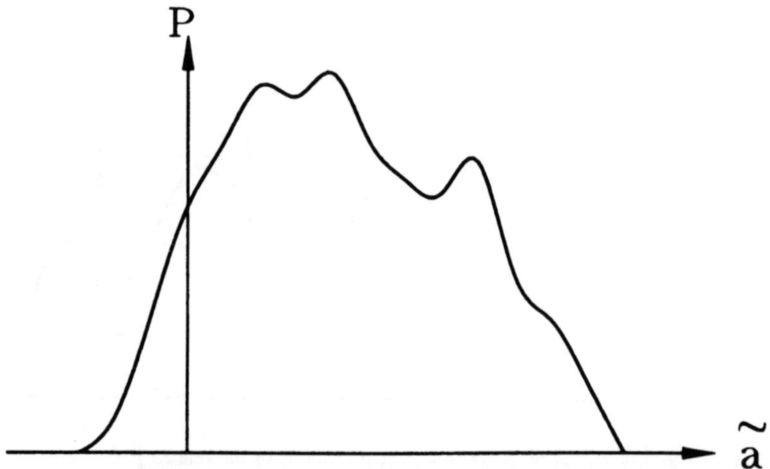

Figure 3.3.2 The histogram and its related probability density function for a variable $\tilde{a}(\mathbf{x}, t)$.

As the subranges of $\tilde{a}(t; \mathbf{x})$ are made smaller and more numerous, the histogram approaches a continuous function. Thus a pdf involves the double limit $\Delta_{\tilde{a}} \rightarrow 0$, $T \rightarrow \infty$, that is,

$$P(\tilde{a}; \mathbf{x}) = \lim_{\Delta_{\tilde{a}} \rightarrow 0} \frac{1}{\Delta_{\tilde{a}}} \left[\lim_{T \rightarrow \infty} \left(\frac{T_{\tilde{a}}}{T} \right) (\mathbf{x}) \right] \tag{3.3.2}$$

where we replace $T_{\tilde{a}_n}$ with the generic value $T_{\tilde{a}}$. In Fig. 3.3.2 we show also the pdf corresponding to the histogram.

An altered perspective is called for if the variable $\tilde{a}(t; \mathbf{x})$ is digitized, i.e., either recorded directly in digital form or converted from analogy to digital form. In this case the time interval is replaced by the number of samples N, $N_{\tilde{a}_n}$ denotes the number of entries in the subrange between $\tilde{a}_n$ and $\tilde{a}_{n+1}$ and replaces $T_{\tilde{a}_n}$, and finally the histogram involves the ratio $N_{\tilde{a}_n}/N$ as $N \rightarrow \infty$. Furthermore, Eq. (3.3.2) becomes

$$P(\tilde{a}; \mathbf{x}) = \lim_{\Delta_{\tilde{a}} \rightarrow 0} \frac{1}{\Delta_{\tilde{a}}} \left[\lim_{N \rightarrow \infty} \left(\frac{N_{\tilde{a}}}{N} \right) (\mathbf{x}) \right] \tag{3.3.3}$$

Our earlier remark concerning the statistical reliability of the entries in the histogram near the edges of the range of $\tilde{a}(t; \mathbf{x})$ clearly applies to pdfs as well, the implication being that such reliability is generally expected to be low.

Returning to Eqs. (3.3.1), we note that the calculation of any additional moments which might be required is clearly suggested. Thus $P(\tilde{a}; \mathbf{x})$ contains all the statistical information on $\tilde{a}$ and its fluctuations a (cf. Section 10.13). We also note that the mean values of various powers of the fluctuations are given by corresponding moments of the related pdf. Thus an intensity, $\overline{a^2}(\mathbf{x})$, is a second moment, while the skewness involves the third moment, $\overline{a^3}(\mathbf{x})$. This nomenclature extends to multivariate correlations; thus $\overline{ab}(\mathbf{x})$ is also a second-moment quantity and $\overline{a^2b}(\mathbf{x})$ is a third-moment quantity.

The physical significance of $P(\tilde{a}; \mathbf{x})$ relates to its determining the fraction of time during which $\tilde{a}$ is between $\tilde{a}_1$ and $\tilde{a}_2 > \tilde{a}_1$, that is, the probability that $\tilde{a}(t; \mathbf{x})$ is within this range, according to

$$\int_{\tilde{a}_1}^{\tilde{a}_2} d\tilde{a}\, P(\tilde{a}; \mathbf{x}) \tag{3.3.4}$$

The most frequently encountered functional form for $P(\tilde{a}; \mathbf{x})$ in the study of probability and statistics is the Gaussian or normal distribution:

$$P(\tilde{a}; \mathbf{x}) = \frac{1}{(2\pi\sigma^2)^{1/2}} \exp\left[\frac{-(\tilde{a} - A)^2}{2\sigma^2} \right] \tag{3.3.5}$$

where $\sigma^2 = \overline{a^2}(\mathbf{x})$ is the variance of $\tilde{a}$ about its mean value $A(\mathbf{x})$. This pdf is symmetric about the mean value, so its skewness $S = 0$ while its kurtosis $K = 3$. From Eq. (3.3.5) we see a feature shared with many other functional forms for pdfs: If $A(\mathbf{x})$ and $\sigma(\mathbf{x})$ are known, then the entire pdf is determined. In the study of turbulence, normal distributions are often compared with experimentally determined pdfs, since the deviations are frequently of interest but non-Gaussian distributions are common (cf. Section 8.5).[†]

We have discussed pdfs of a single variable, but more general, multivariate distributions are also useful. Consider again a statistically stationary system involving two variables, $\tilde{a}(\mathbf{x}, t)$ and $\tilde{b}(\mathbf{x}, t)$, each with unbounded values. Then the bivariate pdf $P(\tilde{a}, \tilde{b}; \mathbf{x})$ has the following properties:

$$
\begin{aligned}
1 &= \int_{-\infty}^{\infty} d\tilde{a} \int_{-\infty}^{\infty} d\tilde{b}\, P(\tilde{a}, \tilde{b}; \mathbf{x}) \\
P(\tilde{a}; \mathbf{x}) &= \int_{-\infty}^{\infty} d\tilde{b}\, P(\tilde{a}, \tilde{b}; \mathbf{x}) \\
P(\tilde{b}; \mathbf{x}) &= \int_{-\infty}^{\infty} d\tilde{a}\, P(\tilde{a}, \tilde{b}; \mathbf{x}) \\
\overline{ab}(\mathbf{x}) &= \int_{-\infty}^{\infty} d\tilde{a}\, (\tilde{a} - A) \int_{-\infty}^{\infty} d\tilde{b}\, (\tilde{b} - B)\, P(\tilde{a}, \tilde{b}; \mathbf{x}) \\
\overline{a^2b}(\mathbf{x}) &= \int_{-\infty}^{\infty} d\tilde{a}\, (\tilde{a} - A)^2 \int_{-\infty}^{\infty} d\tilde{b}\, (\tilde{b} - B)\, P(\tilde{a}, \tilde{b}; \mathbf{x})
\end{aligned}
\tag{3.3.6}
$$

Again the calculation of all the statistical information for each of the two variables individually and for their correlations is clearly suggested.[‡]

We consider joint pdfs further. If the two variables are *independent,* then their joint pdf is the product of their individual pdfs,

$$P(\tilde{a}, \tilde{b}; \mathbf{x}) = P(\tilde{a}; \mathbf{x})\, P(\tilde{b}; \mathbf{x})$$

In this case,

$$\overline{ab}(\mathbf{x}) = \overline{a^2b}(\mathbf{x}) = 0 \tag{3.3.7}$$

and all correlations of their fluctuations are zero. However, two variables which are *not independent* can possess some zero and some nonzero correlations, for example, $\overline{ab} = 0$ but $\overline{a^2b} \neq 0$. To see this, consider for simplicity the pdf of the fluctuations of the two variables and calculate

$$
\begin{aligned}
\overline{ab}(\mathbf{x}) &= \int_{-\infty}^{\infty} da\, a \int_{-\infty}^{\infty} db\, b\, P(a, b; \mathbf{x}) \\
\overline{a^2b}(\mathbf{x}) &= \int_{-\infty}^{\infty} da\, a^2 \int_{-\infty}^{\infty} db\, b\, P(a, b; \mathbf{x})
\end{aligned}
$$

[†]In homogeneous turbulence the pdfs of the velocity components are nearly Guassian but their time derivatives are not, while in turbulent shear flows the pdfs of such components are non-Guassian.

[‡]In current experimental turbulence research, the combination of data recording and data analysis permits ready determination of multivariate pdfs despite the long records required, i.e., large values of T or N, but their presentation, whether graphical or numerical, is difficult.

Now, if the pdf $P(a, b; \mathbf{x})$ is symmetric about the line $a = 0$, then the first quantity is zero while the second is not. With this example it is easy to envisage other circumstances in which some correlations are zero but others involving the same variables are not.

The experimental determination of bivariate pdfs such as $P(\tilde{a},\tilde{b}; \mathbf{x})$, for either continuous or discrete variables, follows closely the procedure discussed earlier; the bins are now two-dimensional, for example, bounded by values $\tilde{a}_n$, $\tilde{a}_{n+1}$ and $\tilde{b}_n$, $\tilde{b}_{n+1}$. Clearly, both the sample time T and number of samples N required to achieve statistical reliability, especially near the boundaries of the ranges of the variables, is greatly increased by the added dimensionality and becomes increasingly so as that dimensionality increases.

As suggested earlier, it is always possible to shift the origin of single and multivariate pdfs so that the distributions of the *fluctuations* about the origin, that is, $P(a; \mathbf{x})$ and $P(a, b; \mathbf{x})$, are considered. We thus have, for example,

$$
\begin{aligned}
0 &= \int_{-\infty}^{\infty} da\ a\ P(a; \mathbf{x}) \\
\overline{ab}(\mathbf{x}) &= \int_{-\infty}^{\infty} da\ a \int_{-\infty}^{\infty} db\ b\ P(a, b; \mathbf{x})
\end{aligned}
\tag{3.3.8}
$$

and contours of $P(a, b; \mathbf{x})$ indicate the sign of the correlation $\overline{ab}(\mathbf{x})$ as shown in Fig. 3.3.3.

A variant of the pdfs discussed so far arises in turbulence. Suppose that our generic variable, $\tilde{a}(\mathbf{x}, t)$, has at the spatial location in question a definite minimum value, denoted a_0, and moreover that it assumes this value a specified fraction of the time $\alpha(\mathbf{x})$. In this case the pdf of $\tilde{a}$ consists of two portions and can be written as

$$
P(\tilde{a}; \mathbf{x}) = \alpha(\mathbf{x})\ \delta(\tilde{a} - a_0) + [1 - \alpha(\mathbf{x})]\ \tilde{P}(\tilde{a}; \mathbf{x})\ H(\tilde{a} - a_0) \tag{3.3.9}
$$

where δ is the delta function, H is the Heaviside function $[H(x) = 0$ if $x < 0$ and $H(x) = 1$ if $x > 0]$, and $\tilde{P}(\tilde{a}; \mathbf{x})$ is the continuous portion of the pdf, normalized so that

$$
1 = \int_{a_0^+}^{\infty} d\tilde{a}\ \tilde{P}(\tilde{a}; \mathbf{x})
$$

In Fig. 3.3.4 we show schematically the pdf of Eq. (3.3.9). In this case the equivalent of Eqs. (3.3.1) yields

$$
\begin{aligned}
A(\mathbf{x}) &= \alpha(\mathbf{x})a_0 + [1 - \alpha(\mathbf{x})] \int_{a_0^+}^{\infty} d\tilde{a}\ \tilde{a}\ \tilde{P}(\tilde{a}; \mathbf{x}) \\
\overline{a^2}(\mathbf{x}) &= \alpha(\mathbf{x})(a_0 - A)^2 + [1 - \alpha(\mathbf{x})] \int_{a_0^+}^{\infty} d\tilde{a}\ (\tilde{a} - A)^2\ \tilde{P}(\tilde{a}; \mathbf{x})
\end{aligned}
\tag{3.3.10}
$$

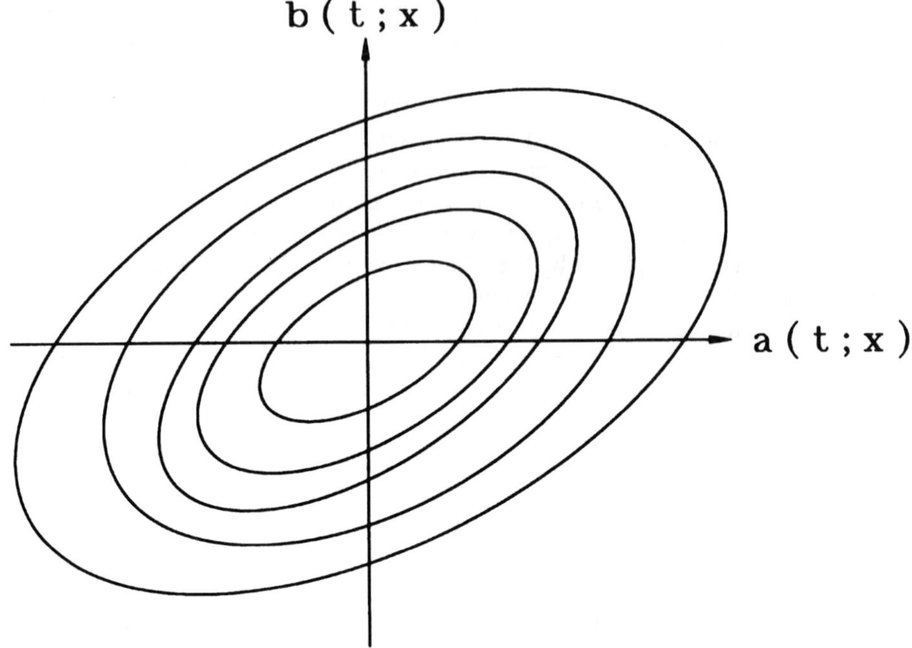

Figure 3.3.3 Contours of a bivariate probability density function $P(a, b; \mathbf{x})$.

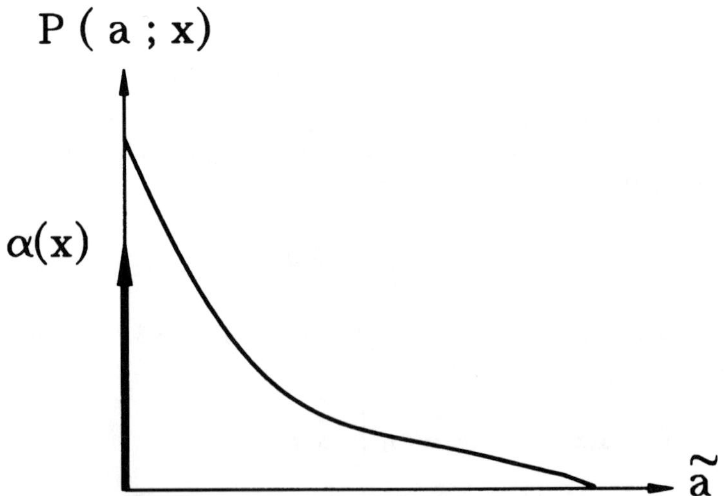

Figure 3.3.4 The probability density function of a variable with a definite minimum value.

Additional statistical quantities can be readily calculated and involve contributions from the delta function relating to the definite value of the variable, i.e., to a_0, and from the continuous pdf relating to the distribution of that variable for $\tilde{a} > a_0$.[†]

Variations and extensions of this discussion are sometimes needed. Thus variables with upper rather than lower limits and variables with both lower and upper limits sometimes arise. Extension to multivariate pdfs with bounds on one or more variables occurs, but we need not give details since the requisite generalizations of Eqs. (3.3.9) and (3.3.10) are clearly suggested.

The pdf description can also be applied to transitory systems. To illustrate, again consider the emptying of a pressurized tank through a tube. In this case each realization provides an entry in a pdf $P(\tilde{a};\, \mathbf{x},\, \tilde{t})$ and we have

$$1 = \int_{-\infty}^{\infty} d\tilde{a}\; P(\tilde{a};\, \mathbf{x},\, \tilde{t})$$

$$A(\mathbf{x},\, \tilde{t}) = \int_{-\infty}^{\infty} d\tilde{a}\; \tilde{a}\; P(\tilde{a};\, \mathbf{x},\, \tilde{t}) \tag{3.3.11}$$

$$\overline{a^2}(\mathbf{x},\, \tilde{t}) = \int_{-\infty}^{\infty} d\tilde{a}\; (\tilde{a} - A)^2\; P(\tilde{a};\, \tilde{t},\, \mathbf{x})$$

with additional statistical quantities following the suggested pattern.

Finally, the generalization of Eqs. (3.3.6) to many variables implies that the statistical properties of individual variables and of their correlations with all other participating variables can be obtained from appropriate integrations of a single multivariate pdf. Thus, for example, in statistically stationary turbulent flows, *all* possible information is contained within the pdf $P(\tilde{u}_1,\, \tilde{u}_2,\, \tilde{u}_3,\, \tilde{p};\, \mathbf{x})$. Indeed, one approach to the treatment of turbulence is based on such a direct attack, but as might be expected, various approximations and considerable computation are required in this approach (cf. Section 10.13).

3.4 SPACE AND TIME CORRELATIONS

Previous discussion focused on the statistical description of a generic variable $\tilde{a}(t;\, \mathbf{x})$. If this is a principal variable, i.e., a velocity component, the pressure, or the temperature at a fixed location, a little reflection establishes that these descriptions contain no information on either the length or time scales associated

[†]Section 8.5 deals with variables having such distributions; the temperature in the thermal wake of a heated cylinder is an example of a function with these characteristics. As a consequence of experimental error, there is always uncertainity in the determination of $\alpha(\mathbf{x})$. Frequently there is employed a threshold value which in the case under discussion involving a minimum value of the variable would be a small positive increment to a_0 (cf. Bilger et al., 1976, for an additional means of dealing rationally with this uncertainty).

with $\tilde{a}(t; \mathbf{x})$. This shortcoming does not prevail if the generic variable is the reciprocal of a space or time derivative of a principal variable. In this case the mean value and various moments clearly provide information on the scales involved. However, in this section we discuss the common problem of obtaining scale information from $\tilde{a}(t; \mathbf{x})$ itself.

In a statistically stationary system, consider the fluctuations of a representative variable, $a(t; \mathbf{x})$, and a second random variable derived therefrom by introduction of a time advance τ, namely, $a(t + \tau; \mathbf{x})$. Figure 3.4.1 shows the fluctuations of this generic variable and the two values associated with t and $t + \tau$.

The experimental determination of the correlation of these two variables depends on whether the variable is continuous or discrete. In the former case we can envisage a tape recorder with two reading heads physically displaced in the tape direction such that the tape speed and separation distance determine the time advance τ. If the output from each head is biased to remove their common mean value $A(\mathbf{x})$ and if the two outputs are multiplied and averaged, there results a quantity that is proportional to the desired correlation. If the variable $\tilde{a}(t; \mathbf{x})$ is in discrete form at uniform time intervals and thus suitable for digital computation, then the multiplication and summing operations needed to calculate their correlation are clear, but the time advance τ is likewise discrete and limited to multiples of those time intervals.

A variant of Eq. (3.2.1) leads to the time autocorrelation coefficient

$$R_a(\tau; \mathbf{x}) = \frac{\overline{a(t; \mathbf{x})\, a(t + \tau; \mathbf{x})}}{\overline{a^2}(\mathbf{x})} \tag{3.4.1}$$

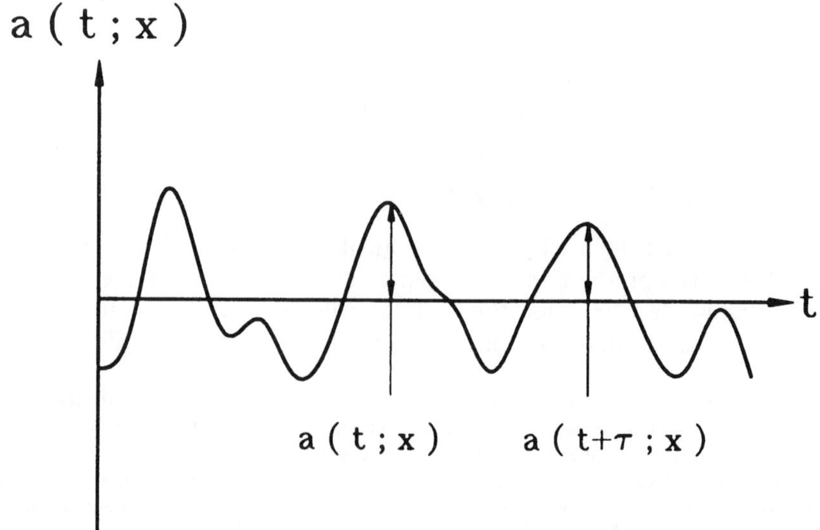

Figure 3.4.1 The distribution of the fluctuations of $a(t; \mathbf{x})$ and its treatment as two random variables.

where the bounds ± 1 cited earlier in connection with Eq. (3.2.1) again apply. Clearly, $R_a(0; \mathbf{x}) = 1$ and for statistically stationary variables $R_a(\tau; \mathbf{x})$ is always an even function of τ. If $a(t; \mathbf{x})$ is periodic with a frequency ω, then $R_a(\tau; \mathbf{x}) = \cos \omega\tau$, that is, periodic, while the combination of a periodic variable and a small random variable leads to a decaying periodic autocorrelation coefficient. When applied to turbulence, Eq. (3.4.1) generally yields $R_a(|\tau| \to \infty; \mathbf{x}) \approx 0$; that is, with sufficient time separation, the two variables become uncorrelated with no or few zero crossings.

A measure of the correlation time, the integral time, is provided by

$$T_{Ia}(\mathbf{x}) = \int_0^\infty d\tau \, R_a(\tau; \mathbf{x}) \tag{3.4.2}$$

where we assume that the integral is convergent. Equation (3.4.2) provides a measure of the large-scale, low-frequency content of the fluctuations of $a(t; \mathbf{x})$.

An alternative time scale is obtained if we restrict attention to a small time advance. Suppose that τ is sufficiently small so that

$$a(t + \tau; \mathbf{x}) \approx a(t; \mathbf{x}) + \frac{\partial a}{\partial t}(t; \mathbf{x}) \, \tau + \frac{1}{2} \frac{\partial^2 a}{\partial t^2}(t; \mathbf{x}) \, \tau^2 + \ldots$$

so that we obtain the fluctuation at an advanced time from the magnitude and derivatives of the fluctuation at any given time.[†] To put this and several similar expansions encountered later in more useful forms, note that

$$a \frac{\partial^2 a}{\partial t^2} = \frac{\partial}{\partial t}\left(a \frac{\partial a}{\partial t}\right) - \left(\frac{\partial a}{\partial t}\right)^2$$

which when averaged yields

$$\overline{a \frac{\partial^2 a}{\partial t^2}} = -\overline{\left(\frac{\partial a}{\partial t}\right)^2} \tag{3.4.3}$$

If we use this result, then it is readily found from Eq. (3.4.1) that

$$R_a(\tau \to 0; \mathbf{x}) \approx 1 - \frac{1}{2} \frac{1}{\overline{a^2}(\mathbf{x})} \overline{\left(\frac{\partial a}{\partial t}\right)^2}(\mathbf{x}) \, \tau^2 + \ldots \approx 1 + \frac{1}{2} R_a''(0; \mathbf{x}) \tau^2 + \ldots$$

and thus that

$$R_a''(0; \mathbf{x}) = -\frac{\overline{(\partial a/\partial t)^2}(\mathbf{x})}{\overline{a^2}(\mathbf{x})} \tag{3.4.4}$$

Thus we can define a time scale

[†]When applied to turbulence, there is a definite limit on the magnitude of τ for the applicability of Eq. (3.4.3). However, since we are interested here only in $\tau \to 0$, we need not be concerned with this restriction.

$$T_{sa} = \left[\frac{2\overline{a^2}(\mathbf{x})}{\overline{(\partial a/\partial t)^2}(\mathbf{x})}\right]^{1/2} = \left[\frac{2}{-R''_a(0;\,\mathbf{x})}\right]^{1/2} \tag{3.4.5}$$

Equation (3.4.5) provides a measure of the rate at which the two variables become uncorrelated. It is possible for a time autocorrelation to decay rapidly with τ for small values of τ, as a consequence of vigorous small-scale fluctuations, but to maintain a significant correlation for large values of τ as a consequence of large-scale fluctuations. Figure 3.4.2 shows schematically a typical $R_a(\tau;\,\mathbf{x})$ and the two time scales suggested thereby.

By utilizing the definition of a second derivative, in this case of the time autocorrelation differentiated with respect to τ, it is found by means of Eq. (3.4.3) that

$$\frac{d^2}{d\tau^2}\,\overline{a(t;\,\mathbf{x})\,a(t+\tau;\,\mathbf{x})} = \overline{a(t;\,\mathbf{x})\,\frac{\partial^2 a}{\partial t^2}(t+\tau;\,\mathbf{x})} = -\overline{\frac{\partial a}{\partial t}(t;\,\mathbf{x})\,\frac{\partial a}{\partial t}(t+\tau;\,\mathbf{x})} \tag{3.4.6}$$

Thus we see that the curvature of the autocorrelation coefficient of $a(t;\,\mathbf{x})$ at any time advance is related to the corresponding coefficient of the *time derivative* of $a(t;\,\mathbf{x})$. Equation (3.4.4) is a special case of Eq. (3.4.6) for $\tau = 0$.

While these notions can be extended to two or more variables, it is sufficient to consider here the correlation between $a(t;\,\mathbf{x})$ and $b(t+\tau;\,\mathbf{x})$, that is, between two variables at the same spatial location but at times differing by τ. Then an extension of Eq. (3.4.1) is

$$R_{ab}(\tau;\,\mathbf{x}) = \frac{\overline{a(t;\,\mathbf{x})\,b(t+\tau;\,\mathbf{x})}}{[\overline{a^2}(\mathbf{x})\,\overline{b^2}(\mathbf{x})]^{1/2}} \tag{3.4.7}$$

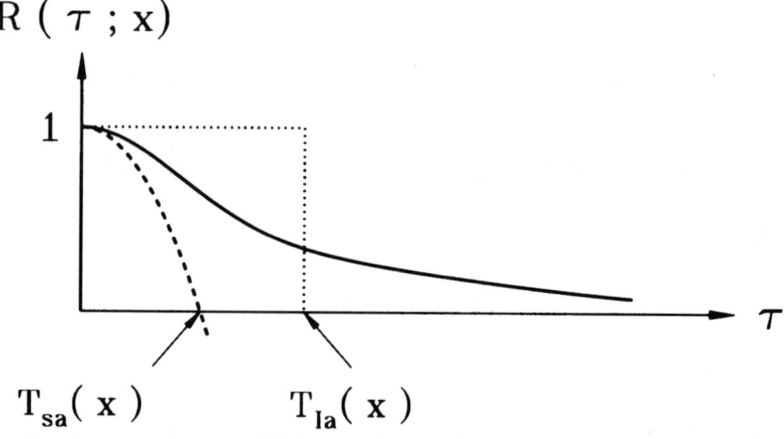

Figure 3.4.2 The time autocorrelation coefficient.

Here $R_{ab}(0; \mathbf{x}) = R_{ab}(\mathbf{x})$ of Eq. (3.2.1), that is, the one-point, one-time correlation coefficient between two variables. When applied to turbulence, $R_{ab}(\tau \to \infty; \mathbf{x}) = 0$. In general, $R_{ab}(\tau; \mathbf{x})$ is *not* an even function of τ, as may be seen as follows: If $b(t + \tau; \mathbf{x})$ is expanded about $\tau = 0$ as indicated by Eq. (3.4.3), then a calculation closely following that resulting in Eq. (3.4.4) leads to

$$R_{ab}(\tau \to 0; \mathbf{x}) \approx R_{ab}(\mathbf{x}) + \frac{\overline{a\, \partial b/\partial t(\mathbf{x})}}{[\overline{a^2}(\mathbf{x})\, \overline{b^2}(\mathbf{x})]^{1/2}} \tau + \ldots \tag{3.4.8}$$

where $\overline{a\, \partial b/\partial t(\mathbf{x})} = -\overline{(\partial a/\partial t) b(\mathbf{x})} \neq 0$ in general. Thus we see from Eq. (3.4.8) that $R_{ab}(\tau; \mathbf{x})$ in the neighborhood of the origin is an odd function of τ.

The notions discussed here can be applied to transitory systems such as the emptying of a pressurized tank provided ensemble averaging of the variables is carried out. Thus, for example, in each realization the values of $a(\tilde{t}; \mathbf{x})$ and $a(\tilde{t} + \tau; \mathbf{x})$ are multiplied to obtain a new variable which is ensemble averaged to obtain the autocorrelation, a function of τ with $\tilde{t}$ and $\mathbf{x}$ as parameters; because of the transitory nature of the flow, the resulting correlation coefficient following Eq. (3.4.1) is in general not an even function of τ.

Returning to statistically stationary variables, note that considerations similar to those involving time separations apply to spatial separations. Suppose that we have two fluctuations, $a(t; \mathbf{x})$ and $a(t; \mathbf{x} + \mathbf{h})$, fluctuations of the same variable, e.g., a velocity component u_i, at the same time but at spatial locations separated by $\mathbf{h}$, some arbitrary direction indicated by a separation vector. Then an extension of Eq. (3.4.1) leads to

$$R_a(\mathbf{h}; \mathbf{x}) = \frac{\overline{a(t; \mathbf{x})\, a(t; \mathbf{x} + \mathbf{h})}}{\overline{a^2}(\mathbf{x})} \tag{3.4.9}$$

Here we have chosen the denominator so that if $\mathbf{h} = 0$, the correlation coefficient is unity. Again, in general, spatial inhomogeneity implies that R_a is not an even function of separation in an arbitrary direction. Figure 3.4.3 shows $R_a(h_1; \mathbf{x})$ schematically for a variable whose statistical quantities vary spatially. Thus, for example, for the separation h_1 in the x_1 direction, we assume that the correlation coefficient $R_a(h_1; \mathbf{x})$ decreases monotonically. For $h_1 < 0$, however, we assume that there is an initial increase in correlation before the two variables become uncorrelated with sufficient separation corresponding to $-h_1$.

Figure 3.4.3 indicates the means for identifying an integral length scale and a length associated with the small-scale fluctuations of $a(t; \mathbf{x})$. To that end let h_{1m} denote the separation length in the x_1 direction for which the correlation is a maximum; in this particular case $h_{1m} < 0$. An integral scale can be defined as

$$L_{Ia}(\mathbf{x}) = \int_{h_{1m}}^{\infty} dh_1\, R_a(h_1; \mathbf{x}) \tag{3.4.10}$$

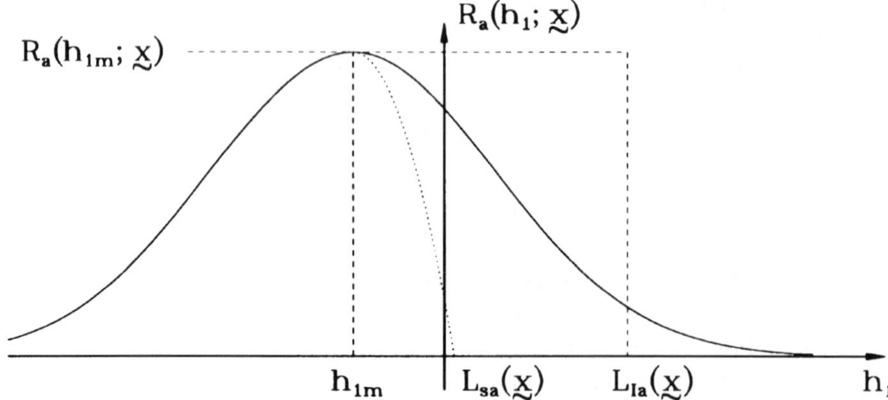

Figure 3.4.3 The spatial correlation coefficient $R_a(h_1; \mathbf{x})$ and the related length scales.

Alternative definitions involving other integration ranges of h_1 are clearly possible. Furthermore, corresponding integral lengths in the other coordinate directions and indeed in arbitrary directions from the reference point $\mathbf{x}$ are evident.

To define a measure of the small scales, expand $R_a(h_1; \mathbf{x})$ about $h_1 = h_{1m}$ and obtain

$$R_a(h_1; \mathbf{x}) = R_a(h_{1m}; \mathbf{x}) + \frac{1}{2}\frac{d^2 R_a}{dh_1^2}(h_{1m}; \mathbf{x})(h_1 - h_{1m})^2 + \ldots \quad (3.4.11)$$

Thus we have a measure of the length scale associated with small fluctuations:

$$L_{sa}(\mathbf{x}) = \left[\frac{2\, R_a(h_{1m}; \mathbf{x})}{-d^2\, R_a/dh_1^2\, (h_{1m}; \mathbf{x})}\right]^{1/2} \quad (3.4.12)$$

Figure 3.4.3 shows the two lengths defined by Eqs. (3.4.10) and (3.4.12). Our earlier remarks concerning the correlations in other coordinate directions apply to these lengths. Note that in spatially homogeneous systems $h_{1m} = 0$ and the definition of Eq. (3.4.12) applies in this case as well.

Although for clarity we have introduced time and length scales on the basis of separate considerations of one-point, two-time and two-point, one-time correlations, more general two-point, two-time correlations are clearly suggested. For example, a combination of the correlation coefficients of Eqs. (3.4.1) and (3.4.9) would lead to

$$R_a(\mathbf{h}, \tau; \mathbf{x}) = \frac{\overline{a(t; \mathbf{x})\, a(t + \tau; \mathbf{x} + \mathbf{h})}}{[\overline{a^2}(\mathbf{x})\, \overline{a^2}(\mathbf{x} + \mathbf{h})]^{1/2}} \quad (3.4.13)$$

Appropriate integrals of $R_a(\mathbf{h}, \hat{\tau}; \mathbf{x})$ with respect to $\mathbf{h}$ and $\hat{\tau}$, jointly or individually, yield length and time scales identifying the correlation scales of the variable $a(t; \mathbf{x})$. Thus length and time scales can be defined in terms of the statistical behavior of variables at multiple points in space and time.

For transitory systems, such as the emptying of a pressurized tank through a tube, two-point, one-time descriptions provide length scale information. Consider the measurement in one realization of $\tilde{a}_n(\mathbf{x}, \hat{t})$ and $\tilde{a}_n(\mathbf{x} + \mathbf{h}, \hat{t})$ and in due course the corresponding fluctuations $a_n(\mathbf{x}, \hat{t})$ and $a_n(\mathbf{x} + \mathbf{h}, \hat{t})$. Such measurements call for two spatially separated probes and the recording of the output from each as the tank is discharged. Then the discussion leading to Eq. (3.4.9) again applies for each observation time $\hat{t}$ so that we get

$$R_a(\mathbf{h}; \mathbf{x}, \hat{t}) = \lim_{N \to \infty} \frac{N^{-1} \sum_{n=1}^{N} a_n(\mathbf{x}, \hat{t})\, a_n(\mathbf{x} + \mathbf{h}, \hat{t})}{\left[N^{-1} \sum_{n=1}^{N} a_n^2(\mathbf{x}, \hat{t})\, N^{-1} \sum_{n=1}^{N} a_n^2(\mathbf{x} + \mathbf{h}), \hat{t}) \right]^{1/2}} \tag{3.4.14}$$

In addition we can define a vector integral length scale $\mathbf{L}_{Ia}(\mathbf{x}, \hat{t})$ by extending Eq. (3.4.9) and a vector small length scale $\mathbf{L}_{sa}(\mathbf{x}, \hat{t})$ by extending Eq. (3.4.12). These considerations arise relative to homogeneous, isotropic turbulence, i.e., to the decay of turbulence in a suitably large container. In this case the correlation coefficient of Eq. (3.4.14) depends only on the magnitude of $\mathbf{h}$, that is, on $|\mathbf{h}|$, and on $\hat{t}$, while the length scales become scalar quantities that depend only on $\hat{t}$.

3.5 SPECTRA

A second but not unrelated means for obtaining time and length scale information concerning the fluctuations involves spectra. Because of their importance in communications, vibrations, acoustics, and turbulence, there are a variety of texts on the theory and practice of spectral analysis. A useful entry to this literature is Bendat and Piersol (1980). Bradshaw (1971) discusses the physics of spectra and correlations with direct application to turbulence.

Our starting point is the definition of the Fourier transform pair,†

$$\begin{aligned} f(x) &= \frac{1}{2} \int_{-\infty}^{\infty} dy\, \hat{f}(y) \exp(-ixy) \\ \hat{f}(y) &= \frac{1}{\pi} \int_{-\infty}^{\infty} dx\, f(x) \exp(ixy) \end{aligned} \tag{3.5.1}$$

where x and y are arbitrary real variables. In our applications $f(x)$ is real; it follows from Eq. (3.5.1) that if $f(x)$ is also even, then $\hat{f}(y)$ is real and even; but

†Factors of 2 and π and/or their roots appear in various places in various definitions of Fourier transforms. The definitions here are standard in the turbulence literature.

if $f(x)$ is asymmetric, then $\hat{f}(y)$ is complex and asymmetric. In the former case Eqs. (3.5.1) and (3.5.2) take on special forms:

$$f(x) = \int_0^\infty dy\, \hat{f}(y) \cos xy$$
$$\hat{f}(y) = \frac{2}{\pi} \int_0^\infty dx\, f(x) \cos xy \tag{3.5.2}$$

As an initial application of Eqs. (3.5.2), we consider $f(x)$ to be the time autocorrelation of $a(t; \mathbf{x})$ and x to be the time advance τ [cf. Eq. (3.4.1)]. In this case y is interpreted as a frequency $\omega, \hat{f}(y)$ is denoted $\phi_a(\omega; \mathbf{x})$, the power spectral density (psd), and we have, from Eqs. (3.5.2),

$$\overline{a(t; \mathbf{x})\, a(t + \tau; \mathbf{x})} = \int_0^\infty d\omega\, \phi_a(\omega; \mathbf{x}) \cos \omega\tau$$
$$\phi_a(\omega; \mathbf{x}) = \frac{2}{\pi} \int_0^\infty d\tau\, \overline{a(t; \mathbf{x})\, a(t + \tau; \mathbf{x})} \cos \omega\tau \tag{3.5.3}$$

A primitive but illuminating perspective on these equations is provided by the observation that there are software programs based on the fast Fourier transform (FFT) (cf. Press et al., 1992) and electronic devices which provide $\phi_a(\omega; \mathbf{x})$ if $a(t; \mathbf{x})$ is input. FFT techniques are so efficient that it is standard practice to compute autocorrelation functions by calculating $\phi_a(\omega; \mathbf{x})$ via FFT and then using the first of Eqs. (3.5.3) to obtain the desired result (cf. Bendat and Piersol, 1980).

An immediate consequence of the first of Eqs. (3.5.3) is that

$$\overline{a^2}(\mathbf{x}) = \int_0^\infty d\omega\, \phi_a(\omega; \mathbf{x}) \tag{3.5.4}$$

i.e., the area under a distribution curve $\phi_a(\omega; \mathbf{x})$ is proportional to the intensity of the fluctuations $a(t; \mathbf{x})$. Furthermore, comparison of the second of Eqs. (3.5.3) and (3.4.2) yields

$$\phi_a(0; \mathbf{x}) = \frac{2}{\pi} \overline{a^2}(\mathbf{x})\, T_{la}(\mathbf{x}) \tag{3.5.5}$$

i.e., the intercept of the psd on the line $\omega = 0$ is proportional to the intensity of a and the integral time scale of a. From both Eqs. (3.5.4) and (3.5.5) we see that the units of $\phi_a(\omega; \mathbf{x})$ are those of a^2 times time. Finally, $\phi_a(\omega; \mathbf{x})$ can be interpreted as yielding the infinitesimal contribution to the intensity $\overline{a^2}(\mathbf{x})$ within the frequency band from ω to $\omega + d\omega$. Note from the first of Eqs. (3.5.3) and (3.5.5) that if two variables have the same psd, they have the same time autocorrelation and the same intensity. Equations (3.5.2)–(3.5.5) expose the intimate connection between the time autocorrelation of a variable and its psd.

A simple but illuminating application of Eqs. (3.5.3) arises from the following consideration. Suppose that[†]

$$a(t; \mathbf{x}) = \alpha(\mathbf{x}) \sin \hat{\omega}\, t$$

Then calculation of the time autocorrelation yields

$$\overline{a(t; \mathbf{x})\, a(t + \tau; \mathbf{x})} = \frac{1}{2} \alpha^2(\mathbf{x}) \cos \hat{\omega}\tau$$

If this is substituted into the left side of the first of Eqs. (3.5.3), this equation implies that

$$\phi_a(\omega; \mathbf{x}) = \frac{1}{2} \alpha^2(\mathbf{x})\, \delta(\omega - \hat{\omega})$$

Thus the psd of a periodic function of frequency $\hat{\omega}$ is a delta function at that frequency. Since the operations leading to this result are linear, a variable $a(t; \mathbf{x})$ involving a series of periodic functions possesses a psd involving a corresponding series of delta functions. More generally, this consideration reinforces our earlier interpretation of the psd as yielding the infinitesimal contribution to the intensity $\overline{a^2}(\mathbf{x})$ from fluctuations with a frequency ω.

We now combine the first of Eqs. (3.5.3) and Eq. (3.4.6) to obtain a useful result. Consider

$$\begin{aligned} \frac{d^2}{d\tau^2} \overline{a(t; \mathbf{x})\, a(t + \tau; \mathbf{x})} &= -\omega^2 \int_0^\infty d\omega\, \phi_a(\omega; \mathbf{x}) \cos \omega\tau \\ &= -\overline{\frac{\partial a}{\partial t}(t; \mathbf{x})\, \frac{\partial a}{\partial t}(t + \tau; \mathbf{x})} \\ &= -\int_0^\infty d\omega \phi_{a_t}(\omega; \mathbf{x}) \cos \omega\tau \end{aligned} \tag{3.5.6}$$

where

$$\phi_{a_t}(\omega; \mathbf{x}) = \frac{2}{\pi} \int_0^\infty d\tau\, \overline{\frac{\partial a}{\partial t}(t; \mathbf{x})\, \frac{\partial a}{\partial t}(t + \tau; \mathbf{x})} \cos \omega\tau$$

is the psd of the time derivative of $a(t; \mathbf{x})$. Comparison of the right sides of the first and third of Eqs. (3.5.6) indicates that

$$\phi_{a_t}(\omega; \mathbf{x}) = \omega^2 \phi_a(\omega; \mathbf{x})$$

Thus the psd of $a(t; \mathbf{x})$ can be used to obtain the psd of its time derivative. As expected intuitively, this result indicates that the peak contribution to the psd of derivatives of $a(t; \mathbf{x})$ is shifted to higher frequencies by the factor ω^2. This

[†]The linearized output from an anemometer in a vortex street is of this nature.

procedure can be employed to relate the psd of the higher time derivatives of $a(t; \mathbf{x})$ to $\phi_a(\omega; \mathbf{x})$. For example,

$$\phi_{a_n}(\omega; \mathbf{x}) = \omega^4 \phi_a(\omega; \mathbf{x})$$

To carry the discussion of spectra further, we return to Eqs. (3.5.1) and consider $f(x)$ to be the real, two-variable time correlation $\overline{a(t; \mathbf{x})\, b(t + \tau; \mathbf{x})}$, which we recall is in general asymmetric in τ so that $\phi_{ab}(\omega; \mathbf{x})$ is complex and asymmetric. In this case,

$$\overline{a(t; \mathbf{x})\, b(t + \tau; \mathbf{x})} = \frac{1}{2}\int_{-\infty}^{\infty} d\omega\, \phi_{ab}(\omega; \mathbf{x}) \exp(-i\omega\tau)$$
$$\phi_{ab}(\omega; \mathbf{x}) = \frac{1}{\pi}\int_{-\infty}^{\infty} d\tau\, \overline{a(t; \mathbf{x})\, b(t + \tau; \mathbf{x})} \exp(i\omega\tau) \tag{3.5.7}$$

where $\phi_{ab}(\omega; \mathbf{x})$, the cross spectral density, with a real part, the cospectrum, and an imaginary part, the quadrature spectrum, respects the restriction

$$\operatorname{Im}\int_{-\infty}^{\infty} d\omega\, \phi_{ab}(\omega; \mathbf{x}) \exp(-i\omega\tau) = 0 \tag{3.5.8}$$

The modulus $|\phi_{ab}(\omega; \mathbf{x})|$ is the coherence which indicates the frequencies dominating the correlation $\overline{ab}(\mathbf{x})$, while the phase is

$$\psi_{ab}(\omega; \mathbf{x}) = \tan^{-1}\frac{\operatorname{Im}[\phi_{ab}(\omega; \mathbf{x})]}{\operatorname{Re}[\phi_{ab}(\omega; \mathbf{x})]} \tag{3.5.9}$$

Clearly, from the first of Eqs. (3.5.7),

$$\overline{a(t; \mathbf{x})\, b(t; \mathbf{x})} = \frac{1}{2}\int_{-\infty}^{\infty} d\omega\, \operatorname{Re}[\phi_{ab}(\omega; \mathbf{x})] \tag{3.5.10}$$

Thus an interpretation similar to that attributed to $\phi_a(\omega; \mathbf{x})$ [cf. Eq. (3.5.4)], modified slightly to account for the complex nature of $\phi_{ab}(\omega; \mathbf{x})$, applies here.

An alternative to the frequency in spectral descriptions is the scalar wave number κ, with units of inverse length. The two variables can be related if κ is multiplied by an appropriate velocity. In flows with a principal direction, such as shear flows, any well-defined mean velocity can be chosen. Thus transforms arising from time correlations are frequently presented in terms of a wave number, which can be made dimensionless by multiplication by an appropriate length so that we have the quantity $\kappa L_0 = \omega L_0 / U_0$.

The corresponding connection between spectra and the *two-point, one-time correlation coefficients* $R_a(\mathbf{h}; \mathbf{x})$ of Eq. (3.4.1) and their generalization $R_{ab}(\mathbf{h}; \mathbf{x})$ of Eq. (3.4.7) is more complicated because $\mathbf{h}$ is a vector with components h_i, $i = 1, 2, 3$. As a consequence, three-dimensional Fourier transforms are involved. Utilization of such transforms in experimental turbulence research requires measurements at two spatial locations, e.g., the use of two anemometers. Since such measurements are not frequently made, three-dimensional transforms do not appear often in the experimental turbulence literature; however, as we shall see in Sections 5.2–5.4, a special, reduced form of such transforms plays an essential role in the study of isotropic, homogeneous turbulence. Moreover, a specialization of the three-dimensional spectra obtained by integration over two wave numbers corresponds to the spectra routinely measured in experimental turbulence. Thus it is worth outlining the generalization of the one-dimensional transforms considered to this point and establishing the connection among the transforms of various dimensions.

In the generalization of Eqs. (3.5.1) the variables x and y are vectors and the single integrals become triple integrals over their three components. In our application x corresponds to the separation length and thus y is a quantity with the dimensions of L^{-1}, namely, *the wave number vector* $\boldsymbol{\kappa}$. Our earlier discussion of the connection between a scalar wave number and frequency applies to each component of $\boldsymbol{\kappa}$. Thus fluctuations associated with large wave numbers correspond to short wavelengths and high frequencies, fluctuations with small wave numbers to long wavelengths and low frequencies.

The Fourier transform pair involving the correlation $\overline{a(t;\, \mathbf{x})a(t;\, \mathbf{x}+\mathbf{h})}$ is

$$\overline{a(t;\, \mathbf{x})a(t;\, \mathbf{x}+\mathbf{h})} = \int_{-\infty}^{\infty} d\kappa_1 \int_{-\infty}^{\infty} d\kappa_2 \int_{-\infty}^{\infty} d\kappa_3\, \Phi_a(\boldsymbol{\kappa};\, \mathbf{x}) \exp(i\boldsymbol{\kappa}\cdot\mathbf{h})$$

$$\Phi_a(\boldsymbol{\kappa};\, \mathbf{x}) = \frac{1}{(2\pi)^3}\int_{-\infty}^{\infty} dh_1 \int_{-\infty}^{\infty} dh_2 \int_{-\infty}^{\infty} dh_3\, \overline{a(t;\, \mathbf{x})a(t;\, \mathbf{x}+\mathbf{h})} \exp(-i\boldsymbol{\kappa}\cdot\mathbf{h})$$

(3.5.11)

where $\Phi_a(\boldsymbol{\kappa};\, \mathbf{x})$ is the wave number spectrum.† In the second of Eqs. (3.5.11) the integrand involves two factors: the two-point, one-time space correlation of the fluctuations $a(t,\, \mathbf{x})$ treated as a surface in three dimensions with coordinates h_1, h_2, and h_3, and an exponential function equal to

$$\cos \kappa_k h_k - i \sin \kappa_k h_k$$

Again, since the correlation is real, $\Phi_a(\boldsymbol{\kappa};\, \mathbf{x})$ is complex and thus respects restraints corresponding to the extension of Eq. (3.5.8). The experimental difficulties of using either of these equations is evident. In particular, to determine $\Phi_a(\boldsymbol{\kappa},\, \mathbf{x})$ from the second of Eqs. (3.5.11), the correlation in the integrand must be measured at a sufficiently large number of spatial points, i.e., a large number of triplets, h_1, h_2, and h_3, surrounding a particular $\mathbf{x}$ so that the three-dimensional integration can be performed with appropriate accuracy.

Our discussion to this point concerns a generic variable $a(t;\, \mathbf{x})$ which can be a velocity component or a scalar, pressure, or temperature. An important variant of Eqs. (3.5.11) involves the fluctuating velocity components $u_i(t;\, \mathbf{x})$ and

†In this case powers of 2π appear in various forms of these equations.

$u_j(t;\, \mathbf{x} + \mathbf{h})$, that is, the two-point, one-time measurement of two, in general different, velocity components, a measurement calling for two anemometers separated by $\mathbf{h}$. In this case we deal with the Fourier transform pair

$$\overline{u_i(t;\, \mathbf{x})u_j(t;\, \mathbf{x} + \mathbf{h})} = \int_{-\infty}^{\infty} d\kappa_1 \int_{-\infty}^{\infty} d\kappa_2 \int_{-\infty}^{\infty} d\kappa_3\, \Phi_{ij}(\boldsymbol{\kappa};\, \mathbf{x}) \exp(i\boldsymbol{\kappa} \cdot \mathbf{h})$$

$$\Phi_{ij}(\boldsymbol{\kappa};\, \mathbf{x}) = \frac{1}{(2\pi)^3} \int_{-\infty}^{\infty} dh_1 \int_{-\infty}^{\infty} dh_2 \int_{-\infty}^{\infty} dh_3\, \overline{u_i(t;\, \mathbf{x})u_j(t;\, \mathbf{x} + \mathbf{h})} \exp(-i\boldsymbol{\kappa} \cdot \mathbf{h})$$

(3.5.12)

The correlation of the two velocities can be converted to a correlation coefficient $R_{ij}(\mathbf{x}, \mathbf{h})$ as in Eq. (3.4.8), so that these equations relate that coefficient with the three-dimensional velocity spectrum tensor $\Phi_{ij}(\boldsymbol{\kappa};\, \mathbf{x})$. Note that as a special case, when $\mathbf{h} = 0$, the first of Eqs. (3.5.12) becomes

$$\overline{u_i(t;\, \mathbf{x})u_j(t;\, \mathbf{x})} = \int_{-\infty}^{\infty} d\kappa_1 \int_{-\infty}^{\infty} d\kappa_2 \int_{-\infty}^{\infty} d\kappa_3\, \Phi_{ij}(\boldsymbol{\kappa};\, \mathbf{x}) \qquad (3.5.13)$$

i.e., the one-point, one-time correlation of two velocity components is given by a three-dimensional integral of $\Phi_{ij}(\boldsymbol{\kappa};\, \mathbf{x})$ over all wave numbers κ_1, κ_2, and κ_3.

One-dimensional counterparts of $\Phi_{ij}(\boldsymbol{\kappa};\, \mathbf{x})$ are obtained by integration. As an example which has special significance in experimental turbulence research, consider $\phi_{ij}(\kappa_1;\, \mathbf{x})$; this is obtained by integrating $\Phi_{ij}(\boldsymbol{\kappa};\, \mathbf{x})$ over all κ_2 and κ_3 for fixed values of κ_1. In this case we have the Fourier transform pair

$$\overline{u_i(t;\, \mathbf{x})u_j(t;\, \mathbf{x} + \mathbf{i}_1\, h_1)} = \int_{-\infty}^{\infty} d\kappa_1 \exp(i\kappa_1 h_1) \left[\int_{-\infty}^{\infty} d\kappa_2 \int_{-\infty}^{\infty} d\kappa_3\, \Phi_{ij}(\boldsymbol{\kappa};\, \mathbf{x})\right]$$

$$= \int_{-\infty}^{\infty} d\kappa_1\, \phi_{ij}(\kappa_1;\, \mathbf{x}) \exp(i\kappa_1 h_1)$$

$$\phi_{ij}(\kappa_1;\, \mathbf{x}) = \frac{1}{2\pi} \int_{-\infty}^{\infty} dh_1\, \overline{u_i(t;\, \mathbf{x})u_j(t;\, \mathbf{x} + \mathbf{i}_1\, h_1)} \exp(-i\kappa_1 h_1) \qquad (3.5.14)$$

The distinction between $\phi_{ij}(\kappa_1;\, \mathbf{x})$, the one-dimensional velocity spectrum tensor, in these equations and $\Phi_{ij}(\kappa_1, 0, 0;\, \mathbf{x})$ from the second of Eqs. (3.5.12) with $\boldsymbol{\kappa} = \mathbf{i}_1\kappa_1$ should be noted. Since the correlation on the left side of the first of these equations is real, $\phi_{ij}(\kappa_1;\, \mathbf{x})$ and $\Phi_{ij}(\boldsymbol{\kappa};\, \mathbf{x})$ are complex.

As suggested earlier, measurement of the correlation on the left side of the first of Eqs. (3.5.14) and thus the determination of $\phi_{ij}(\kappa_1;\, \mathbf{x})$ from the second requires two anemometers, one yielding $u_i(t;\, \mathbf{x})$ and a second displaced in the x_1 direction an amount h_1 and yielding $u_j(t;\, \mathbf{x} + \mathbf{i}_1 h_1)$. An alternative which is frequently and conveniently used in flows with a mean velocity U_1 aligned with the x_1 axis involves recording the signals from *one* anemometer measuring two velocity components $u_i(t; \mathbf{x})$ and $u_j(t; \mathbf{x})$. Now we associate a time delay τ with a spatial separation h_1 by setting $\tau = -h_1/U_1$. The one-point, two-time correlation $\overline{u_i(t; \mathbf{x})u_j(t + h_1/U_1; \mathbf{x})}$ is an approximation to the two-point, one-time correlation appearing in Eqs. (3.5.14), that is, to $\overline{u_i(t, \mathbf{x})u_j(t; \mathbf{x} + \mathbf{i}_1 h_1)}$. It is this approximation which makes the one-dimensional velocity spectrum tensor of special significance in experimental turbulence.

Since this is our first encounter with approximating two-point, one-time correlations with one-point, two-time correlations, we digress to examine its basis. Relating time and space correlations in this fashion is the Taylor hypothesis (Taylor, 1935b), according to which the mean velocity U_1 is imagined to sweep *without change,* i.e., frozen, a pattern of velocity components from x_1 to $x_1 + \mathbf{i}_1 h_1$ in the time h_1/U_1. In the case of grid turbulence, with its uniform mean velocity and low turbulence intensities, this notion is intuitively plausible, but its applicability to more general flows with spatially nonuniform mean velocities and higher intensities is more problematical. Because of its conceptual importance and wide utility in experimental turbulence, the validity of the Taylor hypothesis has been examined by Lin (1953), Lumley (1965), Wyngaard and Clifford (1977), and Gurvich (1980). These studies establish that lateral nonuniformity of the mean velocity is the most important source of error. This is explicable since we expect that lateral fluctuations in a field of nonuniform mean velocity result in streamwise diffusion which distorts the pattern of velocity components as it sweeps downstream, thus vitiating the notion of a *frozen* pattern of fluctuations, the essential ingredient of the hypothesis. Lin (1953) shows that if the hypothesis is to apply in a shear flow, the wave number κ_1 of a fluctuation must satisfy the inequality

$$\kappa_1 U_1 >> \frac{\partial U_1}{\partial x_2}$$

where the usual shear flow coordinates with x_1 in the mainstream direction and x_2 normal thereto are adopted. Thus it is suggested that short-wavelength, high-frequency, large κ fluctuations in a portion of the layer with small shear are convected according to the Taylor hypothesis, but long-wavelength, low-frequency, small κ fluctuations in a portion of the layer subject to large shear are convected with velocities different from the local mean velocity. Put differently, fluctuations of suitably small scale extend laterally relative to the gradient of the mean velocity only a small amount compared with the mean velocity. Despite the uncertainties raised by these considerations as to its quantitative accuracy, the Taylor hypothesis provides a useful connection between space and time correlations and is frequently employed in theoretical and experimental turbulence [cf. Eq. (5.3.17)].

We now return to consideration of two-point, one-time correlations and their spectra. It is useful to consider the implications of the first of Eqs. (3.5.11) and

(3.5.14). Figure 3.5.1 shows schematically the three-dimensional space with κ_1, κ_2, and κ_3 as Cartesian coordinates, the wave number space. Because we deal with a four-dimensional function for each value of $\mathbf{x}$, a value of $\Phi_a(\boldsymbol{\kappa}; \mathbf{x})$ and of any other function of $\boldsymbol{\kappa}$ is assigned to each point in this $\boldsymbol{\kappa}$ space. With the function $\Phi_a(\boldsymbol{\kappa}; \mathbf{x})$ assumed known, we see from the first of Eqs. (3.5.11) that the calculation of $\overline{a(t; \mathbf{x})a(t; \mathbf{x} + \mathbf{h})}$ involves the infinitesimal volume $d\kappa_1\, d\kappa_2\, d\kappa_3$ and integration over all wave number space. However, the calculation of $\phi_{ij}(\kappa_1; \mathbf{x})$, the one-dimensional velocity spectrum tensor, is seen from Eq. (3.5.13) to involve the infinitesimal area $d\kappa_2\, d\kappa_3$ and integration over a surface located at κ_1, i.e., a slice through a related four-dimensional surface in wave number space. In this case integration is more conveniently carried out in a form of cylindrical coordinates, i.e., in terms of κ_1, $r_\kappa \cos\theta$ and $r_\kappa \sin\theta$, where $r_\kappa \equiv (\kappa^2 - \kappa_1^2)^{1/2}$ and $\tan\theta = \kappa_3/\kappa_2$. The equation for $\phi_{ij}(\kappa_1; \mathbf{x})$ then becomes

$$\phi_{ij}(\kappa_1; \mathbf{x}) = 2\int_{\kappa_1}^{\infty} d\kappa\kappa \int_0^{2\pi} d\theta\ \Phi_{ij}(\kappa_1, r_\kappa \cos\theta, r_\kappa \sin\theta; \mathbf{x})$$

The lower limit for the range of κ reflects the restriction that $\kappa \geq \kappa_1$, while the factor 2 arises from the contribution from both positive and negative values of κ_1.

Suppose that the second of Eqs. (3.5.14) is applied to the fluctuations of the streamwise velocity $u_1(t; \mathbf{x})$ so that $i = j = 1$ in a flow with a single mean velocity U_1. Assume that the flow is sufficiently close to homogeneous so that

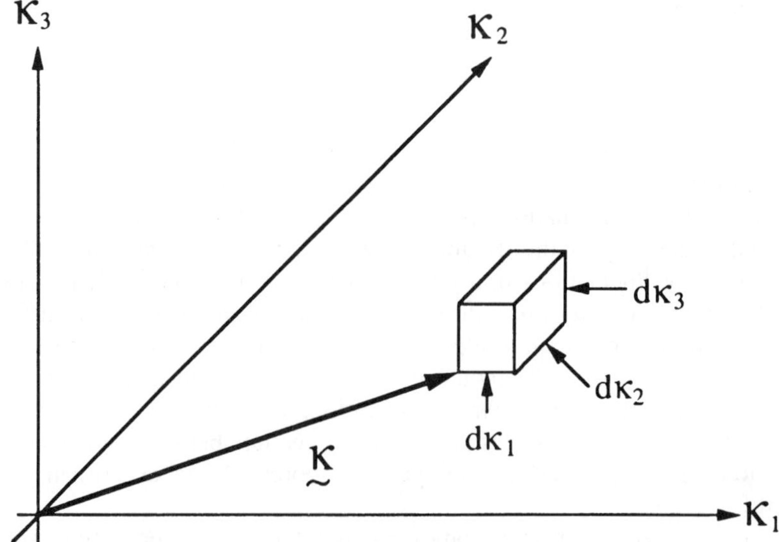

Figure 3.5.1 Wave number space in Cartesian coordinates.

the one-time, two-point correlation $\overline{u_1(t; \mathbf{x})u_1(t; \mathbf{x} + \mathbf{i}_1 h_1)}$ is an even function of h_1, that is, it is the same whether h_1 is positive or negative. In this case,

$$\phi_{11}(\kappa_1; \mathbf{x}) = \frac{2}{\pi} \int_0^\infty dh_1 \, \overline{u_1(t; \mathbf{x})u_1(t; \mathbf{x} + \mathbf{i}_1 h_1)} \cos \kappa_1 h_1 \qquad (3.5.15)$$

Now compare this with the psd given by Eq. (3.5.3) with the generic variable $a(t; \mathbf{x})$ replaced by $u_1(t; \mathbf{x})$; we have

$$\phi_{u_1}(\omega; \mathbf{x}) = \frac{2}{\pi} \int_0^\infty d\tau \, \overline{u_1(\tau; \mathbf{x})u_1(t + \tau; \mathbf{x})} \cos \omega\tau \qquad (3.5.16)$$

If we now invoke the Taylor hypothesis and approximate the one-point, two-time correlation in Eq. (3.5.16) with the two-point, one-time correlation $\overline{u_1(t; \mathbf{x}) \, u_1(t; \mathbf{x} + \mathbf{i}_1 h_1)}$ by interpreting τ as h_1/U_1, we see that the psd and the one-dimensional velocity spectrum tensor are identical. Note that $\omega \, h_1/U_1 = \kappa_1 h_1$.

Although the general three-dimensional Fourier transform pair of Eqs. (3.5.11) is clearly formidable, it becomes more tractable in reduced geometries. In Section 5.2 we need the energy spectrum function $E(\kappa, t)$, which arises in the description of decaying, isotropic, homogeneous turbulence. In this case the space coordinate $\mathbf{x}$ in the present discussion is replaced by the time at which stirring is terminated; the three-dimensional spectrum tensor with this replacement, $\Phi_{ij}(\boldsymbol{\kappa}; t)$, depends only on the magnitude of $\boldsymbol{\kappa}$, that is, on $|\boldsymbol{\kappa}|$, which is a scalar wave number, and not on its direction; and finally, the two-point, one-time velocity correlation depends not on the direction of the separation vector but only on the magnitude of $\mathbf{h}$, that is, on $|\mathbf{h}| \equiv r$, the radius connecting the two points. Thus $E(\kappa, t)$ determines the infinitesimal energy in fluctuations in the wave number range from κ to $\kappa + d\kappa$ and has the units of U^2L, energy per unit mass, per unit wave number $1/L$. In this case integration over all wave numbers and all separations in Cartesian coordinates can be replaced by a more convenient integration in spherical coordinates, where the vector $\boldsymbol{\kappa}$ has components $\kappa_1 = \kappa \sin \phi \cos \theta$, $\kappa_2 = \kappa \sin \phi \sin \theta$ and $\kappa_3 = \kappa \cos \phi$, where ϕ and θ are the spherical polar and azimuthal angles shown in Fig. 3.5.2.

To examine the three-dimensional Fourier transforms for these special circumstances, consider the measurement of velocity components by two anemometers separated r apart in grid turbulence. Here we assume that grid flow adequately approximates decaying, isotropic, and homogeneous turbulence and thus that the required velocity correlations can be measured in that flow as follows. Let r be measured in the x_2 direction and assume that at $\mathbf{x}_1$ and $\mathbf{x}_1 + \mathbf{i}_2 r$ are anemometers that each measure the fluctuations of the two velocity components u_1 and u_2.† With this arrangement the two correlations $\overline{u_1(t; \mathbf{x}_1) \, u_1(t; \mathbf{x}_1 + \mathbf{i}_2 r)}$

†Since the turbulence is assumed isotropic, we could equally as well take the separation along either the x_1 axis or the x_3 axis.

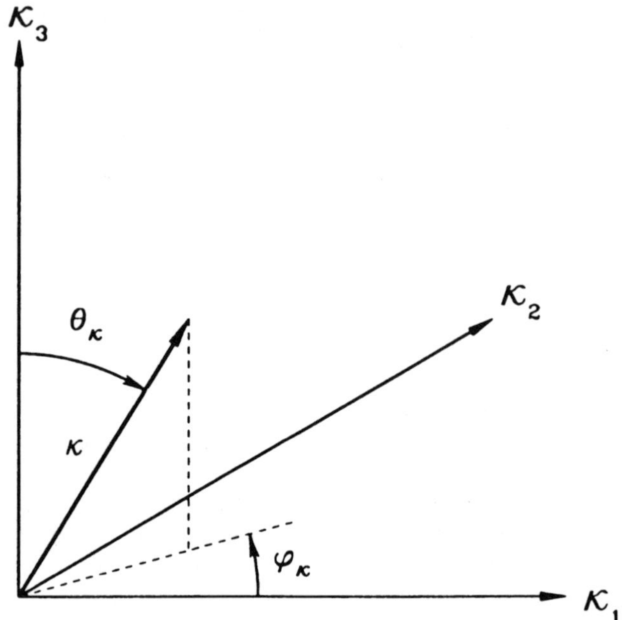

Figure 3.5.2 Wave number space in spherical coordinates.

and $\overline{u_2(t, \mathbf{x}_1)\, u_2(t, \mathbf{x}_1 + \mathbf{i}_2 r)}$ are measured as functions of r. These are the usual two-point, one-time averages. By repeating these measurements at various values of $\mathbf{x}_1$ and *interpreting* these correlations as functions of r and $t = x_1/U_1$, the time of passage from the grid, we obtain $\langle u_1(t)\, u_1(t,r)\rangle$ and $\langle u_2(t)\, u_2(t,r)\rangle$, the ensemble-averaged correlations in a box of *decaying turbulence*. This perspective is adopted repeatedly in Sections 5.2–5.4, which deals largely with ensemble averages. Because of isotropy, the second of these correlations equals $\langle u_3(t)\, u_3(t;\, r)\rangle$.

We convert Eqs. (3.5.12) with $i = j$ into a Fourier transform pair in spherical coordinates as follows. The lack of dependence on the directions of the vectors $\boldsymbol{\kappa}$ and $\mathbf{h}$ permits us to align the κ_3 component with the vector $\mathbf{h}$. As a consequence, $\boldsymbol{\kappa} \cdot \mathbf{h} = \kappa r \cos\theta$, where θ is the polar angle (cf. Fig. 3.5.2). Now consider the integration involved in the first of Eqs. (3.5.12); the infinitesimal volume becomes

$$d\kappa_1\, d\kappa_2\, d\kappa_3 = \kappa^2 \sin\theta\, d\theta\, d\phi\, d\kappa$$

with $d\kappa$ ranging from zero to infinity, $d\phi$ from 0 to π, and $d\theta$ from 0 to 2π (cf. Fig. 3.5.2). Integration with respect to θ yields

$$\int_0^\pi d\theta \sin\theta \exp(i\,\kappa\, r\cos\theta) = \frac{\exp(i\kappa r\cos\phi)}{i\,\kappa\, r}\Big|_0^\pi = -\frac{1}{i\,\kappa\, r}(e^{i\kappa r} - e^{-i\kappa r})$$

$$= 2\,\frac{\sin \kappa r}{\kappa r}$$

The integration with respect to ϕ poses no problem. Moreover, an identical calculation applies to the second of Eqs. (3.5.12), so we obtain the Fourier transform pair,

$$\langle u_i(t)\, u_i(t;\, r)\rangle = 4\,\pi \int_0^\infty d\kappa\, \kappa^2\, \Phi_{ii}(\kappa,\, t)\, \frac{\sin \kappa r}{\kappa r}$$

$$\Phi_{ii}(\kappa,\, t) = \frac{1}{2\pi^2}\int_0^\infty dr\, r^2\, \langle u_i(t)\, u_i(t;\, r)\rangle\, \frac{\sin \kappa r}{\kappa r} \tag{3.5.17}$$

Here both $\langle u_i(t)u_i(t;\, r)\rangle$ and $\Phi_{ii}(\kappa,\, t)$ are real. In contrast with our standard rule, no summation is intended here. Note that as $r \to 0$ we have

$$\langle u_i^2\rangle(t) = 4\pi \int_0^\infty d\kappa\, \kappa^2\, \Phi_{ii}(\kappa,\, t) \tag{3.5.18}$$

i.e., the intensity of the fluctuations of u_i is the integral of $\Phi_{ii}(\kappa,\, t)$ over a spherical shell in wave number space. This equation is the counterpart for ensemble-averaged velocity fluctuations in the isotropic, decaying turbulence of Eq. (3.5.13) which relates to the time average at a fixed spatial location in statistically stationary turbulence.

It will be useful in Section 5.2 to introduce two average quantities,

$$R(r,\, t) \equiv \frac{1}{2}\,\langle u_k(t)\, u_k(t,\, r)\rangle$$

$$E(\kappa,\, t) \equiv 4\pi\, \kappa^2\, \frac{1}{2}\, \Phi_{kk}(\kappa,\, t) \tag{3.5.19}$$

Here $E(\kappa,\, t)$, the energy spectrum function, is the sum of the three components Φ_{11}, Φ_{22}, and Φ_{33}. From Eq. (3.5.18),

$$\frac{1}{2}\,\langle u_k\, u_k\rangle(t) = \int_0^\infty d\kappa\, E(\kappa,\, t) \tag{3.5.20}$$

i.e., the integral of $E(\kappa,\, t)$ with respect to the scalar wave number κ yields the turbulent kinetic energy. The quantity $R(r,\, t)$, the scalar correlation function, is the sum of the correlations associated with each velocity component, each a function of r and t. Thus, if $R(r,\, t)$ is large in some sense, then one or more velocity pairs is correlated over a large distance, while if $R(r, t)$ is small in the same sense, then all velocity pairs are uncorrelated beyond some small distance.

Then, from Eqs. (3.5.17), we have the Fourier transform pair

$$\begin{aligned} R(r, t) &= \int_0^\infty d\kappa\, E(\kappa, t)\, \frac{\sin \kappa r}{\kappa r} \\ E(\kappa, t) &= \frac{2}{\pi} \int_0^\infty dr\, R(r, t)\, r\kappa \sin \kappa r \end{aligned} \qquad (3.5.21)$$

Note that the first of these equations and Eq. (3.5.20) lead to

$$R(0, t) = \int_0^\infty d\kappa\, E(\kappa, t) = \frac{1}{2} \langle u_k\, u_k \rangle(t) \qquad (3.5.22)$$

so that the scalar correlation coefficient at zero separation equals the turbulent kinetic energy.[†]

Because of the isotropy and homogeneity of the flow, the three-dimensional velocity spectrum tensor $\Phi_{ij}(\boldsymbol{\kappa}; t)$ of Eqs. (3.5.12) will obey certain transformation rules, with the consequence that

$$\Phi_{ij}(\kappa, t) = \frac{E(\kappa, t)}{4\pi\, \kappa^2} \left(\delta_{ij} - \frac{\kappa_i\, \kappa_j}{\kappa^2} \right) \qquad (3.5.23)$$

If we calculate $\Phi_{kk}(\kappa, t)$ from this equation, we readily find consistency with the second of Eqs. (3.5.19). Thus, if we know $E(\kappa, t)$, we can determine for the special case of isotropic homogeneous turbulence the components of the three-dimensional velocity spectrum tensor.

In closing this discussion of spectra, several remarks are appropriate. The wave number spectrum can be extended to involve the space–time correlation, e.g., to $a(t; \mathbf{x})\, a(t + \tau; \mathbf{x} + \mathbf{h})$, so as to become the wave number–frequency spectrum. In this case Eqs. (3.5.11) involve further integrations with respect to τ and ω. However, we need not pursue this extension. In terms of *elementary* applications of turbulence theory, those concerned with the determination of the mean velocity components and the mean temperature, the Fourier spectra of various fluid mechanical variables do not play a central role. However, as suggested earlier spectra are conceptually important in understanding the physics of turbulence, e.g., in examining the interaction of fluctuations of different scale and in guiding the modeling in more advanced turbulent theories. Moreover, spectra are important in experimental turbulence in several ways; they provide an efficient means for obtaining space and time correlation coefficients that identify the corresponding turbulence scales. They are also useful in exposing spurious contributions to sensor output from electronic equipment and in establish-

[†]We follow standard practice and denote the scalar correlation function as R but note that in Eqs. (3.4.1), (3.4.9), and (3.4.14) we use R to denote various correlation coefficients.

ing the frequencies beyond which noise dominates. Finally, we shall see in Sections 5.2–5.4 that the evolution of $E(\kappa, t)$ from an initial state plays a central role in the study of isotropic, homogeneous turbulence.

3.6 CONDITIONAL AVERAGING

A special set of the correlations discussed in Sections 3.3–3.5 plays an important role in turbulent flows. Suppose that a fluid mechanical variable $\tilde{a}(t; \mathbf{x})$ is correlated with, or weighted by, a second function denoted by $I(t; \mathbf{x})$ which is a random binary signal in that it possesses only two values, zero and unity. This second function can be thought of, and indeed is sometimes identified as, either a random telegraph signal or an indicator function. The mean value of I, that is, $\bar{I}(\mathbf{x})$, is the fraction of time that $I = 1$; the quantity $1 - \bar{I}(\mathbf{x})$ is the fraction of time that $I = 0$. When $\bar{I}(\mathbf{x}) = 1$, then $I \equiv 1$; and similarly, when $\bar{I}(\mathbf{x}) = 0$, then $I \equiv 0$. If we correlate these two functions, $\tilde{a}(t; \mathbf{x})$ and $I(t; \mathbf{x})$, we can identify separate mean values and separate values for all other statistical characteristics of the former variable depending on whether $I = 0$ or $I = 1$. The resultant *conditional mean values* are *zone averages* defined as follows:

$$
\begin{aligned}
A_1(\mathbf{x}) &= \lim_{T\to\infty} \frac{\int_0^T dt\, \tilde{a}(t; \mathbf{x})\, I(t; \mathbf{x})}{\int_0^T dt\, I(t; \mathbf{x})} \\
A_0(\mathbf{x}) &= \lim_{T\to\infty} \frac{\int_0^T dt\, \tilde{a}(t; \mathbf{x})[1 - I(t; \mathbf{x})]}{\int_0^T dt\, [1 - I(t; \mathbf{x})]}
\end{aligned}
\qquad (3.6.1)
$$

From Eqs. (3.6.1) we see that the zone averages are indeterminant if $I(t; \mathbf{x})$ approaches zero or unity. This situation arises at the outer edge of shear layers, in which case the indeterminacies are finite but with values which are difficult to obtain experimentally; in brief, both the numerator and the denominator approach zero and consequently both are subject to statistical uncertainty.

It is clear from Eqs. (3.6.1) that $I(t; \tilde{x})$ is a conditioning function that selects subsets of $\tilde{a}(t; \mathbf{x})$ depending on whether I is zero or unity. Its role in experimental turbulence is illustrated as follows. Suppose that, in a heated turbulent wake as shown in Fig. 3.6.1, we mount two sensors at the same spatial location, one measuring the x_1 velocity component, $\tilde{u}_1(t; \mathbf{x})$, and a second $\tilde{\theta}(t; \mathbf{x})$, the increment in temperature above that in the external stream. Because the cylinder is responsible for both the turbulence and the elevated temperature, we identify turbulent fluid with $I(t; \mathbf{x}) = 1$ and temperatures $\tilde{\theta}(t; \mathbf{x}) > \tilde{\theta}_t$, where $\tilde{\theta}_t$ denotes a fixed threshold level determined by the accuracy of the thermometry system.

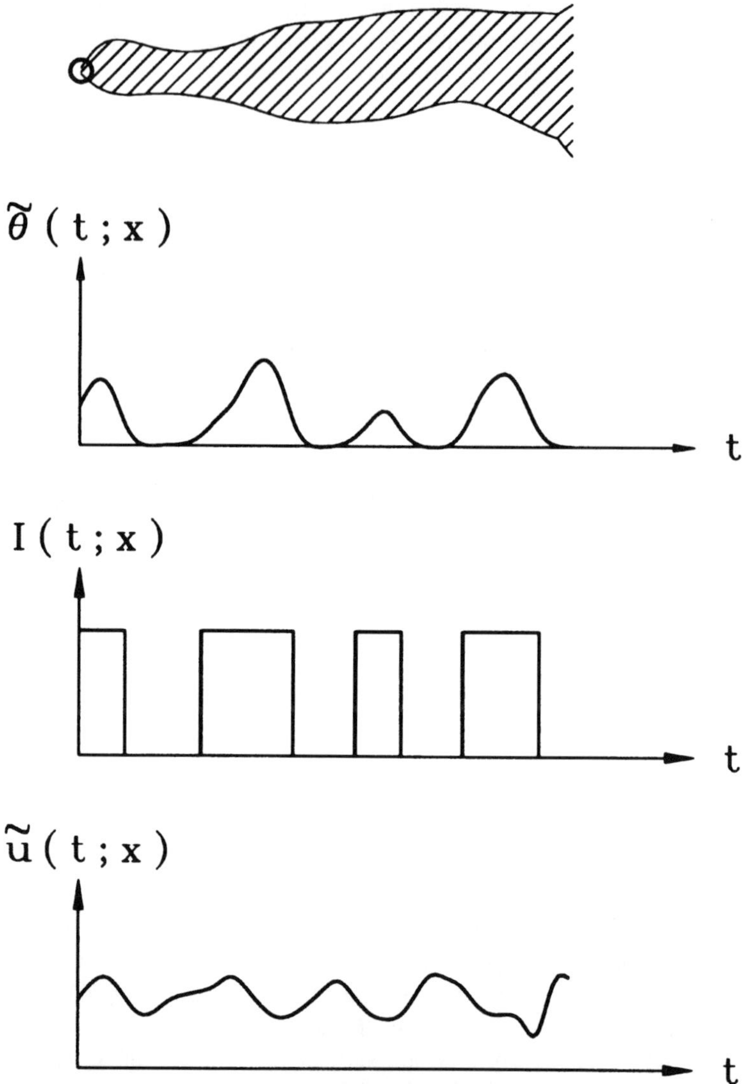

Figure 3.6.1 The wake of a heated cylinder, the signals from an anemometer and thermometer, and the conditioning function.

Similarly, we identify external fluid with $I(t; \mathbf{x}) = 0$ and $0 < \tilde{\theta} < \tilde{\theta}_t$. In this case application of Eqs. (3.6.1) with $\tilde{a}$ replaced by $\tilde{u}_1$ determines the zone averages of the $\tilde{u}_1$ velocity component within the turbulent and external fluids. Knowledge of differences in the two averages provides insight into the mechanics of turbulence in shear flows; for example, in wake flows the zone-averaged streamwise velocity within turbulent fluid is smaller than the external velocity, while the reverse is true for the corresponding velocity within irrotational fluid (cf. LaRue and Libby, 1974).

It should be noted that the variable $\tilde{\theta}(t;\, \mathbf{x})$ corresponds to the variable $\tilde{a}(t;\, \mathbf{x})$ of Section 3.3 and Fig. 3.3.4 although to reflect possible experimental error we have introduced a threshold level $\tilde{\theta}_t$. Had we idealized the situation and set $\tilde{\theta}_t = 0$, the pdf of temperature would involve both a continuous distribution for $\tilde{\theta}(t;\, \mathbf{x}) > 0$ and a delta function at $\tilde{\theta} = 0$ whose strength is $1 - \bar{I}(\mathbf{x})$.

It is possible to impose conditioning on a wide variety of variables. For example, it is sometimes useful to let the output of a sensor at one spatial location and one time determine $I(t;\, \mathbf{x})$ and to correlate the resultant conditioning function with the output of a sensor at another spatial location at another time. In this way the influence of an event at the first location, identified with the conditioning function, on a fluid mechanical variable at the second location can be established (cf. Blackwelder, 1978).

Although we have focused on zone averages, an extension of the idea of conditional averaging applies to fluctuations of variables, for example, to $a(t;\, \mathbf{x})$, to correlations between two variables, etc. Indeed, it is possible and under some circumstances useful to identify fluctuations relative to zone averages and to determine the statistics and correlations of those fluctuations.[†]

Elaboration of the conditioning criteria is also possible. Suppose that in the turbulent wake experiment discussed earlier, the *times* at which $I(t;\, \mathbf{x})$ changes from zero to unity are employed to construct a sequence of delta functions $\delta(t - t_i)$ for $i = 1, 2, 3, \ldots, N$. The conditional mean of the $\tilde{u}_1$ velocity component at the spatial location $\mathbf{x}$ is given by

$$\begin{aligned} \langle \tilde{u}_1 \rangle_{0-1}(\mathbf{x}) &= \lim_{N\to\infty} \frac{1}{N} \sum_{n=1}^{N} \int_{t_n^-}^{t_n^+} dt\, \tilde{u}_1(t;\, \mathbf{x})\, \delta(t - t_n) \\ &= \lim_{N\to\infty} \frac{1}{N} \sum_{n=1}^{N} \tilde{u}_1(t_n;\, \mathbf{x}) \end{aligned} \qquad (3.6.2)$$

This is the mean value of the u_1 velocity component when a front separating nonturbulent and turbulent fluid passes the point $\mathbf{x}$, that is, the *upstream* edge of a region of turbulent fluid. Similar considerations apply to the times when $I(t;\, \mathbf{x})$ changes from unity to zero, i.e., when the corresponding *downstream* edge of a turbulent region passes. In this case we determine $\langle \tilde{u}_1 \rangle_{1-0}(\mathbf{x})$. The two mean values of the velocities at the two sets of crossings need not be the same; indeed, information concerning their differences provides insight into the mechanisms involved in the growth of the turbulent layer (cf. LaRue and Libby, 1974).

[†]This is an especially powerful tool in premixed turbulent combustion, in which the two zones correspond to reactants and products with significantly different velocities and thermochemical states (cf. Bray and Libby, 1994).

As another example of conditional averaging, it is sometimes of interest to study the response of a turbulent flow to external periodic forcing (cf. Ho and Huang, 1982) and to determine the response of the flow at a particular spatial location at a *particular phase angle* of the forcing. This corresponds to phase averaging, clearly a form of conditioning.

The techniques of conditional sampling were developed for, and continue to be widely used in, experimental turbulence (cf. Kovasznay et al., 1970; van Atta, 1974; Smith and Abbott, 1978). However, the notions involved are now commonly applied in direct numerical simulation; for example, in studying the spatial evolution of a laminar shear layer, the flow at the splitter plate is forced at one or more frequencies and the phase averaging discussed earlier is applied to the computed velocities to clarify the spatial evolution of the imposed disturbances. A theory of turbulent shear flows corresponding to conditional sampling and distinguishing between turbulent and external fluid is less developed, although a start is provided by Libby (1976).

3.7 LAGRANGIAN STATISTICS

Throughout the previous discussion in this chapter we considered the statistical description of variables at a fixed spatial location, i.e., we examined $a(t; \mathbf{x})$, $a(t; \mathbf{x} + \mathbf{h})$, etc. There result Eulerian statistics which relate to the technique generally employed in experimental turbulence research, namely, to the measurement of quantities with an instrument fixed in the flow, and to the view generally taken in developing the averaged transport and conservation equations of fluid flow. However, some turbulence problems call for an altered perspective. Suppose, for example, that we are interested for a turbulent flow either in the dispersion of particles from a fixed source or in the burning of fuel particles from such a source injected into a hot oxidizing ambient. In this case we might want to know the probability of finding a particle at a certain location downstream of the source, or the average diameter of fuel particles a specified time following release. To determine these quantities, we are not interested in the flow at a fixed location but rather in the velocity and state variables, e.g., the temperature and composition, *along the trajectory of individual particles.* Formally, we seek to determine the statistical behaviour of an arbitrary flow variable $\tilde{a}[t, \mathbf{x}(t)]$ with $0 < t < \hat{t}$ such that

$$\frac{d^2 x_i}{dt^2} = F_i[\mathbf{u}(\mathbf{x}, t), \ldots] \qquad i = 1, 2, 3 \tag{3.7.1}$$

where F_i denotes the force per unit mass acting on the particle in the ith coordinate direction and where $\hat{t}$ is the time after release from the source. To develop a statistical description of $\tilde{a}(\hat{t})$, ensemble averaging as suggested by Eq. (3.1.3) is clearly called for, with each realization corresponding to a different trajectory. Such averaging defines *Lagrangian statistics.*

With appropriate modification, the Eulerian time autocorrrelation given by Eq. (3.4.1) applies to Lagrangian statistics, but the spatial coordinate appearing there requires modification. To appreciate this, consider the analogous Lagrangian correlation coefficient,

$$R_{\mathrm{La}}(\tau;\, \hat{t}) = \frac{\langle \tilde{a}(\hat{t})\; \tilde{a}(\hat{t} + \tau)\rangle}{\langle \tilde{a}^2(\hat{t})\rangle} \tag{3.7.2}$$

where the variable $\hat{t}$ was defined earlier, $\tilde{a}$ denotes a property associated with the particle, e.g., its temperature, velocity, or lateral displacement from a mean path, and τ is a time increment beyond $\hat{t}$. A Lagrangian integral time scale follows from the analog of Eq. (3.4.2), i.e.,

$$T_{\mathrm{ILa}}(\hat{t}) = \int_0^\infty d\tau\; R_{\mathrm{La}}(\tau;\, \hat{t}) \tag{3.7.3}$$

Similarly, a short time scale follows from Eq. (3.4.5),

$$T_{\mathrm{sLa}}(\hat{t}) = \left[\frac{2}{-R''_{\mathrm{La}}(0;\, \hat{t})}\right]^{1/2} \tag{3.7.4}$$

Although the difference between Eulerian and Lagrangian statistics is implicit in these definitions, the difficulties of obtaining the latter experimentally has led to considerable efforts devoted to relating the two. A crude but illuminating approximation applies to a turbulent flow with a single mean velocity U_1. In this case we can invoke the Taylor hypothesis discussed in Section 3.5 to obtain the two-point, two-time Eulerian correlation according to

$$R_a(h_1 = U_1\,\tau,\, \tau;\, \mathbf{x}) = \frac{\overline{\tilde{a}(t;\, \mathbf{x})\; \tilde{a}(t + \tau;\, \mathbf{x} + \mathbf{i}_1\, U_1\, \tau)}}{\overline{\tilde{a}^2(t;\, \mathbf{x})}} \tag{3.7.5}$$

To determine this quantity we must select τ and record measurements of $\tilde{a}$ at the two spatial locations separated by $U_1\ \tau$ in the x_1 direction. We can then construct the correlation in the numerator of Eq. (3.7.5); if this procedure is followed for a suitable range of τ, there results an approximation to the Lagrangian correlation $R_{\mathrm{La}}(\tau;\, \hat{t})$ of Eq. (3.7.2), provided we identify $\hat{t}$ with $\mathbf{x}$. The principal shortcoming of this approximation relates to its dependence on the Taylor hypothesis and to its neglect of particle inertia if the particles are not neutrally buoyant.

In one of the seminal contributions to turbulence theory, Taylor (1921) provided an analysis of particle dispersion under idealized conditions. Following his analysis, we consider a grid flow with a single mean velocity U_1 and a suitably short distance downstream such that the decay of turbulence may be neglected; i.e., we consider a statistically homogeneous, isotropic flow. A further idealization is based on the particles being small and neutrally buoyant so that their velocity is identical with the fluid velocity at each location in space and time. Under these circumstances the radial position of a particle $r \equiv (x_2^2 +$

$x_3^2)^{1/2}$ at a particular time $\hat{t}$ after its release from the source is an obvious example of a random variable $\tilde{a}$ discussed earlier. At a specified $\hat{t}$ the location of a particle in a single realization can be identified with the two coordinates, x_1 and r. The mean value of the former is $U_1\ \hat{t}$, which in the context of particle dispersion is not particularly interesting. Clearly, the mean value of r is zero, but its variance is a measure of the extent of dispersion from a fixed source.

If a particle has the same velocity as the surrounding fluid in each realization, an assumption consistent with the assumption of suitably small particles, its radial position is given by the equation

$$\frac{dr_n}{dt} = v_n(t) \tag{3.7.6}$$

and its radial position at time $\hat{t}$ is given by

$$r_n(\hat{t}) = \int_0^{\hat{t}} dt\ v(t) \tag{3.7.7}$$

Thus application of ensemble averaging to many such particles yields

$$\left\langle r_n \frac{dr_n}{d\hat{t}} \right\rangle = \frac{1}{2} \frac{d}{d\hat{t}} \langle r_n^2 \rangle = \left\langle v_n(\hat{t}) \int_0^{\hat{t}} dt\ v_n(t) \right\rangle \tag{3.7.8}$$

where the average on the extreme right side involves the product of an instantaneous velocity $v_n(\hat{t})$ and an *average velocity* $\int_0^{\hat{t}} dt\ v_n(t)$. Under the assumption of stationarity and homogeneity the radial velocity component is statistically uniform in space and time, and it is convenient to specialize Eq. (3.7.2) to the Lagrangian velocity correlation coefficient,

$$R_L(\tau \equiv \hat{t} - t) = \frac{\langle v_n(\hat{t})\ v_n(t) \rangle}{\overline{v^2}} \tag{3.7.9}$$

The dependence of R_L solely on the difference $\hat{t} - t$ and not on $\hat{t}$ and t individually reflects the idealization of nondecaying turbulence. With the introduction of Eq. (3.7.9), we can integrate Eq. (3.7.8) to obtain

$$\overline{r^2}(\hat{t}) = 2\ \overline{v^2} \int_0^{\hat{t}} dt \int_0^{t} d\tau\ R_L(\tau) = 2\ \overline{v^2} \int_0^{\hat{t}} (\hat{t} - \tau)\ R_L(\tau)\ d\tau \tag{3.7.10}$$

Consider the implications of Eq. (3.7.10). Since $R_L(\tau \to 0) \to 1$, we see that initially, i.e., for small $\hat{t}$,

$$\overline{r^2}(\hat{t}) \approx \overline{v^2}\ \hat{t}^2 \tag{3.7.11}$$

while for large $\hat{t}$

$$\overline{r^2}(\hat{t}) = 2\,\overline{v^2}\,T_L\,\hat{t} \tag{3.7.12}$$

where

$$T_L \equiv \lim_{t\to\infty} \int_0^t d\tau\; R_L(\tau)$$

defines a Lagrangian integral time. For values of τ that are significantly greater than T_L, the radial velocities at $\hat{t}$ and t are independent.

Equations (3.7.11) and (3.7.12) show that particles initially disperse from a point source at a parabolic rate, but for $\hat{t} >> T_L$ they disperse linearly with time. This long time behavior is equivalent to a random-walk process with uncorrelated random steps. Brownian diffusion in a static fluid is another example of such a process. Note that in Eq. (3.7.12) the product $\overline{v^2}\,T_L$ has the dimensions of UL, that is, a diffusion coefficient, and may be considered as the diffusivity of our idealized particles in a turbulent flow.

It is useful to consider the implications of this analysis in terms of the probability density function $P(r;\,\hat{t})$ for the particle position r. A function closely resembling this can be obtained experimentally as follows. Suppose that a continous source of idealized particles is located in a stream of grid turbulence and that we measure the rate at which these particles pass various radial positions in planes displaced downstream of the source. By converting the rate measured at each radial position into the fraction of emitted particles passing through an annulus with that radius, such measurements yield a pdf of radial position $P(r;\,x_1)$. The analysis given here implies that the most probable location of the particles is on the axis $r = 0$; that the variance of the pdf for small downstream separations increases quadratically with x_1; and finally, that for greater separations the variance increases linearly with x_1.

Although an idealized experiment of this sort represents an example of Lagrangian statistics, some reflection establishes that the experimental determination of the statistics of an arbitrary variable $\tilde{a}[t,\,\mathbf{x}(t)]$ in a more complicated situation, e.g., when particle inertia is appreciable, is a formidable task (cf. Snyder and Lumley, 1971). Recent large-scale computations (cf. Pope 1988; Elgobashi & Truesdell, 1992), in which the velocity field in space and time is obtained from DNS, yield Lagrangian statistics in flows of low turbulence Reynolds numbers.

3.8 VOLUME AVERAGING

A different perspective on averaging is called for in large eddy simulation (cf. Section 2.7 and Reynolds, 1990). The underlying notion is that large-scale turbulent fluctuations are specific to the flow under consideration, while the small scales exhibit universal characteristics. These ideas regarding the various scales of turbulence are developed in more detail in Section 5.3, but here we introduce the concept of volume averaging, the operation that allows us to exploit this point of view. Volume averaging corresponds to a filtering procedure implying that fluctuations with scales greater than a length defined by the filter are resolved computationally, while the influence of those with scales less than that length is described in an average sense.

To proceed, consider the averaging operation defined by

$$A(\mathbf{x}, t) = \int d\mathbf{x}' \, \tilde{a}_i(\mathbf{x}', t) \, G(\mathbf{x} - \mathbf{x}') \tag{3.8.1}$$

where $G(\mathbf{x} - \mathbf{x}')$ is a specified filter function. To be general and in keeping with LES averaging, we consider the mean, i.e., the left side of Eq. (3.8.1), to be space and time dependent. Although various functions are used, the most common are the top-hat filter defined by

$$\begin{aligned} G(\mathbf{x} - \mathbf{x}') &= \frac{3}{4\pi\Delta^3} \qquad |\mathbf{x} - \mathbf{x}'| < \Delta \\ &= 0 \qquad\qquad |\mathbf{x} - \mathbf{x}'| > \Delta \end{aligned} \tag{3.8.2}$$

and the Gaussian filter defined by

$$G(\mathbf{x} - \mathbf{x}') = \left(\frac{\gamma}{\pi\Delta}\right)^{3/2} \exp\left[-\gamma \frac{(|\mathbf{x} - \mathbf{x}'|)^2}{\Delta^2}\right] \tag{3.8.3}$$

Here Δ determines the volume over which the average of the generic variable $\tilde{a}(\mathbf{x}, t)$ is taken. If $\mathbf{x}$ is considered to be the location of a grid point in a calculation, the integration with respect to $\mathbf{x}'$ leads to the average of $\tilde{a}$ in the neighborhood of that point. The multiplicative factor on the right side of Eqs. (3.8.2) and (3.8.3) is chosen so that the average of a constant preserves the constant.[†] The parameter γ in Eq. (3.8.3) is arbitrary, but an assigned value of 6 is sufficient to ensure that Δ is representative of the smallest resolved scales (cf. Reynolds, 1990). Fluctuations are again determined by the difference between the value $\tilde{a}(\mathbf{x}, t)$ and $A(\mathbf{x}, t)$, but their calculus differs from that discussed earlier, e.g., in Eqs. (3.1.4). In particular, the repeated volume average of an averaged quantity is not equal to the initial average, and the volume average of a fluctuation is not zero; these characteristics arise because filtering corresponds to a smoothing operation whose repeated application results in yet further smoothing. Thus, when applied to the conservation and transport equations of fluid motion, volume averaging results in special space- and time-dependent correlations of fluctuations, the Leonard stresses (Leonard, 1974), and while the resulting equations resemble those developed in Chapter 4, their interpretation, closure, and utilization are unique to LES. Although the filters indicated by Eqs. (3.8.2) and

[†]To show this, replace the integration variable $d\mathbf{x}'$ with $4\pi r^2 \, dr$ and integrate over the sphere of radius Δ in the case of the top-hat filter and from zero to infinity in the case of the Gaussian filter.

(3.8.3) are isotropic, different widths in the three coordinate directions can be chosen so that, for example, Δ_1 in the streamwise direction is greater than Δ_2 and Δ_3 in the cross-stream directions.

3.9 SUMMARY

The various procedures used to obtain statistical information regarding the velocity, temperature, and indeed any fluid mechanical variable in a turbulent flow have been described in this chapter. The most useful for developments in Chapter 4 are those concerned with one-point, one-time correlations of velocity components, those obtained from a hot-wire anemometer with one or more sensors, and of a velocity component and temperature. More elaborate averaging procedures involving one-point, two-time correlations and their associated spectra are needed in Chapter 5 to describe the length and time scales of turbulence. Modifications of the more common averaging methods lead to conditional and Lagrangian statistics which have special applications. For example, the latter is relevant to the determination of the dispersion of particles in turbulent flows. Finally, we discussed volume averaging, which finds application in LES.

CHAPTER
FOUR

AVERAGED CONSERVATION AND TRANSPORT EQUATIONS

We now decompose the flow variables, velocity components, pressure, and temperature, as they appear in the conservation and transport equations, Eqs. (2.1.1), (2.2.1), and (2.3.1), into mean and fluctuating quantities. Although we shall apply the time averaging of Eq. (3.1.1) in order to facilitate the exposition, we could equally apply the ensemble averaging of Eq. (3.1.2). We consider only statistically stationary flows so that we have, for example, $\tilde{u}_i(\mathbf{x}, t) = U_i(\mathbf{x}) + u_i(\mathbf{x}, t)$, $\tilde{p}(\mathbf{x}, t) = P(\mathbf{x}) + p(\mathbf{x}, t)$, and $\tilde{\theta}(\mathbf{x}, t) = \Theta(\mathbf{x}) + \theta(\mathbf{x}, t)$.

4.1 CONSERVATION OF MASS

Averaging Eq. (2.1.1) yields a kinematic condition on the mean velocity components,

$$\frac{\partial U_k}{\partial x_k} = 0 \qquad (4.1.1)$$

An important subsidiary result is obtained by subtracting Eq. (4.1.1) from the originating equation, Eq. (2.1.1), to find a similar restraint on the instantaneous fluctuating velocity components,

$$\frac{\partial u_k}{\partial x_k} = 0 \tag{4.1.2}$$

Application of this equation will simplify many later equations.

The simple and direct results given by Eqs. (4.1.1) and (4.1.2) are a consequence of the linearity of Eq. (2.1.1). In the next section we shall see the complications which result from the averaging of nonlinear terms.

4.2 TRANSPORT OF MOMENTUM AND REYNOLDS STRESSES

Averaging the transport equations of Eq. (2.2.1) leads to

$$\rho \frac{\partial}{\partial x_k} (U_k\, U_i) = -\frac{\partial}{\partial x_i} (P + \rho \overline{u_i^2}) + F_i + \frac{\partial}{\partial x_k} [T_{ik} - \rho \overline{u_i u_k}(1 - \delta_{ik})] \tag{4.2.1}$$

where F_i and T_{ij} are the mean values of the ith component of the body force and of the shear stress, respectively. More specifically,

$$T_{ij} = \mu \left(\frac{\partial U_i}{\partial x_j} + \frac{\partial U_j}{\partial x_i} \right) \tag{4.2.2}$$

is the mean viscous stress due to the mean rate of strain. Combined with these mean stresses due to viscosity and with the mean pressure gradient are terms of the form $\rho \overline{u_i u_j}$. These are the Reynolds stresses; in a general three-dimensional flow there are six such stresses, three normal stresses $\rho \overline{u_i^2}$, $i = 1, 2, 3$, and three shear stresses, $\rho \overline{u_i u_j}$, $i \neq j$.

The terms on the left side of Eq. (4.2.1) involve the product of the density and the mean convective derivative,

$$U_k \frac{\partial}{\partial x_k}$$

which appears repeatedly in our considerations. Accordingly, the terms on the right side describe processes which alter the variable being convected. For example, in Eq. (4.2.1) the change in the mean value of the $\tilde{u}_i$ velocity component of a fluid element moving along the mean streamline is altered by gradients of the mean pressure P, of the mean molecular stresses T_{ij}, and of the Reynolds stresses $\rho \overline{u_i u_j}$.

The Reynolds stresses are written in Eq. (4.2.1) on the right side of the equation to reflect their contribution to the forces acting on a fluid element, but they arise from the nonlinearity of the convection terms on the left side. Since these stresses are new unknowns, we see that averaging applied to the equations of conservation of momentum leads to a significant complication. Note that the

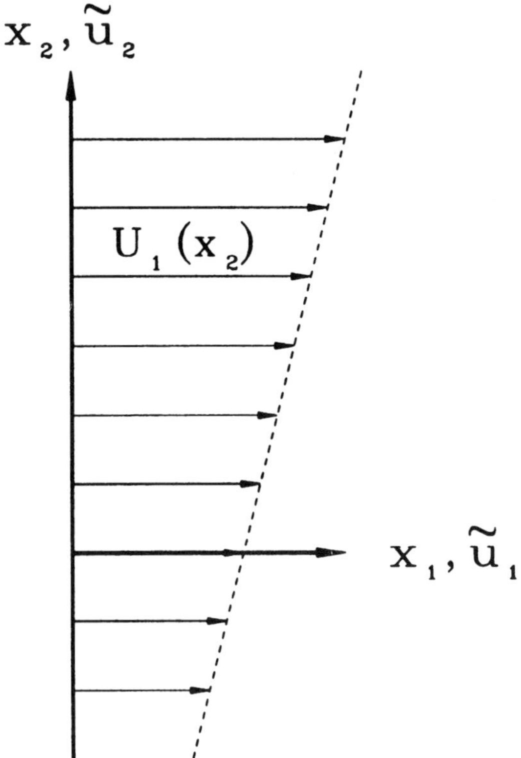

Figure 4.2.1 A simple shear flow.

Reynolds stresses involve examples of the one-point, one-time correlations discussed in Section 3.2.

The physical significance of the Reynolds shear stresses can be understood in terms of a simple shear flow as shown in Fig. 4.2.1, that is, a flow with only one mean velocity component, U_1, considered a function of the x_2 coordinate alone, and with $\partial/\partial x_1 \equiv \partial/\partial x_3 \equiv U_2 \equiv U_3 \equiv 0$.† This flow is well represented by the central region between two plates, one fixed and one moving as in turbulent Couette flow (cf. Sections 6.3 and 6.4). We also neglect body forces, so that Eqs. (4.2.1) specialized to this situation yield after integration

$$\mu \frac{dU_1}{dx_2} - \rho \overline{u_1 u_2} = \text{constant} \tag{4.2.3}$$

Since the first term on the left side of this equation is the shear stress due to

†We shall use simple shear flow repeatedly to illustrate in a direct fashion fundamental concepts of fluid mechanics in general and of turbulence in particular.

the shear rate of strain, the implication is that the term $\rho\overline{u_1u_2}$ is effectively a shear stress.

Further insight into the nature of the Reynolds shear stresses is provided if we consider a specific plane identified with the coordinate $x_2 = \hat{x}_2$ as indicated in Fig. 4.2.1. Suppose that at a particular instant of time the fluctuation of the x_2 velocity component u_2 is positive. At a somewhat earlier time the fluid associated with this fluctuation must have been at an x_2 location less than $\hat{x}_2$ and is thus likely to have had a velocity component in the x_1 direction less than U_1 $(\hat{x}_2)$. According to this reasoning, the positive fluctuation in u_2 is likely to be accompanied by a negative fluctuation in u_1 and thus the contribution to the correlation $\overline{u_1u_2}$ from this event is likely to be negative. Of course, the averaging indicated by the overbar implies that many such events are taken into account and not all events need yield negative contributions in order for the mean value of the product to be negative. These considerations imply that the $\overline{u_1u_2}$ correlation relates to the bivariate pdf of u_1 and u_2 as shown schematically in Fig. 3.3.3.

Similar arguments can be made for a negative fluctuation in the u_2 velocity and for simple shear flows with $dU_1/dx_2 < 0$. The implication of these considerations is that the Reynolds shear stresses add to the molecular shear stress; that is, with $dU_1/dx_2 > 0$, then $\rho\overline{u_1u_2} < 0$. These qualitative arguments must be generalized to more complex flows and be made quantitative before Eq. (4.2.1) is a useful statement of mean transport of momentum. Although we have put forth these arguments to support the interpretation of the correlations $\overline{u_iu_j}$ multiplied by ρ as stresses, these considerations underlie the gradient transport theories of turbulent exchange discussed in Chapter 7, in particular in the mixing length theory in Section 7.4.

A further appreciation of the stresslike nature of the terms $\rho\overline{u_iu_j}$ is gained by examining the influence of a rotation of the coordinate system. In Fig. 4.2.2 we consider a plane that is normal to the x_3 axis with two sets of coordinates, x_1, x_2 and $\hat{x}_1$, $\hat{x}_2$, one set rotated with respect to the other by an angle ϕ but with a common origin. Associated with these coordinates are the fluctuating velocity components u_1, u_2 and $\hat{u}_1$, $\hat{u}_2$, respectively. At any instant of time the two sets of velocity components are related according to the equations

$$\begin{aligned} u_1 &= \hat{u}_1 \cos\phi - \hat{u}_2 \sin\phi \\ u_2 &= \hat{u}_1 \sin\phi + \hat{u}_2 \cos\phi \end{aligned} \tag{4.2.4}$$

so that

$$\rho\overline{u_1u_2} = \frac{1}{2}\rho(\overline{\hat{u}_1^2} - \overline{\hat{u}_2^2}) \sin 2\phi + \rho\overline{\hat{u}_1\hat{u}_2} \cos 2\phi \tag{4.2.5}$$

Thus we see that a shear stress in one coordinate system corresponds to a combination of normal and shear stresses in another system. We also see that if

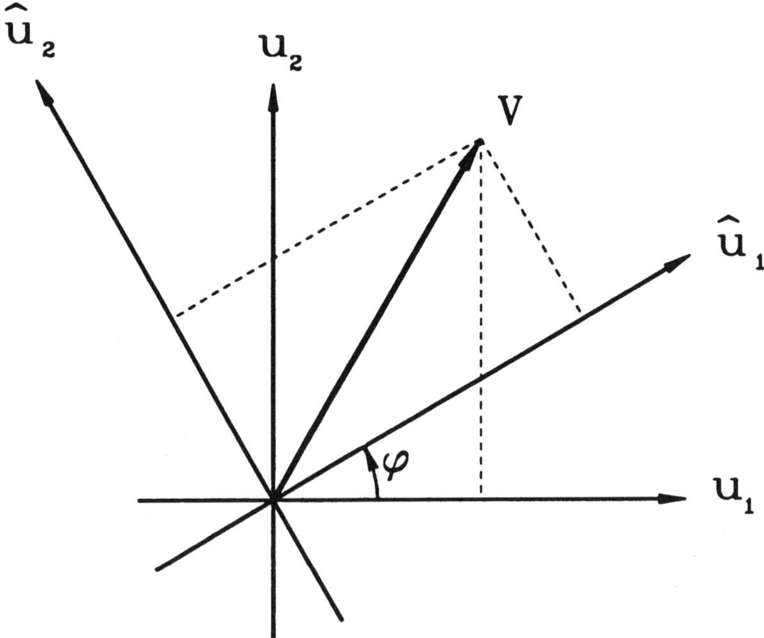

Figure 4.2.2 The velocity components in two coordinate systems at an angle ϕ.

$\phi = \pi/4$, the stress $\rho\overline{u_1u_2} = \frac{1}{2}\rho(\overline{\hat{u}_1^2} - \overline{\hat{u}_2^2})$, that is, one-half the difference in the two normal stresses in the rotated coordinates. In an isotropic flow the normal stresses are all equal and *independent* of the coordinate system; if that is the case, then $\overline{u_1u_2} = \overline{\hat{u}_1\hat{u}_2} = 0$ and from Eq. (4.2.5) we infer that $\overline{u_iu_j} = \delta_{ij}\overline{u_i^2}$, that is, the Reynolds shear stresses are zero in all coordinate systems. On the contrary, if the normal stresses are unequal in one coordinate system, there are certainly Reynolds shear stresses in some other coordinate systems. Note also that if $\overline{\hat{u}_1^2}$, $\overline{\hat{u}_2^2}$, and $\overline{\hat{u}_1\hat{u}_2}$ are known, Eq. (4.2.5) can be used to define two principal directions, i.e., two orthogonal planes on which the shear stress is zero. These directions are given by

$$\tan 2\phi_p = 2\frac{\overline{\hat{u}_1\hat{u}_2}}{\overline{\hat{u}_1^2} - \overline{\hat{u}_2^2}} \tag{4.2.6}$$

It is sometimes convenient to orient the coordinates in the analysis of turbulent flows with the principal directions, but in a general flow field such directions rotate and the need to incorporate that rotation in the tranport equations vitiates the advantages of dealing with principal directions. These considerations reinforce the identification of the product of density and the various velocity correlations, that is, $\rho\overline{u_iu_j}$, as *stresses.*

4.3 TRANSPORT OF THERMAL ENERGY AND REYNOLDS FLUXES

New unknown quantities also arise when the thermal energy equation, Eq. (2.3.1), is averaged. We have

$$\rho \frac{\partial}{\partial x_k}(U_k\Theta) = \frac{\partial}{\partial x_k}\left(\frac{\mu}{N_\sigma}\frac{\partial \Theta}{\partial x_k} - \rho\overline{u_k\theta}\right) \tag{4.3.1}$$

Here we see that the heat flux due to thermal conductivity is supplemented by a Reynolds flux $\rho\overline{u_i\theta}$. The physical significance and sign of these fluxes can be appreciated by arguments closely following those used in the previous section if we add to the simple shear flow of Fig. 4.2.1 a constant mean temperature gradient $d\Theta/dx_2$. Thus, if $d\Theta/dx_2 > 0$ and if at a reference plane at a particular instant of time $u_2 > 0$, then it is likely that $\theta < 0$ and thus that on average $\overline{u_2\theta} < 0$. The conclusion from these considerations is that the sign of the Reynolds flux is opposite to that of the mean temperature gradient and thus that the Reynolds flux adds to the flux due to the product of thermal conductivity and temperature gradient. Again, this line of reasoning is the basis of gradient transport theory of thermal energy exchange in turbulent flows as discussed in Section 7.2.

The appearance of the Reynolds flux terms in Eq. (4.3.1) implies that averaging the thermal energy equation or its various analogs, such as the transport equation for species, also results in a closure problem. Thus, even if the mean velocity distribution $U_i(\mathbf{x})$ were somehow known, we would still have to deal with the mean fluxes $\rho\overline{u_i\theta}$ in the scalar transport problem.

4.4 SECOND-MOMENT EQUATIONS FOR VELOCITY

Further insight into the closure problem and the mechanics of turbulence is achieved if we perform additional averages of the conservation and transport equations with appropriate weighting factors (cf. Section 1.3). To be specific, consider Eq. (2.2.1) for the ith velocity component and again for the jth component. Now cross-multiply these two equations by u_j and u_i, respectively, add, and average. This procedure constitutes taking moments of the momentum equations with weighting factors equal to the fluctuations of the various velocity components. Thus, while the original averaging leads to *first-moment equations,* the present procedure leads to the *second-moment equations* (cf. Section 3.3). Clearly, arbitrary higher moments can be introduced, although third- and higher-moment equations are infrequently considered in theoretical turbulence studies. The second-moment equation for the Reynolds stress $\rho\overline{u_iu_j}$ is

$$\rho \frac{\partial}{\partial x_k}(U_k\overline{u_iu_j}) = -\left(\rho\overline{u_ju_k}\frac{\partial U_i}{\partial x_k} + \rho\overline{u_iu_k}\frac{\partial U_j}{\partial x_k}\right) - \rho\frac{\partial}{\partial x_k}\overline{u_iu_ju_k}$$
$$-\left(\frac{\partial}{\partial x_i}\overline{pu_j} + \frac{\partial}{\partial x_j}\overline{pu_i}\right) + \overline{p\left(\frac{\partial u_i}{\partial x_j} + \frac{\partial u_j}{\partial x_i}\right)}$$
$$+ \overline{f_iu_j} + \overline{f_ju_i} + \left(\overline{u_j\frac{\partial \tau_{ik}}{\partial x_k}} + \overline{u_i\frac{\partial \tau_{jk}}{\partial x_k}}\right) \tag{4.4.1}$$

This equation calls for several comments. It is important to recall that the terms on the left side of this equation involve the convective operator applied to the Reynolds stress $\rho\overline{u_iu_j}$ and all the terms on the right side relate to processes which alter that stress as a fluid element follows the mean streamline. The first two sets of terms on the right side describe the influence of the interaction between fluctuating quantities appearing in the second-moment correlations and gradients of the mean velocities. These are production terms which can be either positive or negative in sign and can thus either increase or decrease the stresses. The last terms on the right side represent contributions from the viscous stresses, termed dissipative effects, since in many cases they tend to diminish the Reynolds stresses.

We also see that the right side of Eq. (4.4.1) involves many additional terms which are unknown. In a general three-dimensional flow there are 10 third-moment correlations of the form $\overline{u_iu_ju_m}$, which, by analogy with the second-moment correlations discussed earlier, can be loosely interpreted as describing the mean transport of the instantaneous stresses ρu_iu_j in the mth coordinate direction. There also arise pressure–velocity correlations, $\overline{pu_i}$, pressure-rate-of-strain correlations, $\overline{p(\partial u_i/\partial x_j + \partial u_j/\partial x_i)}$, and contributions from fluctuations in the body forces. The appearance of additional correlations involving pressure fluctuations implies that the closure problem arises not only from the nonlinearity of the convective forces but also from the global nature of the pressure field in fluid flows. We see this nature from Eq. (2.4.1), which shows that $p(\mathbf{x}, t)$ depends on the velocity field throughout the space occupied by the fluid.

Despite this situation and the number of additional terms which must be dealt with before Eq. (4.4.1) is quantitatively useful, it contains important physical implications. Thus, in Chapter 10 where we deal with second-moment methods, these equations with the new unknowns arising from the averaging replaced by approximations play a central role.

4.5 DISSIPATIVE EFFECTS

Because of their importance, it is worth considering the viscous terms in Eq. (4.4.1) in more detail. From Eq. (2.2.3),

$$\overline{u_j \frac{\partial \tau_{ik}}{\partial x_k}} + \overline{u_i \frac{\partial \tau_{jk}}{\partial x_k}} = \mu \frac{\partial^2}{\partial x_k \partial x_k} \overline{u_i u_j} + \mu \frac{\partial}{\partial x_k} \left(\overline{u_j \frac{\partial u_k}{\partial x_i}} + \overline{u_i \frac{\partial u_k}{\partial x_j}} \right)$$

$$- \mu \left[2 \overline{\frac{\partial u_i}{\partial x_k} \frac{\partial u_j}{\partial x_k}} + \overline{\frac{\partial u_k}{\partial x_i} \frac{\partial u_j}{\partial x_k}} + \overline{\frac{\partial u_k}{\partial x_j} \frac{\partial u_i}{\partial x_k}} \right] \tag{4.5.1}$$

We thus see that the dissipation terms involve three types of contribution: gradients of mean quantities consisting of second moments; mean values of products of velocity fluctuations and gradients of those fluctuations; and mean values of products of such gradients. In many circumstances the first two contributions are negligible compared to the third. Although this will be confirmed later by more precise estimates for the relative sizes of these contributions, it is sufficient for present purposes to consider the following argument. Suppose that the derivatives of mean quantities are inversely proportional to a length scale characterizing the large-scale motion L_g, for example, the width of a channel or the thickness of a boundary layer; all velocity fluctuations are roughly equal to a measure of the rms of the velocity fluctuations U_f; and finally, spatial derivatives of the fluctuating velocities are inversely proportional to the small-scale motion L_s, where $L_s << L_g$. The three contributions in Eq. (4.5.1) are thus estimated to have magnitudes of $\mu U_f^2/L_g^2$, $\mu U_f^2/L_s L_g$, and $\mu U_f^2/L_s^2$, respectively, and thus to be in proportion to $(L_s/L_g)^2{:}(L_s/L_g){:}1$. If $L_s/L_g << 1$, as is the case in many turbulent flows, then the dominant contribution to the viscous term in the balance equation for the Reynolds stress is given by the mean value of products of gradients. Accordingly, for future use we define

$$\varepsilon_{ij} \equiv \nu \left(2 \overline{\frac{\partial u_i}{\partial x_k} \frac{\partial u_j}{\partial x_k}} + \overline{\frac{\partial u_k}{\partial x_i} \frac{\partial u_j}{\partial x_k}} + \overline{\frac{\partial u_k}{\partial x_j} \frac{\partial u_i}{\partial x_k}} \right) \tag{4.5.2}$$

For the typical diagonal term, e.g., $i = j = 1$, this takes the simpler form

$$\varepsilon_{11} \equiv 2\nu \overline{\left(\frac{\partial u_1}{\partial x_k} + \frac{\partial u_k}{\partial x_1} \right) \frac{\partial u_1}{\partial x_k}} \tag{4.5.3}$$

with similar equations for $i = 2, 3$.

Several comments are appropriate. In some applications to be discussed later, models for ε_{ij} are required; but for the moment it is sufficient to note that these quantities have the units of U^3/L or alternatively $U^2/(L/U)$, i.e., energy per time (cf. Section 4.7 for a further development of this topic). Note that although these dissipation quantities are averaged quantities, there is no overbar indicating explicitly that this is the case. We shall refer to *mean* dissipation to emphasize this averaging, because the instantaneous dissipation is sometimes of interest.

The experimental determination of quantities such as $\varepsilon_{ij}(t; \mathbf{x})$ with adequate time and the requisite spatial resolution is clearly difficult, requiring the measurement of *gradients* such as $\partial u_i/\partial x_j$, $i, j = 1, 2, 3$, as functions of time. As a consequence, measurement strategies yielding approximate values for $\varepsilon_{ij}(\mathbf{x})$ are generally adopted (cf. Browne et al., 1987, and Section 4.7). In turbulence modeling such as discussed in Chapter 10, the mean viscous dissipation is generally assumed to be isotropic, so

$$\varepsilon_{ij} = \tfrac{2}{3}\delta_{ij}\varepsilon \tag{4.5.4}$$

and $\bar{\varepsilon}(\mathbf{x})$ is assumed to be given with sufficient accuracy by Eq. (5.3.17).

4.6 SECOND-MOMENT EQUATIONS FOR TEMPERATURE

Two useful equations arise from second moments involving the temperature. The first is the transport equation for the Reynolds fluxes, $\rho\overline{u_i\theta}$; if Eq. (2.2.1) is multiplied by θ, Eq. (2.3.1) by u_i, and the resulting equations added and averaged,

$$\rho\frac{\partial}{\partial x_k}(U_k\overline{u_i\theta}) = \left(\rho\overline{u_k\theta}\frac{\partial U_i}{\partial x_k} + \rho\overline{u_iu_k}\frac{\partial\Theta}{\partial x_k}\right) - \rho\frac{\partial}{\partial x_k}\overline{u_iu_k\theta}$$
$$-\frac{\partial}{\partial x_i}\overline{p\theta} + \overline{p\frac{\partial\theta}{\partial x_i}} + \overline{f_i\theta} + \left[\overline{\theta\frac{\partial\tau_{ik}}{\partial x_k}} + \overline{u_i\frac{\partial}{\partial x_k}\left(\frac{\mu}{N_\sigma}\frac{\partial\theta}{\partial x_k}\right)}\right] \tag{4.6.1}$$

The left side describes the convection of the Reynolds flux by the mean velocity, i.e., the convective derivative based on the mean velocity components, while the first two terms on the right side represent the alteration of that flux by an interaction between fluctuating quantities and gradients of mean flow variables. The $\overline{f_i\theta}$ term on the right side describes the influence of the body force and is frequently a deterministic quantity. As in the case of weighted averaging of the momentum equation, new unknowns arise: Six third-moment correlations $\overline{u_iu_j\theta}$; the temperature–pressure correlations; and terms describing viscous transport. The latter can be rearranged to make their physical significance more apparent. Using Eq. (2.2.3),

$$\overline{\theta\frac{\partial\tau_{ik}}{\partial x_k}} + \overline{u_i\frac{\partial}{\partial x_k}\left(\frac{\mu}{N_\sigma}\frac{\partial\theta}{\partial x_k}\right)} = \left(1+\frac{1}{N_\sigma}\right)\mu\frac{\partial^2}{\partial x_k\,\partial x_k}\overline{u_i\theta} - \frac{\mu}{N_\sigma}\frac{\partial}{\partial x_k}\overline{\theta\frac{\partial u_i}{\partial x_k}}$$
$$-\mu\frac{\partial}{\partial x_k}\overline{u_i\frac{\partial\theta}{\partial x_k}} - \left(1+\frac{1}{N_\sigma}\right)\mu\overline{\frac{\partial\theta}{\partial x_k}\frac{\partial u_i}{\partial x_k}} \tag{4.6.2}$$

If we supplement our earlier argument leading to Eq. (4.5.3) by introducing a measure of the temperature fluctuations and follow the same line of reasoning, we readily determine that the last term on the right side of Eq. (4.6.2) is dominant. It is thus useful for later purposes to define

$$\varepsilon_{i\theta} \equiv \left(1 + \frac{1}{N_\sigma}\right) \nu \overline{\frac{\partial \theta}{\partial x_k} \frac{\partial u_i}{\partial x_k}} \tag{4.6.3}$$

Another second-moment equation involving the temperature is obtained by weighting Eq. (2.3.1) by θ and averaging. There results a transport equation for the intensity of the temperature fluctuations,

$$\frac{1}{2}\rho \frac{\partial}{\partial x_k}(U_k \overline{\theta^2}) = -\frac{1}{2}\, \rho \overline{u_k \theta} \frac{\partial \Theta}{\partial x_k} - \rho \frac{\partial}{\partial x_k} \overline{u_k \theta^2}$$

$$+ \frac{1}{2} \frac{\mu}{N_\sigma} \frac{\partial^2 \overline{\theta^2}}{\partial x_k \partial x_k} - \frac{\mu}{N_\sigma} \overline{\frac{\partial \theta}{\partial x_k} \frac{\partial \theta}{\partial x_k}} \tag{4.6.4}$$

Our earlier identification of the various terms in transport equations applies here as well. Especially noteworthy is the last term on the right side, which represents the dominant viscous influence on the intensity $\overline{\theta^2}(\mathbf{x})$. We thus define the mean scalar dissipation,

$$\varepsilon_\theta \equiv \frac{\nu}{N_\sigma} \overline{\frac{\partial \theta}{\partial x_k} \frac{\partial \theta}{\partial x_k}} \tag{4.6.5}$$

Clearly, ε_θ is a positive definite quantity.

4.7 TURBULENT KINETIC ENERGY AND THE ANISOTROPY TENSOR

An additional second-moment equation which plays an important role in developing insight as well as in formulating predictive methods can be derived from Eq. (4.4.1). We refer to a transport equation for the turbulent kinetic energy, $k \equiv \frac{1}{2}\overline{u_l u_l}$, which provides a convenient measure of the intensity of turbulence at any spatial location. A more useful measure is the relative intensity $(k/U_l U_l)^{1/2}$, with typical values of the order of 10^{-2} in grid turbulence and 10^{-1} in turbulent shear flows. If at a particular spatial location in the flow this quantity is small compared to unity, the instantaneous velocity vector $\tilde{\mathbf{u}}$ varies only slightly about its mean $\mathbf{U}$. On the contrary, values of this quotient on the order of unity imply large fluctuations in the magnitude of the instantaneous velocity vector. Note that knowledge of $k(\mathbf{x})$ does not distinguish among the individual contributions from $\overline{u_1^2}$, $\overline{u_2^2}$, etc., without supplemental information, but frequently estimates of the relative contributions can be made; for example, in flows with significant mean shear and x_1 in the direction of the principal velocity component, we can assume that $\overline{u_1^2} = 2\overline{u_2^2} = 2\overline{u_3^2}$, so $\overline{u_1^2} = k$, $\overline{u_2^2} = \overline{u_3^2} = \frac{1}{2}k$. We shall see later that gradient transport methods permit the $\overline{u_i^2}$ intensities, indeed all Reynolds stresses, to be calculated from knowledge of $\mathbf{U}(\mathbf{x})$ and $k(\mathbf{x})$ [cf. Eqs. (10.1.3)]. In many flows of applied interest, however, the resulting distribution of the contributions to the turbulent kinetic energy is not accurate.

To obtain a transport equation for $k(\mathbf{x})$, we set $i = j$ in Eq. (4.4.1) and then sum over i, with the result

$$\rho \frac{\partial}{\partial x_k}(U_k k) \approx -\rho \overline{u_k u_l} \frac{\partial U_l}{\partial x_k} - \frac{\partial}{\partial x_k}\left(\overline{p u_k} + \frac{1}{2}\rho \overline{u_k u_l u_l}\right) + \overline{f_k u_k} - \rho \varepsilon \quad (4.7.1)$$

where we take advantage of our earlier discussion of molecular effects to retain only the dominant contributions and where we introduce the mean viscous dissipation (cf. Eq. 4.5.3).

$$\varepsilon \equiv \frac{1}{2}\varepsilon_{kk} = \nu \overline{\left(\frac{\partial u_l}{\partial x_k} + \frac{\partial u_k}{\partial x_l}\right)\frac{\partial u_l}{\partial x_k}} = \frac{\nu}{2}\overline{\left(\frac{\partial u_l}{\partial x_k} + \frac{\partial u_k}{\partial x_l}\right)\left(\frac{\partial u_l}{\partial x_k} + \frac{\partial u_k}{\partial x_l}\right)} \quad (4.7.2)$$

The second form clearly shows the positive-definite nature of ε.† It also shows that the mean viscous dissipation is the mean of the product of the instantaneous viscous stresses and the instantaneous rate of strain.

We see that the left side of Eq. (4.7.1) consists of the familiar convection terms, while the right side involves turbulent production, pressure–velocity, and third-moment terms, a body force contribution, and molecular effects. Note that the pressure–rate-of-strain correlations vanish as a consequence of the restraint on the velocity fluctuations [cf. Eq. (4.1.2)]; the implication from this behavior is that these terms describe a redistribution of the contributions to the turbulent kinetic energy but do not alter that energy. This matter is discussed in more detail in Section 6.3.

If body forces are absent and the flow is spatially homogeneous, i.e., the one-point statistical quantities do not vary spatially (a situation closely approximated by grid turbulence) and furthermore, if the flow is observed in a frame of reference moving with the mean velocity U_1, then Eq. (4.7.1) becomes

$$\frac{Dk}{Dt} = \varepsilon \quad (4.7.3)$$

where for present purposes we include the $\partial/\partial t$ contribution to the convective derivative. Equation (4.7.3) indicates that the rate of decrease of the turbulent kinetic energy under these circumstances is equal to the mean viscous dissipation. Also implied is that the influences inoperative under these special circumstances, in particular the production associated with the $\overline{u_k u_l}\ \partial U_l/\partial x_k$ terms, are required if the turbulent kinetic energy is to be maintained or increased.

Recall from Section 4.5 that the main contribution to the mean viscous dissipation is due to small-scale velocity fluctuations. It will become evident in Section 5.5 that in high-Reynolds-number flows these small scales tend to be locally isotropic and locally homogeneous. If this is indeed the case, ε can be obtained experimentally with relative ease. Although the validity of this as-

†This second form is found by noting that indices k and l can be interchanged, so addition of the identical term and division by 2 leads to the desired result.

sumption is discussed in Section 5.5, it is sufficient to assert here that in flows with suitably small mean rates of strain and with suitably high Reynolds numbers, the assumptions of local isotropy and homogeneity are valid. In such flows, statistical quantities involving the velocity fluctuations and their gradients respect certain kinematic relations reflecting their independence of the coordinate system, i.e., the rotation and reflection of the coordinates (cf. Taylor, 1935b). Thus we have, e.g.,

$$\begin{aligned}
&\overline{\left(\frac{\partial u_1}{\partial x_1}\right)^2} = \overline{\left(\frac{\partial u_2}{\partial x_2}\right)^2} = \overline{\left(\frac{\partial u_3}{\partial x_3}\right)^2} \\
&\overline{\left(\frac{\partial u_1}{\partial x_2}\right)^2} = \overline{\left(\frac{\partial u_1}{\partial x_3}\right)^2} = \overline{\left(\frac{\partial u_2}{\partial x_1}\right)^2} = \overline{\left(\frac{\partial u_2}{\partial x_3}\right)^2} \\
&\qquad = \overline{\left(\frac{\partial u_3}{\partial x_1}\right)^2} = \overline{\left(\frac{\partial u_3}{\partial x_2}\right)^2} = 2\,\overline{\left(\frac{\partial u_1}{\partial x_1}\right)^2} \\
&\overline{\frac{\partial u_1}{\partial x_2}\frac{\partial u_2}{\partial x_1}} = \overline{\frac{\partial u_2}{\partial x_3}\frac{\partial u_3}{\partial x_2}} = \overline{\frac{\partial u_3}{\partial x_1}\frac{\partial u_1}{\partial x_3}} = -\frac{1}{2}\overline{\left(\frac{\partial u_1}{\partial x_1}\right)^2}
\end{aligned} \tag{4.7.4}$$

When these and similar relations are used in Eq. (4.7.2), we obtain as an intermediate result from strictly kinematic considerations,

$$\varepsilon = 6\nu\left[\overline{\left(\frac{\partial u_1}{\partial x_1}\right)^2} + \overline{\left(\frac{\partial u_1}{\partial x_2}\right)^2} + \overline{\frac{\partial u_1}{\partial x_2}\frac{\partial u_2}{\partial x_1}}\right] \tag{4.7.5}$$

A further reduction is achieved when these derivatives are associated with fluid flow by incorporation of the conservation and transport equations, i.e., the continuity equation and the mean of Eq. (2.4.1) for the pressure, to yield the simple result

$$\varepsilon = 15\nu\,\overline{\left(\frac{\partial u_1}{\partial x_1}\right)^2} \tag{4.7.6}$$

If the flow involves a dominant mean velocity U_1 aligned with the x_1 coordinate such that the Taylor hypothesis can be invoked (cf. Section 3.5), this becomes

$$\varepsilon = 15\,\frac{\nu}{U_1^2}\,\overline{\left(\frac{\partial u_1}{\partial t}\right)^2} \tag{4.7.7}$$

a quantity which can be readily measured with a single anemometer. As stated earlier, the validity of Eq. (4.7.6) and its consequent Eq. (4.7.7) are discussed in Section 5.5.

As an immediate consequence of the second of Eqs. (4.7.4), we obtain relations prevailing in isotropic, homogeneous turbulence, namely,

$$\overline{\left(\frac{\partial u_1}{\partial x_1}\right)^2} = \frac{1}{2}\overline{\left(\frac{\partial u_2}{\partial x_1}\right)^2} = \frac{1}{2}\overline{\left(\frac{\partial u_3}{\partial x_1}\right)^2} \tag{4.7.8}$$

which are used in Section 5.5.

A convenient means for characterizing various states of turbulence and changes in those states in terms of the turbulent kinetic energy is provided by the *anisotropy tensor*,

$$b_{ij} \equiv \frac{\overline{u_i u_j}}{2k} - \frac{1}{3}\delta_{ij} \tag{4.7.9}$$

We see that b_{ij} scales the Reynolds stresses on the turbulent kinetic energy. If we set $i = j$ and sum over i, we readily find that

$$b_{kk} = 0 \tag{4.7.10}$$

Moreover, since $\overline{u_i^2} \geq 0$, $i = 1, 2, 3$, we have

$$-\frac{1}{3} \leq b_{11} \leq \frac{2}{3} \tag{4.7.11}$$

with identical inequalities for b_{22} and b_{33}. When the inequality on the left applies, $\overline{u_i^2} = 0$; while when that on the right applies, e.g., with $i = 1$, then $\overline{u_2^2} = \overline{u_3^2} = 0$ and $\overline{u_1^2} = 2k$.

The constraint implied by Eq. (4.7.10) permits us to represent the components of the turbulent kinetic energy in any flow in a two-dimensional plane with coordinates consisting of two of the three quantities b_{11}, b_{22}, and b_{33}. For example, in Fig. 4.7.1 we select the first two of these quantities as coordinates; as a consequence of Eq. (4.7.10), the value of b_{33} at each point in this plane is known. Since turbulence is isotropic whenever all the normal Reynolds stresses are equal, this state lies at the origin in this figure. Axisymmetric turbulence corresponds to two equal intensities, e.g., to $\overline{u_2^2} = \overline{u_3^2}$, so that $\overline{u_1^2} = 2(k - \overline{u_2^2})$. In this case $b_{22} = b_{33}$ and $b_{22} = -\frac{1}{2}b_{11}$ and the various states lie on a line in Fig. 4.7.1 through the origin and the vertex $(\frac{2}{3}, -\frac{1}{3}, -\frac{1}{3})$. When other pairs of intensities are equal, other lines through the origin to the vertices describe possible states of axisymmetric turbulence. These are shown in Fig. 4.7.1 as dashed lines through the origin. The idealized state of two-dimensional turbulence (cf. Section 6.6) involves the total suppression of fluctuations in one coordinate direction. For example, if $\overline{u_3^2} = 0$, then $b_{33} = -\frac{1}{3}$ and $b_{22} = \frac{1}{3} - b_{11}$. These states are represented in Fig. 4.7.1 by the slanted edge of the triangle. The corresponding states for $\overline{u_1^2} = 0$ and $\overline{u_2^2} = 0$ are readily seen to the other edges of the triangle, i.e., $b_{11} = -\frac{1}{3}$ and $b_{22} = -\frac{1}{3}$, respectively. Finally, one-dimensional

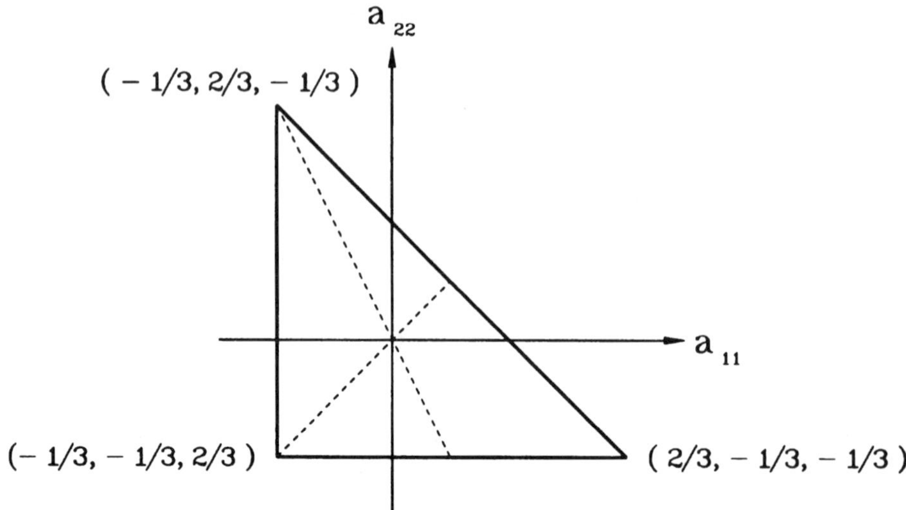

Figure 4.7.1 The space of the anisotropy tensor.

turbulence corresponds to a further idealization, e.g., to $\overline{u_2^2} = \overline{u_3^2} = 0$, $\overline{u_1^2} = 2k$, and thus to $b_{11} = \frac{2}{3}$, $b_{22} = b_{33} = -\frac{1}{3}$, i.e., to the lower right vertex of the triangle in Fig. 4.7.1. The other vertices correspond to other states of one-dimensional turbulence. Thus, from these considerations we see that all possible turbulent states must lie on the edges or within the triangle of Fig. 4.7.1.

In idealized grid turbulence the flow remains isotropic as the turbulence intensity decays, so the state is always at the origin of Fig. 4.7.1; in more general flows, however, the turbulence evolves and thus traces out a path in that figure. For example, we shall see later (cf. Section 6.5) that if isotropic turbulence is subjected to a high rate of extensional strain in one coordinate direction, e.g., in the x_1 direction, and corresponding rates of compressional strain in the other two directions, the $\overline{u_1^2}$ intensity is reduced while the other two intensities are increased. A limiting case of such a flow involves a path from the origin to a point $(-\frac{1}{3}, \frac{1}{6}, \frac{1}{6})$, i.e., to a point on the left boundary of the triangle and to two-dimensional turbulence. This situation corresponds roughly to an axisymmetric contraction cone with a large area change (cf. Section 6.7).

Other circumstances also lead to nearly two-dimensional turbulence; we discuss in Section 10.5 that the turbulence near a wall involves fluctuations of the velocity components parallel to the wall and none of the velocity component normal to the wall, with the consequence that a two-dimensional turbulent state prevails. Finally, in Section 6.4 and again in Section 10.13 it is shown that in a *stably stratified* medium the intensity of the vertical velocity component is strongly suppressed by buoyancy effects so that most, if not all, of the turbulent kinetic energy is contributed by the horizontal velocity components.

Within the context of this discussion, consider the application of one of the methods dealt with in either Chapter 7 or Chapter 10 to determine the evolution of turbulence along a mean streamline from an initial state within the triangle of Fig. 4.7.1. Since the intensities of the velocity fluctuations in a physically acceptable flow must lie either on the edges or within that triangle, it would be highly desirable if calculation methods inherently preclude unacceptable states. Such a consideration leads to the notion of realizability (Schumann, 1977; Lumley, 1983), i.e., the formulation of closure models which *assure* that pathological turbulent states are never predicted. As might be expected, the most difficult flows to preclude prediction of physically unacceptable states are those corresponding to points in Fig. 4.7.1 near the edges of the triangle, e.g., two-dimensional turbulence. A critique of realizability and a discussion of the significant cost of its imposition in moment methods are provided by Speziale et al. (1993). The theories we discuss do not guarantee the prediction of unphysical states, but this shortcoming is not significant for the flows we consider.

Because the invariance properties of the anisotropy tensor b_{ij} are invoked in the most recent modeling of the pressure–rate-of-strain terms (cf. Section 10.10), it is appropriate to discuss those properties briefly. A tensor such as b_{ij} can be represented as a 3×3 matrix with row and column entries $b_{11}, b_{12}, \ldots, b_{21}$ etc. The eigenvalues of such a matrix are obtained by replacing the entries on the principal diagonal by $b_{11} - \lambda$, $b_{22} - \lambda$, and $b_{33} - \lambda$, by treating the resultant matrix as a determinant, and by setting that determinant to zero. By using minors, the cubic equation determining the eigenvalues can be obtained by straightforward but tedious algebra. There results

$$\lambda^3 + A_{\mathrm{I}}\lambda^2 + A_{\mathrm{II}}\lambda + A_{\mathrm{III}} = 0 \tag{4.7.12}$$

where

$$A_{\mathrm{I}} = -(b_{11} + b_{22} + b_{33}) = 0$$

$$A_{\mathrm{II}} = \mathrm{b}_{11}(b_{22} + b_{33}) + b_{22}b_{33} - b_{12}b_{21} - b_{13}b_{31} - b_{23}b_{32}$$

$$A_{\mathrm{III}} = -b_{11}(b_{22}b_{33} - b_{23}b_{32}) + b_{12}(b_{21}b_{33} - b_{31}b_{23}) - b_{13}(b_{21}b_{32} - b_{31}b_{22})$$

and where $A_{\mathrm{I}} = 0$ in the present calculation is a consequence of Eq. (4.7.10). Some simplification of A_{II} and A_{III} occurs if the symmetries of the b_{ij} components are taken into account, but we need not give details. Now, if a different coordinate system is used to describe the velocity components, then the correlations $\overline{u_i u_j}$ entering the definition of b_{ij} of Eq. (4.7.9) and the entries of the matrix are altered but the eigenvalues given by Eq. (4.7.12) *are not changed.* The coefficients A_{I}–A_{III} are the coordinate invariants of the tensor b_{ij}, a result well known from linear algebra (cf. Strang, 1976). Thus, turbulence models utilizing these invariants are independent of the coordinate system employed and hence invariant under rotations and translations.

4.8 THE CLOSURE PROBLEM

In this chapter we establish that the expected benefits of restricting attention to a statistical description of turbulent flows, i.e., to averaged forms of the conservation equations, are vitiated by a fundamental problem. In particular, the equations for mean velocity components $U_i(\mathbf{x})$, mean pressure $P(\mathbf{x})$, and mean temperature $\Theta(\mathbf{x})$ contain new unknowns, the Reynolds stresses $\rho\overline{u_i u_j}$ and the Reynolds fluxes $\rho\overline{u_i\theta}$. Moreover, if we systematically develop transport equations for these stresses and fluxes by averaging weighted transport equations appropriately, we find equations with a large number of *new unknowns.* In particular, the momentum equations treated in this fashion involve third-order moments and correlations involving pressure fluctuations. This situation constitutes *the closure problem,* which arises in any statistical theory involving nonlinear effects. Although the nature of turbulence, its many manifestations, and its numerous applications render attempts to narrowly define "the turbulence problem" an activity of dubious value, if such identification is demanded, then closure must be regarded as *the central problem.*

Within the context of the discussion in Section 2.7 concerning the limitation of exact numerical solutions, the Navier-Stokes equations with respect to Reynolds number, we note that the computing power needed to obtain solutions of the averaged equations closed in some suitable fashion, for flows with the complex geometries and Reynolds numbers of interest in many applications is significantly less than that required for DNS and large eddy simulation of those same flows. Thus the statistical properties of such flows are routinely calculated. However, as discussed in Chapter 10, the large systems of partial differential equations which arise in more advanced methods of analysis of complex flows and the requirement of adequate spatial resolution, i.e., small grid sizes, call for present-day supercomputers. However, this situation is expected to improve as computing speeds increase. Thus the averaged equations which owe their genesis to Osborne Reynolds a century ago, with appropriate approximations to achieve closure, will continue to provide a useful strategy for the future treatment of many turbulent flows.

Given that approximation techniques, some systematic and others ad hoc, are always required, our aim in the present work is henceforth to develop physical insight and understanding based on the physics and mechanics of turbulence and to illustrate their application to various flows. At this juncture it is appropriate to introduce the notion of a "model," since the words, "model," "modeling," etc., frequently appear in the turbulence literature and in subsequent portions of this text. These words are applied in various senses; for example, the basic conservation and transport equations of fluid mechanics can be considered "models" describing the behavior of real fluids. However, in the turbulence literature, "models" generally imply an approximation of either higher-order statistical quantities in terms of lower-order quantities, or correlations involving pressure fluctuations or dissipative effects. For example, we shall see that a Reynolds shear stress is frequently "modeled" by the substitution

$$-\rho\overline{u_1 u_2} \propto \frac{\partial U_1}{\partial x_2}$$

where the proportionality is replaced by an equality with the introduction of an appropriate coefficient or function of proportionality involving at most other principal dependent variables so that the Reynolds shear stress is eliminated as an unknown.

4.9 SUMMARY

By applying the simplest averaging procedure set forth in Chapter 3, in this chapter we developed the averaged conservation and transport equations of fluid motion. There result from the momentum equations correlations of the velocity components which are interpreted as Reynolds stresses. Similarly, averaging of the thermal energy equation results in correlations of the velocity components and temperature which are interpreted as Reynolds fluxes. Additionally, by appropriate weighting of the original transport equations before averaging, equations for these stresses and fluxes are obtained, the second-moment equations. A special and important example of a second-moment equation is that for the transport of turbulent kinetic energy, an equation which involves the mean viscous dissipation. The anisotropy tensor provides a convenient means to characterize the various states of turbulence. We concluded with a discussion of the fundamental problem of turbulence, namely, closure.

CHAPTER

FIVE

LENGTH AND TIME SCALES OF TURBULENCE

As suggested in Chapter 3, turbulent flows involve a variety of length and time scales whose understanding is important to gaining insights into various turbulence phenomena. We begin our consideration of turbulent scales by identifying certain global features of the flow that are not associated directly with turbulence. In channel flow, for example, the mean centerline velocity U_g and the channel height L_g clearly characterize the global features of the flow and can be used to form a Reynolds number $U_g L_g / \nu$. An associated time scale would be L_g / U_g, the time a fluid element with velocity U_g requires to travel the distance L_g. Similarly, for boundary layers and wakes, the mean external velocity and the shear-layer thickness at a particular streamwise station serve to identify the global flow characteristics.

We next consider the length and time scales of turbulence of turbulent flows involving a principal flow direction which, without loss of generality, can be taken to be aligned with the x_1 coordinate axis such that U_2, $U_3 << U_1$, a situation which prevails in many flows of fundamental and applied interest. For example, grid turbulence involving U_1 = constant while $U_2 = U_3 = 0$ is the subject of extensive theoretical study, as discussed in Section 5.2–5.4. If viewed in a coordinate system moving with the mean velocity U_1, grid turbulence is nearly spatially homogeneous and decays with time.

5.1 THE INTEGRAL LENGTH AND TIME SCALES

The general time and space correlations discussed in Section 3.4 [cf. Eq. (3.4.13)] can be applied directly to turbulent flows to characterize some of their scales. The experimental measurement of velocity space–time correlation components requires multiple anemometers, each permitting the determination of either one or two velocity components. One anemometer may be considered to be at a fixed location **x**, while the second is at various locations **x** + **h**. From continuous records of the signals from these anemometers, various arbitrary time delays can be subsequently imposed. Similar techniques are needed for the space–time correlations of temperature, using multiple thermometers. In the discussion which follows we focus on anemometer signals, but our considerations apply with little change to thermometer signals.

First we consider grid flow experiments. Since, as suggested earlier, this flow plays an important role in turbulence theory and experiment, it is appropriate to describe its essential features. A uniform mesh of rods, usually circular in cross section, in a square grid pattern with a charactistic spacing of L_M is placed normal to a uniform stream, e.g., in a wind tunnel. In flows with suitably high velocity, the grid generates turbulence which becomes nearly isotropic at some distance downstream of the grid, i.e., it becomes such that $\overline{u_i u_j} \approx \delta_{ij}\overline{u^2}$. With the x_1 coordinate aligned in the downstream direction, the flow involves one mean velocity U_1, a constant, and $U_2 \equiv U_3 \equiv 0$. All statistical quantities depend only on x_1 and are independent of the cross-stream coordinates. Since it is free of external influences, the turbulence decays in the downstream direction. In the following discussion we are concerned with an intermediate range of downstream distances where the turbulence is essentially isotropic and suitably intense. There exists a final period of decay with a different decay mechanism and low turbulence intensity, a period leading to the complete demise of the velocity fluctuations and the return of the flow to a uniform velocity $\tilde{U}_1 \equiv U_1$.

We now imagine that in this grid flow one anemometer is located at x_1 and the second at $\mathbf{i}_1\ x_1 + \mathbf{h}$. An origin for the x_1 coordinate is selected, either at the grid on the tunnel centerline or at a virtual origin on that centerline. The output from each anemometer is continuously recorded so that it can be played back with any desired time delay τ imposed on the measured velocities. We discuss here the essential findings from the comprehensive correlation results reviewed by Favre (1965).

Figure 5.1.1 shows the distributions of the correlation coefficient $R_{u_1}(\tau, h_1; x_1)$ defined by Eq. (3.5.14) for fixed values of h_1 made dimensionless with L_M and with the time advance τ made dimensionless with L_M/U_1. Consider first the autocorrelation coefficient corresponding to the special case of $h_1/L_M = 0$, that is, of no spatial separation; this coefficient decreases, becoming nearly zero at $U_1\ \tau/L_M \approx 2.5$ and approaching zero from below as τ increases to infinity. This

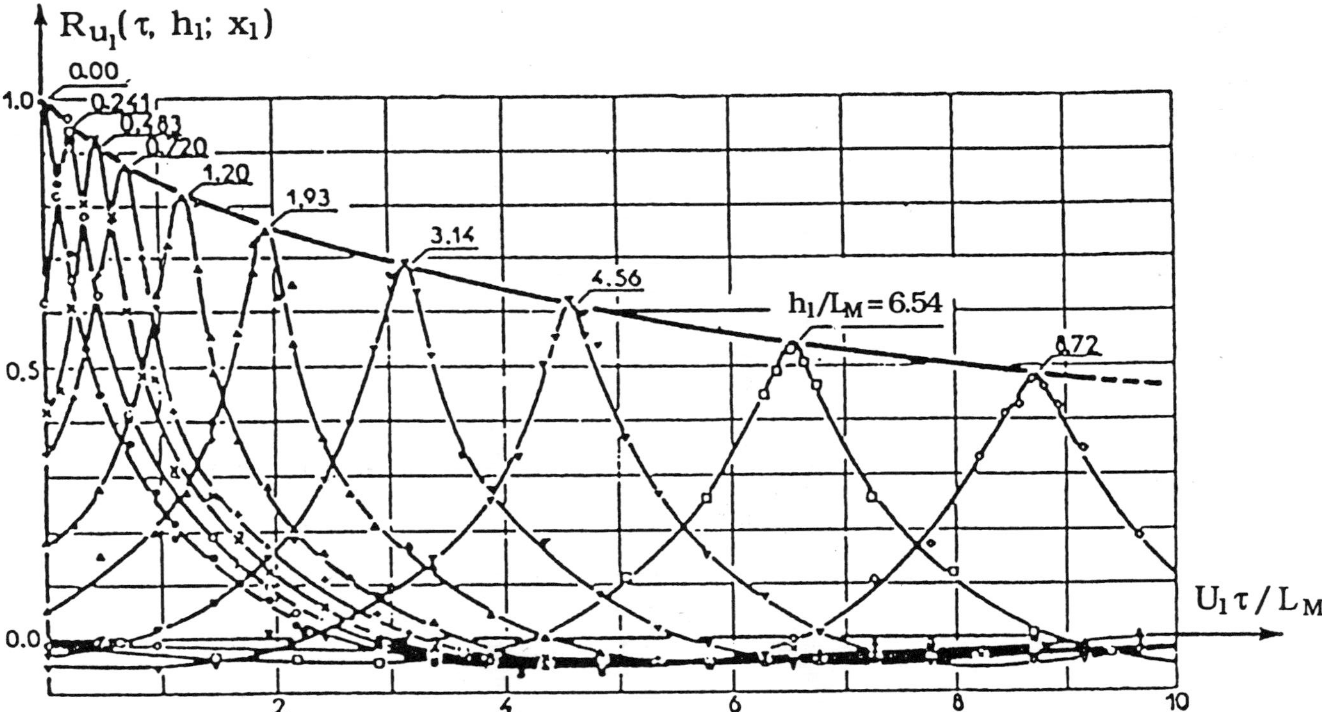

Figure 5.1.1 The space–time correlation of the u_1 velocity component in grid flow. (From Favre, 1965.)

leads to a correlation time as defined by Eq. (3.4.2) of about L_M/U_1, implying an integral length of roughly L_M. This inference is supported by considering the purely spatial correlations, that is, for $\tau = 0$. For example, when $h_1/L_M = 1$ and $\tau = 0$, we see that $R_{u_1} \approx 0.2$, that is, the two velocities are poorly correlated when the two anemometers are separated by L_M. Furthermore, for $h_1/L_M = 2$ we have $R_{u_1} \approx 5\ (10^{-2})$ and thus, from Eq. (3.4.9) with $\mathbf{i}_h = \mathbf{i}_1$, the integral length is again about equal to L_M.

Consider next the variation of the correlation coefficient with τ for a fixed, nonzero separation, e.g., for $h_1/L_M = 3$. We see that if $\tau = 0$, R_{u_1} takes on a small negative value, but that with increasing τ the coefficient increases to a maximum of approximately $7\ (10^{-1})$ at $U_1\ \tau/L_M \approx h_1/L_M$ and then decreases with further increases in τ. This implies that the fluctuations in the u_1 velocity at the fixed location x_1 are transported by the mean velocity U_1, but that during the transit from x_1 to $x_1 + h_1$ are changed so that the correlation coefficient is reduced. The same considerations apply qualitatively to the other values of the separation length h_1, but as that length increases the maximum value of the correlation coefficient decreases.

It is informative to compare the results in Fig. 5.1.1 with the Taylor hypothesis (cf. Section 3.5), that the time and space correlations in the mean streamline direction, i.e., in the x_1 direction, with mean velocity U_1 are equal if the time advance and spatial separation are related by $\tau = h_1/U_1$. The hypothesis implies that the maximum of the time correlation for a given h_1 should occur at a time advance of h_1/U_1 and the correlation coefficient should equal unity at the maximum. We see from Fig. 5.1.1 that this assumption is partially confirmed; the maximum does occur at roughly the estimated time advance, but the value of the coefficient at the maximum is less than unity. Later we shall find that this reduction in the correlation coefficient is a consequence of the relatively rapid decay of small-scale fluctuations, i.e., their relatively short lifetime.

Further, with respect to the Taylor hypothesis, note that in Section 3.7 a rough connection between Eulerian and Lagrangian correlations is discussed, one in which a particular two-point, one-time Eulerian correlation serves as an approximation to a Lagrangian correlation. Consider $R_{u_1}(\tau, h_1 = U_1\tau; x_1) = R_{u_1}(\tau; U_1, x_1)$, that is, the space–time correlation with the spatial separation given by the Taylor hypothesis. This correlation closely approximates the peaks in the space–time correlations for fixed h_1 in Fig. 5.1.1, so an extended version of the hypothesis appears to be useful.

Although not presented in Favre (1965), we know from other sources and can expect on intuitive grounds that a combination of streamwise and radial separations h_1 and $r = (h_2^2 + h_3^2)^{1/2}$, respectively, yield $R_{u_1}(\tau, h_1, r; x_1)$ values similar to those shown in Fig. 5.1.1 but with reduced peaks.

As a second set of experiments for turbulent length and time scales, we consider results for a turbulent boundary layer, again from Favre (1965). The measurement of space–time correlations in turbulent boundary layers involves techniques similar to those described earlier. The output from two anemometers, one at a fixed location within the boundary layer, the second at a location with a separation **h**, is continuously recorded so that with playback various time delays can be imposed on one variable relative to the other. However, in a boundary layer the inhomogeneity of the mean flow makes the space–time correlation coefficient more complex. In our discussion a standard orientation of coordinates for boundary layers is adopted; namely, the x_1 coordinate is directed along the plate in the streamwise direction, x_2 normal to the plate, and the x_3 in the spanwise direction. The separation vector **h** possesses corresponding components. Although Favre gives extensive results for turbulent boundary layers, the main points to be made here are shown in Fig. 5.1.2. One anemometer is fixed at $x_2/\delta = 3.0\ (10^{-2})$, where δ is the boundary-layer thickness ($\delta = 3.3$ cm), i.e., relatively close to the wall, and the second is placed downstream at $h_1/\delta = 7.7\ (10^{-1})$ and displaced farther from the wall at $h_2/\delta = 9.1 \times 10^{-2}$. The various correlation coefficients correspond to displacements in the spanwise direction h_3/δ and to time delays $U_\infty \tau/\delta$, where $U_\infty = 16$ m/s is the velocity in the external stream. The results are symmetric with respect to h_3/δ, as is to be expected in a two-dimensional boundary layer. With this arrangement of the anemometers in mind, consider the results shown in Fig. 5.1.2 for $h_3/\delta = 7.6\ (10^{-2})$; again we see that as the time delay is increased, the correlation $R_{u_1}(\tau;\ \mathbf{h}, \mathbf{x})$ increases from small values, approximately 10^{-1}, to a maximum of approximately $2.0\ (10^{-1})$ when $U_\infty \tau/\delta = -1.5$. Further changes in time delay lead to a reduction in the correlation until for $U_\infty \tau/\delta \approx -6$ the velocities become uncorrelated. When the sign of the time delay is changed, we see that the correlation again vanishes when $U_\infty \tau/\delta \approx 3$. From the value of the $U_\infty \tau/\delta$ corresponding to the maximum correlation and from the streamwise separation $h_1/\delta = 7.7\ (10^{-1})$, it is easy to estimate that the convection velocity of the

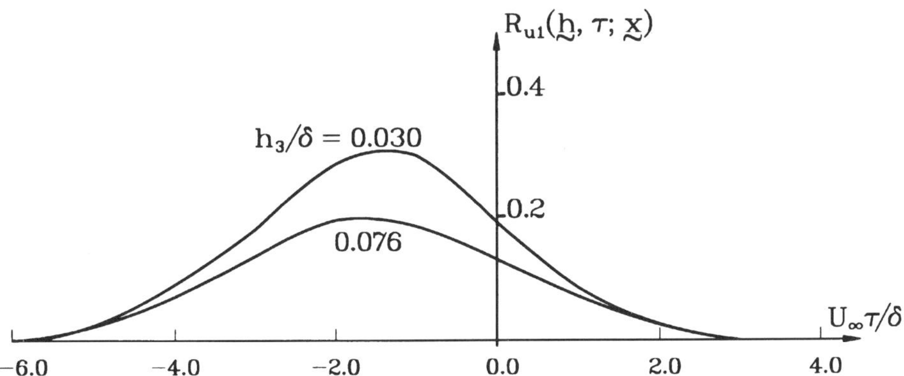

Figure 5.1.2 The space–time correlation of the u_1 velocity component in a turbulent boundary layer. (From Favre, 1965.)

fluctuations is 0.5 U_∞. With this convection velocity the correlation length in the streamwise direction is roughly one-quarter of the boundary-layer thickness. When the cross-stream displacement of the movable anemometer is decreased so that $h_3/\delta = 3.0\ (10^{-2})$, the peak correlation is increased to roughly 3 (10^{-1}) but at the same time delay and with the same correlation limits in terms of $U_\infty \tau/\delta$.

Favre (1965) gives extensive results for a variety of locations of the two anemometers, with particular emphasis on the distributions of the correlation coefficient with optimum time advance, i.e., corresponding to the maxima in Fig. 5.1.2. The surfaces of constant correlation coefficient with such τ values are roughly ellipsoids of revolution having major axes aligned in the x_1 direction. The maximum radius of the ellipsoids for a fixed value of the coefficient occurs when $h_1 = 0$ and is typically an order of magnitude smaller than the semimajor axis of the ellipsoid. This characteristic reflects the downstream transport of fluid noted earlier in our discussion of grid flow.

We can conclude from these considerations that the integral length and time scales in turbulent shear flows are of the order of magnitude of the global length and time, respectively, of the flows in which the turbulence arises. Put simply, in shear flows the largest scales of turbulence fill the flow.

In recent years extensive space–time correlations in turbulent boundary layers have been measured with the purpose of clarifying interactions between the wall and outer flows and the role of the large turbulent structures in such layers (cf. Smith and Abbott, 1978).

5.2 THE EVOLUTION OF SPECTRA

In the previous section attention focused largely on the low-frequency, relatively large-scale velocity fluctuations. To expose the role of high-frequency fluctuations and their interaction with these low frequencies, it is useful to adopt an altered perspective and to consider the idealization associated with isotropic homogeneous turbulence.[†] As noted earlier, the turbulent flow downstream of a grid in a wind tunnel observed in a frame of reference moving with the mean velocity U_1 closely approximates this idealization. In this frame the mean velocities U_i, $i = 1, 2, 3$ are all zero, and we deal only with the fluctuations u_i, $i = 1, 2, 3$. The spatial decay of the turbulence in a wind tunnel implies a slight spatial inhomogeneity in this moving frame, but this is not an essential shortcoming according to the following argument. The size of the imaginary box moving with the velocity U_1 is considered to be bounded between the lengths equal on the one hand to a *fraction* of the length characterizing the decay of the

[†]For a full discussion of the theory of isotropic homogeneous turbulence, see Hinze, 1975; Batchelor, 1967.

turbulence $k(dk/dx_1)^{-1}$ and on the other to a *multiple* of the integral scale of the turbulence L_M. With a judicious choice of the size of the volume under consideration, a size within these limits, the assumption of homogeneity is closely satisfied. The usual slight anisotropy of grid turbulence must also be ignored.

In Section 3.5 we discussed the two-point, one-time velocity correlations in grid turbulence as functions of downstream distance x_1 and separation r and their interpretation in terms of ensemble averages as functions of time equal to x_1/U_1 and separation r. In the present section we take up the notion of turbulence in a box and introduce time as an independent variable after some preliminaries concerned with measurements in grid turbulence.

We start by considering two distinct two-point, one-time correlations, $\overline{u_2(t; x_1)u_2(t; x_1 + h_2)}$ and $\overline{u_1(t; x_1)u_1(t; x_1 + h_2)}$ measured in grid flow with anemometers placed as shown in Fig. 5.2.1. Although, as we shall see, other orientations of the two points could be chosen, a separation along the x_2 coordinate direction is especially convenient in an experiment because the two probes do not interfere with one another. Central to the theory of isotropic, homogeneous turbulence is the independence of these correlations on the direction of separation and their dependence only on the separation distance r, which in the case of these correlations is h in the x_2 direction. The first is the parallel correlation, since the velocity components are aligned with the separation vector. From the assumption of isotropy, we know that this correlation is the same as

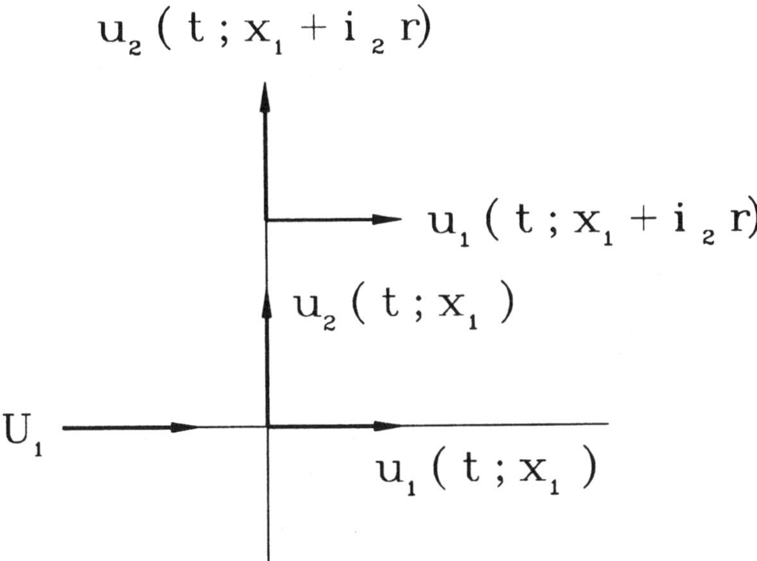

Figure 5.2.1 Measurement of the parallel and normal correlations in grid turbulence.

$\overline{u_1(t; x_1)u_1(t; x_1 + h_1)}$ and $\overline{u_3(t; x_1)u_3(t; x_1 + h_3)}$. The second is the lateral correlation $\overline{u_1(t; x_1)u_1(t; x_1 + h_2)}$, which is the same as $\overline{u_1(t; x_1)u_1(t; x_1 + h_3)}$, $\overline{u_3(t; \mathbf{x}_1)u_3(t; \mathbf{x}_1 + h_1)}$, etc. Thus we introduce *two scalar functions,*

$$\begin{aligned} f(r; x_1) &= \frac{\overline{u_2(t; x_1)\, u_2(t; x_1 + h_2)}}{\overline{u^2}(x_1)} \\ g(r; x_1) &= \frac{\overline{u_1(t; x_1)\, u_1(t; x_1 + h_2)}}{\overline{u^2}(x_1)} \end{aligned} \tag{5.2.1}$$

where these functions, the parallel and lateral correlation coefficients, respectively, are dependent on the arbitrary separation length r, here conveniently taken to be h_2, and where $\overline{u^2}(x_1) = \frac{1}{3}\, \overline{u_k u_k}(x_1) = \frac{2}{3}\, k(x_1)$ is related to the turbulent kinetic energy at station x_1. Although in Eqs. (5.2.1) we have defined these scalar functions in terms of the velocity components suggested by Fig. 5.2.1, we could equally well use other velocity components and other separation directions, provided the distinction between the parallel and lateral correlations is kept in mind. Clearly, $f(0; x_1) = g(0; x_1) = 1$; since we deal with turbulence $f(r \to \infty; x_1) = g(r \to \infty; x_1) = 0$, that is, with sufficient spatial separation, the velocity fluctuations are uncorrelated.

The functions $f(r; x_1)$ and $g(r; x_1)$ permit the correlation between any two velocity components at any separation $\mathbf{h} = \mathbf{i}_k h_k$ to be calculated; we sketch the argument. Consider a generalization of the transformation of Eqs. (4.2.4) by introducing a second coordinate system $\hat{\mathbf{x}}$ with associated velocity components $\hat{\mathbf{u}}$ and unit vector $\hat{\mathbf{i}}$. Align the $\hat{x}_1$ axis with an arbitrary separation vector $\mathbf{h}$ so that $\mathbf{h} = \hat{\mathbf{i}}_1 r$. The velocity components in the two coordinate systems are related according to $u_i(t; \mathbf{x}) = e_{ik}\, \hat{u}_k(t; \mathbf{x})$, where e_{ij} is the cosine of the angle between the ith coordinate in the $\mathbf{x}$ system and the jth coordinate in the $\hat{\mathbf{x}}$ system.† Thus we can calculate that

$$\overline{u_i(t; \mathbf{x}_1)\, u_j(t; \mathbf{x}_1 + \mathbf{h})} = e_{il}\, e_{jk}\, \overline{\hat{u}_k(t; \mathbf{x}_1)\, \hat{u}_l(t; \mathbf{x}_1 + \hat{\mathbf{i}}_1\, r)}$$

But we know that because of isotropy the six entries arising from the double sum over k and l on the right side consist of only two nonzero contributions, the longitudinal correlation $\overline{\hat{u}_1(t; \mathbf{x}_1)\hat{u}_1(t; \mathbf{x}_1 + \hat{\mathbf{i}}_1\, r)} = \overline{u^2}(x_1)f(r; x_1)$ and the lateral correlation $\overline{\hat{u}_2(t; x_1)\, \hat{u}_2(t; \mathbf{x}_1 + \hat{\mathbf{i}}_1\, r)} = \overline{u^2}(x_1)\, g(r; x_1)$. Now we can see that measurement of the velocity components $\hat{u}_1(t; \mathbf{x}_1)$, $\hat{u}_1(t; \mathbf{x}_1 + \hat{\mathbf{i}}_1\, r)$ determines the longitudinal correlation while $\hat{u}_2(t; \mathbf{x}_1)$ and $\hat{u}_2(t; \mathbf{x} + \mathbf{i}_1\, r)$ determine the lateral correlation. Such measurements call for two anemometers, *each* sensing two orthogonal velocities so that $\hat{u}_1$ and $\hat{u}_2$ can be determined at each instant of time. Now for isotropic, homogeneous flow, the second-order tensor

†The reduction to Eqs. (4.2.4) is achieved if we set $u_3 \equiv \hat{u}_3 \equiv 0$ and if we write, e.g., $u_1 = e_{11}\, \hat{u}_1 + e_{12}\, \hat{u}_2 = \hat{u}_1 \cos\phi + \hat{u}_2 \cos[(\pi/2) + \phi] = \hat{u}_1 \cos\phi - \hat{u}_2 \sin\phi$.

$\overline{u_i(t;\, \mathbf{x}_1)u_j(t;\, \mathbf{x}_1 + \mathbf{h})}$ must depend only on the scalar separation r and must also obey certain coordinate transformation rules implying that it must equal $Ah_ih_j + B\delta_{ij}$, where A and B are *even functions* of r. Thus we find that

$$\begin{aligned}\overline{u_i(t;\, \mathbf{x}_1)u_j(t;\, \mathbf{i}_1\, x_1 + \mathbf{h})} &= e_{i1}e_{j1}\overline{u^2}(\mathbf{x}_1)[f(r;\, x_1) - g(r;\, x_1)] \\ &\quad + \overline{u^2}(x_1)g(r;\, x_1)\delta_{ij} \\ &= \overline{u^2}(x_1)\left[\frac{f(r;\, x_1) - g(r;\, x_1)}{r^2}\, h_ih_j + g(r;\, x_1)\delta_{ij}\right]\end{aligned} \tag{5.2.2}$$

where i and j and $\mathbf{h}$ are arbitrary and where we use the relations $h_i = e_{i1}r$ resulting from the special orientation of the $\hat{x}_1$ coordinate. To see that Eq. (5.2.2) is consistent with the first member of Eqs. (5.2.1), set $i = j = 2$ and $\mathbf{h} = \mathbf{i}_2\, r$ so that $h_i = h_j = r$. To do likewise for the second member, set $i = j = 1$ and $h_1 = 0$. To determine the correlation $\overline{u_1(t;\, \mathbf{x}_1)u_2(t;\, \mathbf{x}_1 + \mathbf{i}_2\, r)}$, set $i = 1, j = 2$, $h_1 = 0$ and find $\overline{u_1(t;\, \mathbf{x}_1)u_2(t;\, \mathbf{x}_1 + \mathbf{i}_2\, \mathrm{r})} = 0$. Finally, alter this last calculation by taking $h_1 = h_2 = 2^{-1/2}\, r$ so that we obtain the correlation of the two velocities with a separation in the first quadrant. In this case

$$\overline{u_1(t;\, x_1)\, u_2(t;\, x_1 + h_1,\, h_2,\, 0)} = \frac{1}{2}\,\overline{u^2}(x_1)[f(r;\, x_1) - g(r;\, x_1)]$$

We see that this correlation is not zero if $\mathrm{r} \neq 0$, but as $r \to 0$ the correlation related to the Reynolds shear stress goes to zero as expected.

Equation (5.2.2) gives the desired result, namely, the general two-point, one-time correlation of velocity components in isotropic, homogeneous turbulence, provided only that the two correlation functions $f(r;\, \mathbf{x})$ and $g(r;\, \mathbf{x})$ are known. Of significance to our discussion is the resulting relation between those two functions. Consider

$$\frac{\partial}{\partial h_k}\, \overline{u_i(t;\, x_1)u_k(t;\, \mathbf{i}_1x_1 + \mathbf{h})} = \overline{u_i(t;\, x_1)(\partial u_k/\partial h_k)(t;\, \mathbf{i}_1x_1 + \mathbf{h})} = 0$$

where from the continuity equation we have that $\partial u_k/\partial h_k \equiv 0$. Thus if we differentiate both sides of Eq. (5.2.2), we obtain the result due to von Karman and Howarth (1938):

$$g = f + \frac{1}{2}\, r\, \frac{\partial f}{\partial r} \tag{5.2.3}$$

We thus see that measurement of one correlation coefficient permits determination of the other. Despite the imperfect simulation of isotropic homogeneous turbulence provided by grid flow, Eq. (5.2.3) is well confirmed by experiments.

Equation (5.2.3) can be employed to infer certain information regarding the behavior of $g(r;\, x_1)$ as follows. Consider

$$\int_0^\infty dr\; r\; g(r; x_1) = \int_0^\infty dr\; r\; f(r; x_1) + \frac{1}{2}\int_0^\infty dr\; r^2\,\frac{\partial f}{\partial r} = \frac{1}{2}\, r^2 f|_0^\infty = 0 \qquad (5.2.4)$$

where we have integrated the second integral by parts and have assumed that f decays sufficiently rapidly as $r \rightarrow \infty$ so that the end-point contribution vanishes. The implication of Eq. (5.2.4) is that $g(r; x_1)$ must be negative for some range of r values.

Equations (5.2.1) can be used to calculate large and small length scales associated with spatial correlations in accord with Eqs. (3.4.8). The parallel and lateral integral scales are defined by

$$L_{Ip}(x_1) = \int_0^\infty dr\; f(r; x_1)$$
$$L_{Il}(x_1) = \int_0^\infty dr\; g(r; x_1) = \frac{1}{2}\, L_{Ip}(x_1) \qquad (5.2.5)$$

respectively, where in the second equation we follow the proceedure employed in Eq. (5.2.4). Recall from the previous section that in the initial period of decay L_{Ip} is roughly equal to the mesh size of the grid generating the turbulence.

Applying the notions used to establish Eq. (3.4.11), we define the Taylor microscale L_λ as

$$\frac{\partial^2 f}{\partial r^2}(r \rightarrow 0; x_1) = -\frac{1}{L_\lambda^2(x_1)}$$

which provides a measure of the rate at which the spatial correlation initially decreases. Thus, from Eq. (5.2.3) we have

$$f(r \rightarrow 0; x_1) \approx 1 - \frac{r^2}{2\, L_\lambda^2(x_1)} \cdots + g(r \rightarrow 0; x_1) \approx 1 - \frac{r^2}{L_\lambda^2(x_1)} + \ldots \qquad (5.2.6)$$

Figure 5.2.2 shows schematically typical distributions of the two correlation functions in terms of r/L_λ.

With this preliminary discussion suggesting the means for experimental determination of $f(r; x_1)$ and $g(r; x_1)$ and establishing certain of their mathematical properties, we change our emphasis and, as discussed earlier, consider these quantities to be functions of the scalar separation r measured from an arbitrary origin of a Cartesian coordinate system and of the time t, that is, we take $t = x_1/U_1$. The means for interpreting two-point, one-time averages at arbitrary distances downstream of the grid in terms of the *ensemble averages* required henceforth are discussed in Section 3.5. From this altered perspective the longitudinal and lateral correlation functions of Eqs. (5.2.1) can be written as

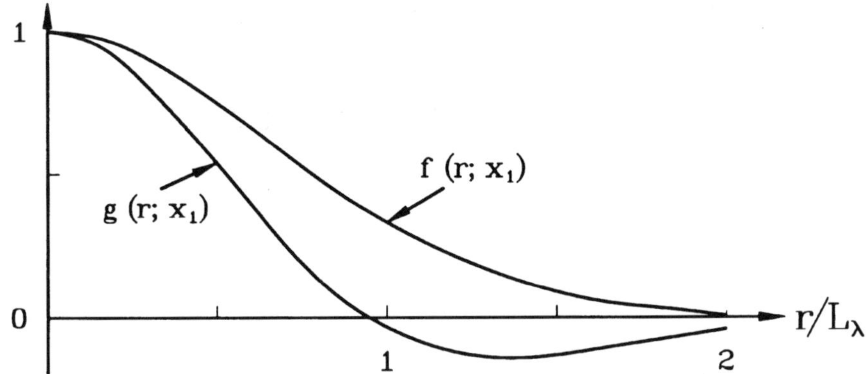

Figure 5.2.2 Distributions of the longitudinal and lateral correlation coefficients.

$$f(r, t) = \frac{\langle u_1(t, 0)u_1(t, r)\rangle}{\langle u_1^2(t)\rangle}$$

$$g(r, t) = \frac{\langle u_2(t, 0)u_2(t, r)\rangle}{\langle u_1^2(t)\rangle}$$

where the u_1 and u_2 velocity components are taken along and normal to the separation vector of length r connecting the two points, respectively. This reinterpretation requires that the time t in Eqs. (5.2.1) and in these equations have different meanings; in the former x_1 is fixed and t denotes the time associated with conventional time averaging at that spatial location, while in these equations the averaging involves repeated sampling of the velocities at a location associated with the mean flow time elapsed since passage through the grid.

To this point we have considered statistical information obtained experimentally. However, to proceed we need *an equation* for the correlation $u_i(t; \mathbf{x}_1)u_j(t; \mathbf{x}_1 + \mathbf{h})$ of Eq. (5.2.2) based on the Navier-Stokes equations. Such an equation can be derived as follows. Consider the x_i momentum equation at a point $\mathbf{x}$ and the x_j momentum equation at a second point $\mathbf{x} + \mathbf{h}$. The independent variables in this second equation are considered to be $\mathbf{x}$, $\mathbf{h}$, and t, and the spatial derivatives are changed accordingly. The two equations are then cross-multiplied, averaged, and added, which results in the general equation for the two-point, one-time correlation in question. The conditions of isotropy and homogeneity are then imposed; homogeneity eliminates all derivatives with respect to $\mathbf{x}$, so only those with respect to $\mathbf{h}$ and time remain. Finally, when Eqs. (5.2.2) and (5.2.3) are introduced to account for isotropy, there results a partial differential equation for $f(r; t)$, the von Karman-Howarth equation (cf. von Karman and Howarth, 1938),

$$\frac{\partial}{\partial t}[\langle u^2\rangle(t)f(r,t)] = \langle u^3\rangle(t)\left(\frac{\partial}{\partial r}+\frac{4}{r}\right)k(r,t) + 2\nu\langle u^2\rangle(t)\left(\frac{\partial^2}{\partial r^2}+\frac{4}{r}\frac{\partial}{\partial r}\right)f(r,t) \tag{5.2.7}$$

where $\langle u^2\rangle$ is the intensity of any of the velocity components and $k(r, t)$, not to be confused with the turbulent kinetic energy $k(t) = \frac{3}{2}\langle u^2\rangle$, is a triple correlation arising from the third-moment terms in the intensity equations [cf. Eq. (4.4.1)] and is the manifestation of the closure problem in this formulation.

Equation (5.2.7) is the starting point for several developments. If we accept that $k(r \to 0, t) \propto r^3$ and if we recall the behavior of $f(r \to 0, t)$ from Eq. (5.2.6) to evaluate the second term on the right side, Eq. (5.2.7) applied at $r = 0$ yields

$$\frac{d}{dt}\left[\frac{3}{2}\langle u^2\rangle(t)\right] = -\langle\varepsilon\rangle(t) = -\frac{15\,\nu\,\langle u^2\rangle(t)}{L_\lambda^2(t)} \tag{5.2.8}$$

where $\langle\varepsilon\rangle$ is the mean viscous dissipation, directly identified here with the rate of decay of the turbulent kinetic energy. This equation provides a connection between that rate and the Taylor microscale, a connection which prevails despite a finding later that the dissipation mechanism in intense turbulence arises from fluctuations with length scales *significantly smaller* than $L_\lambda(t)$.

Assume that $k(r, t)$ is represented in terms of $f(r, t)$ or otherwise given as a function of r and t. Then at any time t, the Taylor microscale $L_\lambda(t)$ is known from the nature of $f(r \to 0; t)$ [cf. Eq. (5.2.6) with x_1 replaced by t]. In this case Eqs. (5.2.7) and (5.2.8) determine $\langle u^2\rangle(t)$ and $f(r, t)$. Finally, an initial distribution, $f(r, 0)$, and the boundary conditions $f(0\,,\,t) = 1$ and $f(r \to \infty, t) = 0$ determine the evolution of $f(r, t)$.

Given a solution for $f(r, t)$, we can calculate the scalar correlation function $R(r, t)$ and the energy spectrum function $E(\kappa, t)$ developed in Section 3.5 via a special form of three-dimensional Fourier transforms. The former function determines the correlation length for the velocity components and the latter yields the turbulent kinetic energy $k(t)$ when integrated over all wave numbers. From Eqs. (5.2.2) and (5.2.3), the scalar correlation function defined by Eqs. (3.5.17) is related to $f(r, t)$ by

$$\begin{aligned} R(r,t) &= \frac{1}{2}\langle u_k(t)u_k(t;r)\rangle = \frac{1}{2}\langle u^2\rangle(t)[f(r,t) + 2\,g(r,t)] \\ &= \frac{1}{2}\langle u^2\rangle(t)\left[3f(r,t) + r\frac{\partial f}{\partial r}(r,t)\right] \end{aligned} \tag{5.2.9}$$

and is connected with the energy spectrum function $E(\kappa, t)$ via the Fourier transform pair of Eqs. (3.5.19),

$$R(r,t) = \int_0^\infty d\kappa\, E(\kappa,t)\,\frac{\sin \kappa r}{\kappa r}$$
$$E(\kappa,t) = \frac{2}{\pi}\int_0^\infty dr\, R(r,t)\,\kappa r \sin \kappa r \tag{5.2.10}$$

Note that the dimensions of R and E are U^2 and U^2L, respectively. It will be recalled that the wave number κ has the dimensions of L^{-1} and thus the product of velocity and wave number has the dimensions of frequency. Thus E can be considered an energy per unit mass U^2 per wave number $1/L$.

From Eq. (5.2.9) we see that the Karman-Howarth equation explicitly involves the scalar correlation function, which via the second of Eqs. (5.2.10) yields the energy spectrum function. However, if we take the Fourier transform of each of the momentum equations, invoke isotropy and homogeneity, and introduce the energy spectrum function, we obtain (cf. Batchelor, 1967) the spectral transfer equation which determines $E(\kappa, t)$ *directly,* although we must again expect a manifestation of the closure problem. We have

$$\frac{\partial}{\partial t} E(\kappa,t) = T(\kappa,t) - 2\,\nu\,\kappa^2 E(\kappa,t) \tag{5.2.11}$$

If $E(\kappa, t)$ is known from a solution to Eq. (5.2.11), closed in some appropriate fashion, then the first of Eqs. (5.2.10) determines $R(\kappa, t)$. The term on the left of Eq. (5.2.11) describes the rate of change with time of the energy at a particular wave number κ, while the second on the right accounts for the decrease in that energy due to viscosity at a rate which *increases quadratically* with wave number. Of course this term can *decrease* with increasing wave number, provided $E(k, t)$ decreases at a rate faster than κ^{-2}.

The first term on the right side, the spectral transfer function, describes the change in $E(\kappa, t)$ by exchange among wave numbers; for $T(\kappa, t) > 0$ there is a net energy transfer into wave number κ, leading, e.g., to an increase with time in $E(\kappa, t)$ and vice versa for $T(\kappa, t) < 0$. The unknown spectral transfer function is the manifestation of the closure problem in this formulation and, as expected, arises from the convection terms, i.e., from inertial effects, with their nonlinearity. It is related to the two-point, one-time correlation between one velocity component at one spatial location and two velocity components at another location.† We discuss models for this function in Section 5.4.

Since $T(\kappa, t)$ describes the transfer of energy among wave numbers, it possesses the property that

$$\int_0^\infty d\kappa\, T(\kappa,t) = 0$$

†See van Atta and Chen (1969) for a discussion of the means for *measuring* $T(\kappa, t)$ in grid turbulence under the assumption of local isotropy.

Thus integration of Eq. (5.2.11) over all wave numbers yields the result consistent with Eq. (5.2.8) that

$$\frac{d}{dt}\int_0^\infty d\kappa\, E(\kappa, t) = -2\nu \int_0^\infty d\kappa\, \kappa^2 E(\kappa, t)$$
$$= \frac{3}{2}\frac{d\langle u^2\rangle}{dt}(t) \equiv -\langle\varepsilon\rangle(t) \qquad (5.2.12)$$

We again identify the mean dissipation with the rate of change of the turbulent kinetic energy, but now make the connection between dissipation and the integral of $\kappa^2 E$. Later we shall need the dissipation at time zero,

$$\varepsilon_0 \equiv \langle\varepsilon\rangle(0) = 2\nu \int_0^\infty d\kappa\, \kappa^2 E(\kappa, 0) \qquad (5.2.13)$$

which implies that

$$\lim_{\kappa\to\infty} \kappa^2 E(\kappa, 0) = 0$$

i.e., that $E \propto \kappa^{-n}$ where $n > 2$.

If an appropriate form for $T(\kappa, t)$ is substituted into Eq. (5.2.11), we have in effect an ordinary differential equation for $E(\kappa, t)$ to be solved subject to an initial distribution $E(\kappa, 0) = E_0(\kappa)$. Such a solution determines the evolution with time of the energy spectrum function and the relative rates of decay of the high- and low-wave number contributions to the turbulent kinetic energy. This provides the connection for isotropic homogeneous turbulence between the large- and small-scale fluctuations we seek in this section. In addition, certain global information—for example, the instantaneous dissipation and, equivalently, the time rate of change of the turbulent kinetic energy—are obtained by appropriate integrals of $E(\kappa, t)$ [cf. Eqs. (5.2.12)]. Finally, from the first of Eqs. (5.2.10), the scalar correlation function $R(r, t)$ can be calculated so that the change with time of the correlation coefficients $f(r, t)$ and $g(r, t)$ is obtained from Eqs. (5.2.9) and (5.2.3), respectively.

Since time does not appear explicitly in Eq. (5.2.11), the assumed initial distribution $E_0(\kappa)$ may be associated with any phase of the evolution of the energy spectrum function. Accordingly, when we use $E_0(\kappa)$ to characterize the length and velocity scales of the turbulence, we are concerned with the evolution of $E(\kappa, t)$ *from that particular state* and not necessarily from its true initial state immediately after passage through the grid or after termination of stirring.

We examine next the behavior of the energy spectrum function as $\kappa \to 0$. In the second of Eqs. (5.2.10) with Eq. (5.2.9) substituted for $R(r, t)$ we let $\kappa \to 0$; to obtain a nontrivial result we must take two terms in the expansion of the sine function. There results

$$E(\kappa \to 0, t) = \frac{1}{3\pi} \langle u^2 \rangle(t)\, \kappa^4 \int_0^\infty dr\, r^4 f(r, t) \tag{5.2.14}$$

Now, if Eq. (5.2.7) is multiplied by $r^4\, dr$ and integrated from zero to infinity, and if both terms on the right side are integrated by parts, the first once and the second twice, it is found that

$$\frac{d}{dt}\left[\langle u^2 \rangle(t) \int_0^\infty dr\, r^4 f(r, t)\right] = r^4 k(r, t)\Big|_0^\infty \tag{5.2.15}$$

The implication from this equation is that $i_f k(r, t)$ decays sufficiently fast with increasing r so that the end-point contribution vanishes, e.g., if $\lim_{r\to\infty} k(r, t) \propto r^{-5}$, then

$$\left[\langle u^2 \rangle(t) \int_0^\infty dr\, r^4 f(r, t)\right] = \text{constant} = \left[\langle u^2 \rangle(0) \int_0^\infty dr\, r^4 f(r, 0)\right] \tag{5.2.16}$$

We thus see from Eq. (5.2.14) that under these circumstances $E(\kappa \to 0, t) = C\kappa^4$, where C is the Loitsianskii invariant (Loitsianskii, 1939). However, Batchelor and Proudman (1956) show that C is invariant only in the final stages of turbulent decay, i.e., in weak turbulence, and more generally is a slowly varying function of time. The implication is that the end-point contribution in Eq. (5.2.15) is not zero under these circumstances. Recently, Chasnov (1993) reported DNS results showing that $C \propto t^{0.25}$. For the purposes of our discussion we require only that C vary slowly with time.

This behavior of $E(\kappa, t)$ for small κ implies that large-scale fluctuations, i.e., those of small wave numbers, are fixed by the initial spectral distribution, hence largely dependent on the details of the initial turbulence generation, and possess a permanence not shared by the smaller-scale fluctuations. A second implication arises from the near invariance of C, namely, that as $\langle u^2 \rangle(t)$ decreases with time under the influence of dissipation, the longitudinal correlation $f(r, t)$ *increases with time* such that the combination reflected in Eq. (5.2.14) is invariant. We shall see the reason for this increase later, but note briefly here that the small fluctuations have a shorter lifetime than the larger-scale fluctuations, with the consequence that $f(r, t)$ *increases* as the flow evolves. These considerations lead to the schematic representation shown in Fig. 5.2.3 of the evolution of the energy spectrum function $E(\kappa, t)$ from an initial state $E_0(\kappa)$.

5.3 THE DIMENSIONLESS SPECTRAL TRANSFER EQUATION AND ITS IMPLICATIONS

Additional developments based on Eq. (5.2.11) are now taken up. Referring to Fig. 5.2.3, we see that the dominant contribution to the integral of $E(\kappa, t)$, that

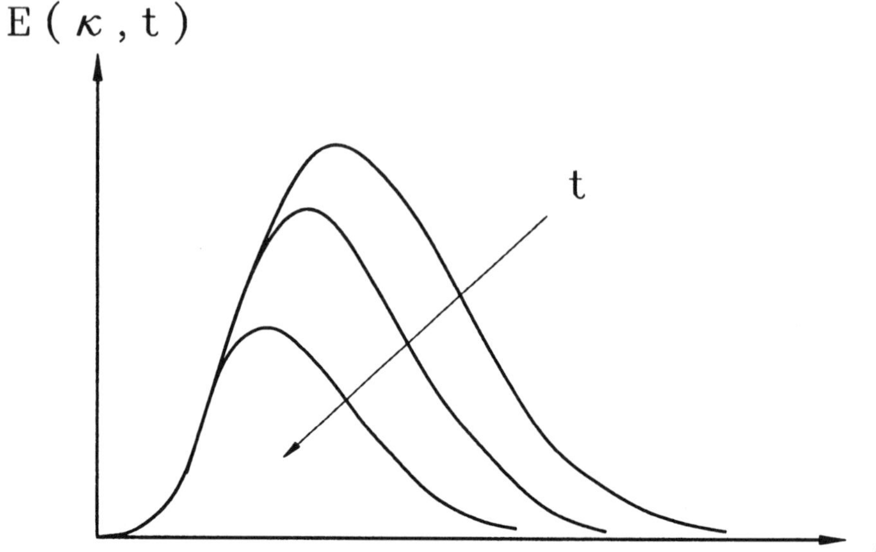

Figure 5.2.3 Evolution with time of the energy spectrum function $E(\kappa, t)$.

is, to the turbulent kinetic energy at any time, arises from wave numbers in the neighborhood of the maximum in E. Accordingly, we define a reference wave number κ_e based on the distribution $E_0(\kappa)$, where subscript e is indicative of the energy-containing fluctuations, and defined by

$$\frac{dE_0}{d\kappa}(\kappa = \kappa_e) = 0 \tag{5.3.1}$$

A reference velocity U and a corresponding Reynolds number characterizing the initial state of the turbulence are defined by

$$U \equiv \left[\frac{2}{3}\int_0^\infty d\kappa\, E_0(\kappa)\right]^{1/2} = [\langle u^2\rangle(0)]^{1/2} \qquad N_{R\kappa} \equiv \frac{U}{\kappa_e \nu} \tag{5.3.2}$$

In addition, we introduce the dimensionless variables

$$\tilde{t} = t\,U\kappa_e \qquad \tilde{\kappa} = \frac{\kappa}{\kappa_e} \qquad \tilde{E} = \frac{\kappa_e E}{U^2} \qquad \tilde{T} = \frac{T}{U^3} \tag{5.3.3}$$

so that Eq. (5.2.11) becomes

$$\frac{\partial \tilde{E}}{\partial \tilde{t}} = \tilde{T} - \frac{2\,\tilde{\kappa}^2\tilde{E}}{N_{R\kappa}} \tag{5.3.4}$$

Note that $(U\kappa_e)^{-1}$ is a time characterizing the spectral transfer rate. An equivalent notion is the *turnover time*; if we associate U with the turbulent kinetic energy and κ_e^{-1} with the large-scale fluctuations, then $L/k^{1/2}$ defines a turnover time. From this perspective $N_{R\kappa}$ is the ratio of a diffusion time $(\kappa_e^2\nu)^{-1}$ to the spectral transfer or turnover time $(U\ \kappa_e)^{-1}$.

We now assume that the initial conditions determining $N_{R\kappa}$ are such that $N_{R\kappa} << 1$. This implies that we begin our calculation of the evolution of the turbulence at a suitably long time after stirring has ceased, e.g., at the onset of the final period of decay in grid turbulence. Under these circumstances the second term on the right side of Eq. (5.3.4) dominates, and the rate of change of the energy spectrum function for arbitrary wave number is determined by dissipation. Thus we have the simple equation

$$\frac{\partial \tilde{E}}{\partial \tilde{t}} \approx -\frac{2\,\tilde{\kappa}^2\tilde{E}}{N_{R\kappa}}$$

which has the solution

$$\tilde{E}(\tilde{\kappa},\tilde{t}) = \tilde{E}_0(\tilde{\kappa})\exp\left(-\frac{2\,\tilde{\kappa}^2\tilde{t}}{N_{R\kappa}}\right) = \tilde{E}_0\exp(-\kappa^2\nu\ t) \tag{5.3.5}$$

We see clearly from Eq. (5.3.5) that the rate of decay of the energy spectrum function increases with increasing wave number, so small-scale, high-wave number fluctuations are short-lived. Conversely, we also see that $\tilde{E}(\tilde{\kappa}\to 0,\tilde{t}) \approx \tilde{E}_0(\tilde{\kappa}\to 0)$, so large-scale, low-wave number fluctuations exhibit permanence. These are the same properties deduced earlier from a different point of view but demonstrated explicitly here for $N_{R\kappa} << 1$. More specifically, from the second form of Eq. (5.3.5) we can define a characteristic decay time for fluctuations of wave number κ,

$$T_{l\kappa} = \frac{1}{2\,\kappa^2\nu}$$

the time in which the energy in a fluctuation of wave number κ is reduced to $1/e$ of its initial value. Clearly, it is this variation in the characteristic times of fluctuations of various scales which leads to the initial reduction in space and time correlations discussed in general terms in Chapter 3.

With the dimensionless quantities defined by Eqs. (5.3.3), the mean dissipation rate at the initial time is given by

$$\langle\varepsilon\rangle(0) \equiv \langle\varepsilon_0\rangle = 2\,\nu\,\kappa_e^2U^2\int_0^\infty d\tilde{\kappa}\ \tilde{\kappa}^2\tilde{E}_0$$

and we see that when $N_{R\kappa} << 1$ the dissipation takes a form involving the kinematic viscosity, that is, $\langle\varepsilon\rangle \propto \nu\kappa_e^2U^2$.[†]

[†]As will be seen later, this behavior contrasts with the dissipation rate in turbulence of high Reynolds numbers, dissipation which is independent of ν but rather proportional to U^3/L, where L is the scale of large fluctuations.

Although the solution given by Eq. (5.3.5) relates specifically to turbulence of low intensity, it provides a vehicle for developing the consequences of knowing the energy spectrum function $E(\kappa, t)$. For example, the evolution of the turbulent kinetic energy from its initial value is obtained by quadrature; from Eq. (5.2.11) we have

$$\frac{3}{2} \langle u^2 \rangle(t) = \int_0^\infty d\kappa \; E_0(\kappa) \exp(-\kappa^2 \nu t) \tag{5.3.6}$$

Similarly, from Eq. (5.2.10) we obtain the scalar correlation function as

$$R(r, t) = \int_{0t}^\infty d\kappa \; E_0(\kappa) \exp(-\kappa^2 \nu t) \frac{\sin \kappa r}{\kappa r} \tag{5.3.7}$$

and from Eq. (5.2.9) we obtain the longitudinal correlation function $f(r, t)$ as

$$f(r, t) = \frac{2}{r^3 \langle u^2 \rangle(t)} \int_0^r dr' \; r'^2 \; R(r', t) \tag{5.3.8}$$

Note that $f(r \rightarrow 0, t) = 1$ as required. The lateral correlation coefficient is obtained from Eq. (5.2.3). Finally, the one-dimensional velocity spectrum tensor is obtained from Eq. (3.5.23). Thus we see that the turbulence characteristics follow from the solution for $E(\kappa, t)$.

Suppose we now wish to associate initial conditions with the turbulent state either near the grid or close to termination of stirring, where the turbulence is vigorous. Here it is tempting to seek an approximation for $N_{R\kappa} >> 1$ and to consider only the $\tilde{T}$ term on the right side of Eq. (5.3.4). However, the dimensionless form of Eq. (5.2.12) obtained by use of the new variables given by Eqs. (5.3.3) and by a convenient change of sign indicates the dangers of such an approach; we have

$$-\frac{d}{d\tilde{t}} \int_0^\infty d\tilde{\kappa} \; \tilde{E}(\tilde{\kappa}, \tilde{t}) = \frac{2}{N_{R\kappa}} \int_0^\infty d\tilde{\kappa} \; \tilde{\kappa}^2 \tilde{E}(\tilde{\kappa}, \tilde{t}) = \frac{\langle \varepsilon \rangle}{\kappa_e u^3} \equiv \tilde{\varepsilon}(\tilde{t}) \tag{5.3.9}$$

Since the integral on the left side is positive, we see that no matter how large the Reynolds number $N_{R\kappa}$ is, the dissipation cannot be neglected; i.e., as $N_{R\kappa}$ increases, so too must the integral, and we must deal with all three terms in Eq. (3.5.4). This is an equation rich in physical detail which can be exposed without actually developing complete solutions, as we shall do later for a restricted range of κ by introducing a model for the spectral transfer function.

To proceed, recall that the wave numbers in the range of $\tilde{\kappa} \approx 1$ are associated with the energy-containing fluctuations and introduce a second wave number κ_d associated with the dissipative range making the greatest contribution to the integral on the left of Eq. (5.3.10). Thus we have

$$\frac{d}{d\kappa} (\kappa^2 E_0)(\kappa = \kappa_d) = 0 \tag{5.3.10}$$

as depicted in Fig. 5.3.1. We thus encounter a two-scale problem associated with κ_e and κ_d. Clearly, we are led to a new wave number scaling based on κ_d, leading to a new independent variable $\hat{\kappa} \equiv \kappa/\kappa_d$, but less obvious is the scaling for the energy spectrum function. Without a priori justification, consider

$$\hat{E} \equiv \kappa_e^2 \frac{U}{\langle \varepsilon_0 \rangle} \left(\frac{\kappa_d}{\kappa_e} \right)^{5/3} E \qquad \hat{T} = \frac{\kappa_d T}{\langle \varepsilon_0 \rangle} \tag{5.3.11}$$

where $\langle \varepsilon_0 \rangle$ is the mean dissipation at $t = 0$ defined earlier. In this case the connection between the dissipation and the integral of $\kappa^2 E$ [cf. Eq. (5.3.10)] applied at time zero yields

$$\left(\frac{\kappa_d}{\kappa_e} \right)^{4/3} \frac{1}{N_{R\kappa}} \int_0^\infty d\hat{\kappa}\; \hat{\kappa}^2 \hat{E}_0(\hat{\kappa}) = \frac{1}{2}$$

which implies that as $N_{R\kappa} \to \infty$,

$$\left(\frac{\kappa_d}{\kappa_e} \right)^{4/3} \frac{1}{N_{R\kappa}} = O(1) \tag{5.3.12}$$

and hence that the ratio κ_d/κ_e between the wave numbers associated with the turbulent kinetic energy and dissipation increases with $N_{R\kappa}$. This behavior was first noted by Taylor (1938), who used a spectrum determined from grid turbulence to show that $E(\kappa, t)$ and $\kappa^2 E(\kappa, t)$ involve essentially nonoverlapping ranges of κ, as we show schematically in Fig. 5.3.1.

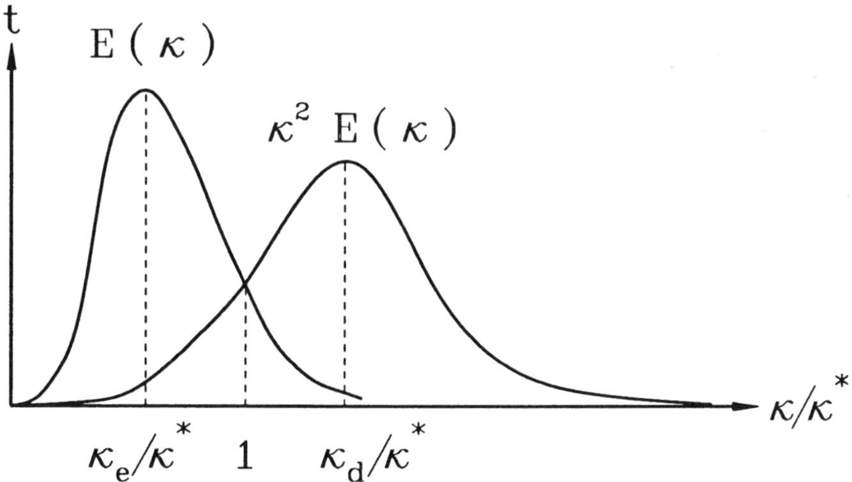

Figure 5.3.1 The nonoverlapping of the energy and dissipation energy spectrum function for high Reynolds numbers.

Proceeding further, we substitute Eqs. (5.3.11) into Eq. (5.2.11) *without rescaling the time* and find

$$\frac{1}{2}\left(\frac{\kappa_e}{\kappa_d}\right)^{2/3}\frac{\partial \hat{E}}{\partial \tilde{t}}(\hat{\kappa}, \tilde{t}) = \frac{1}{2}\hat{T}(\hat{\kappa}, \tilde{t}) - \left(\frac{\kappa_d}{\kappa_e}\right)^{4/3}\frac{1}{N_{R\kappa}}\hat{\kappa}^2\hat{E}(\hat{\kappa}, \tilde{t}) \tag{5.3.13}$$

As $N_{R\kappa}$ and the wave number ratio (κ_d/κ_e) increase indefinitely but subject to the constraint of Eq. (5.3.12), the right side of Eq. (5.3.13) is well behaved whereas the left side approaches zero. It is this result which justifies a posteriori the scaling of Eq. (5.3.11). The implication is that on a time scale associated with the spectral transfer time there is a balance or equilibrium between dissipation associated with fluctuations having wave numbers in the neighborhood of $\hat{\kappa} = 1$, the dissipative range, and spectral transfer associated with inertia. Thus $\hat{E}$ as determined by Eq. (5.3.13) depends primarily on $\hat{\kappa}$ and weakly on time. Moreover, the absence of κ_e except in the quotient $\kappa_d/\kappa_e >> 1$ implies that the energy spectrum in terms of $\hat{E}(\hat{\kappa})$ is independent of the low-wave number, energy-containing fluctuations, except as $\hat{\kappa} \rightarrow 0$, a limit which implies that $\kappa \rightarrow \kappa_e$. This defines the range of equilibrium wave numbers. It should be recognized that the slow evolution of the energy spectrum function could be described by a new time $\tilde{t}$ rescaled by multiplication by a large quantity $(\kappa_d/\kappa_e)^{2/3}$, so that the left side of Eq. (5.3.13) in the new time variable is no longer small. This development, of no interest in the present discussion, is not pursued.

On the basis of these arguments we write, as an approximation to Eq. (5.3.13),

$$\frac{1}{2}\hat{T}(\hat{\kappa}) - \left(\frac{\kappa_d}{\kappa_e}\right)^{4/3}\frac{1}{N_{R\kappa}}\hat{\kappa}^2\hat{E}(\hat{\kappa}) = 0 \tag{5.3.14}$$

where, since time is simply a parameter, we drop its explicit functional dependence. In so doing we emphasize the balance for high wave numbers between spectral transfer and dissipation when $N_{R\kappa} >> 1$.

Several comments are appropriate. The short-time approximation leading to Eq. (5.3.14) is consistent with our earlier finding of the short lifetimes of the small-scale, high-wave number fluctuations. The influence of the large-scale fluctuations on the small scales arises from the spectral transfer term in Eq. (5.3.14), but the absence of a time dependence of this influence implies that the behavior of the small scales must be dependent only on parameters characterizing those scales and must be independent of parameters characterizing the large scales. Kolmogorov hypothesizes that there are only two such parameters: the mean dissipation $\langle\varepsilon\rangle$ and the kinematic viscosity ν. Accepting this hypothesis, we can use dimensional arguments to define various scales at these wave numbers:

$$L_K = \left[\frac{\nu^3}{\langle\varepsilon\rangle}\right]^{1/4} \qquad U_K = (\nu\langle\varepsilon\rangle)^{1/4} \qquad T_K = \left[\frac{\nu}{\langle\varepsilon\rangle}\right]^{1/2} \tag{5.3.15}$$

These are the Kolmogorov parameters associated with the high wave numbers in turbulence. Note that a consequence of these definitions is that $U_K L_K/\nu = 1$, that is, the Reynolds number based on U_K and L_K is unity while the Kolmogorov time $T_K = L_K/U_K$ is a turnover time for the smallest fluctuations. The most important of these quantities is the Kolmogorov length L_K, which for the flow under consideration is a function of time via $\langle\varepsilon\rangle(t)$. Empiricism establishes that κ_d is almost a decade smaller than L_K^{-1}, an indication of the limitations of dimensional analysis; i.e., we could readily introduce a constant into the definition of L_K so that $L_K \approx \kappa_d^{-1}$, but this is not standard.† Since L_K relates to scales smaller than those associated with dissipation and thus to lengths over which viscous stresses are certainly dominant, we can make the following inferences relative to the Kolmogorov length. Suppose that at a given instant of time the velocity vectors at two spatially adjacent points in a turbulent flow are known either from experiment or from calculation. Then, if these two points are within L_K apart, we can interpolate so as to make meaningful estimates of the distribution of velocity vectors along the line connecting the two points. On the other hand, if the points are separated by more than L_K, such interpolation can be either greatly in error or meaningless. In a similar sense we can estimate the velocity at a fixed location but at an advanced time from the velocity and its time derivatives at an arbitrary time τ [cf. Eq. (3.4.4)], provided $\tau < T_K$.

It is this perspective which leads to the restriction on the maximum grid size in direct numerical solutions of the conservation equations, i.e., to Eq. (2.7.1); if finite difference approximations of the partial derivatives are to be accurate, it can be argued that the grid spacing must be roughly *equal to* the Kolmogorov length. However, experience with such solutions indicates that an accuracy acceptable for many purposes is obtained if the spacing is some small multiple of L_K, for example, four to six L_K. This finding tends to set the grid spacing at about 0.5 κ_d^{-1}. In experimental turbulence, however, significant efforts are made to use probes whose spatial resolution equals a cube with sides equal to the anticipated L_K.

We now return to developments connected with Eq. (5.3.4) for $N_{R\kappa} >> 1$. An extension of the dimensional analysis leading to Eqs. (5.3.15) based on the Kolmogorov hypothesis that the energy spectrum function depends only on κ, the kinematic viscosity ν, and the mean viscous dissipation $\langle\varepsilon\rangle$ yields

$$E = \beta\,(\langle\varepsilon\rangle\nu)^{1/2}L_K\,(\kappa L_K)^{-(5/3)+(4/3)m} \tag{5.3.16}$$

where β is an empirical coefficient and where the power law form of the right side anticipates later developments and m is an empirical exponent that is unknown at this point. An alternative form for two of the factors on the right side found from the first of Eqs. (5.3.15) is

$$(\langle\varepsilon\rangle\nu)^{1/4}L_K = \frac{\nu^2}{L_K}$$

The importance of the Kolmogorov length suggests that estimates of the

†Hinze (1975) states that $\kappa_d = 0.2\ L_K^{-1}$, but currently $\kappa_d = 0.1\ L_K^{-1}$ is believed to be a more appropriate estimate.

mean viscous dissipation in various flows are valuable. Two perspectives are possible. If a rough estimate for L_K prior to any experiment or computation is needed, a global value can be obtained for various flows as follows. Consider a turbulent boundary layer with an external velocity U_1 and a thickness L_δ at a particular streamwise station; then $\langle\varepsilon\rangle \approx U_1^3/L_\delta$. We shall see in Section 10.8 that the dissipation rate varies widely over the thickness of the boundary layer, but global estimates cannot take such nonuniformity into account. In the case of a mixing layer with velocity U_1 in the high-speed stream, U_0 in the low-speed stream, and thickness L_δ at a particular streamwise station, we can again estimate the mean viscous dissipation as $(U_1 - U_0)^3/L_\delta$. In fully developed pipe flow, the rate of work by the mean pressure drop per unit length per unit mass of fluid equals the mean viscous dissipation; thus a straightforward calculation yields $\langle\varepsilon\rangle \approx (\overline{U}/\rho)(dP/dx_1)$, where $\overline{U}$ is the average velocity across the cross section. By using one of the various laws relating the pressure drop to the friction factor (cf., e.g., Bird et al., 1960), this formula leads to a relation between $\langle\varepsilon\rangle$ and the pipe Reynolds number. Again the mean viscous dissipation varies considerably across the radius of the pipe. Finally, previous experiments in turbulent grid flow for a particular grid geometry permit estimates to be made of the rate of decay of the turbulent kinetic energy so that $\langle\varepsilon\rangle$ can be estimated from Eq. (5.2.12). The adequacy of all of these global estimates relative to L_K clearly benefits from the dependence $L_K \propto \varepsilon^{-1/4}$.

An alternative perspective for determining the mean viscous dissipation involves making a preliminary measurement of the fluctuations of the principal velocity in a turbulent shear flow $u_1(t; \mathbf{x})$, where the mean velocity is U_1. Then from Eq. (4.7.6), which is based on the assumptions of local isotropy and the Taylor hypothesis, we have

$$\langle\varepsilon\rangle \approx 15 \frac{\nu}{U_1} \overline{\left(\frac{\partial u_1}{\partial t}\right)^2} \tag{5.3.17}$$

By measuring $\partial u_1/\partial t$ at appropriate points in the flow, the needed values of $\langle\varepsilon\rangle$ and thus of L_K can be determined. Equation (5.3.17) provides the most widely used means for the experimental determination of mean viscous dissipation. Thus, when presentation of experimental results calls for a value of $\langle\varepsilon\rangle$ (cf., e.g., Fig. 5.5.1), this equation is usually adopted.

Returning to Eq. (5.3.14), as $\hat{\kappa} \to 0$, that is, as $\kappa \to \kappa_e$ so that we consider the low-wave number end of the equilibrium range, we see that the dissipation becomes less important and spectral transfer becomes zero. That is, at any given wave number in this range, the energy transferred from fluctuations of lower wave numbers equals that transferred to fluctuations of higher wave numbers, so the *net energy transfer* is zero. This is the inertial subrange of the wave number spectrum, subrange because it is a portion of the equilibrium range. As suggested by our earlier considerations, in this range E does not depend on the kinematic viscosity, so the exponent m in Eq. (5.3.13) must be set to zero and we have

$$E = \beta\,(\langle\varepsilon\rangle\nu)^{1/2}L_K\,(\kappa L_K)^{-5/3} = \beta\,\frac{\nu^2}{L_K}\,(\kappa\,L_K)^{-5/3} \tag{5.3.18}$$

where β is an empirical constant with a value in the range 1.44–1.50 (cf., e.g., Grant et al., 1962). This theoretical deduction is a central result from the study of isotropic, homogeneous turbulence, namely, that in the inertial subrange the spectral transfer function is zero and the energy spectrum function decays as $\kappa^{-5/3}$. The significance of this result in the presentation and interpretation of experimental data from a variety of turbulent flows is discussed in Section 5.5.

These considerations result in a spectral transfer function $T(\kappa, t)$ shown schematically in Fig. 5.3.2. At low wave numbers there is a net energy transfer *out of* fluctuations of large scale into those of smaller scale; a spectral range in which such transfer is zero, the inertial subrange; and a final range of high wave numbers involving a net gain of energy which is balanced by viscous dissipation. This is the energy cascade process of turbulence. The extent of the wave number range for which $T \approx 0$ depends on the turbulence Reynolds number $N_{R\kappa}$ in the present analysis. The magnitude of the energy transfer through the inertial subrange of wave numbers to the dissipative range is determined by the turbulent energy, i.e., by fluctuations associated with wave numbers in the neighborhood of κ_e, is independent of the kinematic viscosity and equals the mean viscous dissipation. Provided we accept the existence of a mechanism for transferring energy among fluctuations of various scales, i.e., a mechanism described by

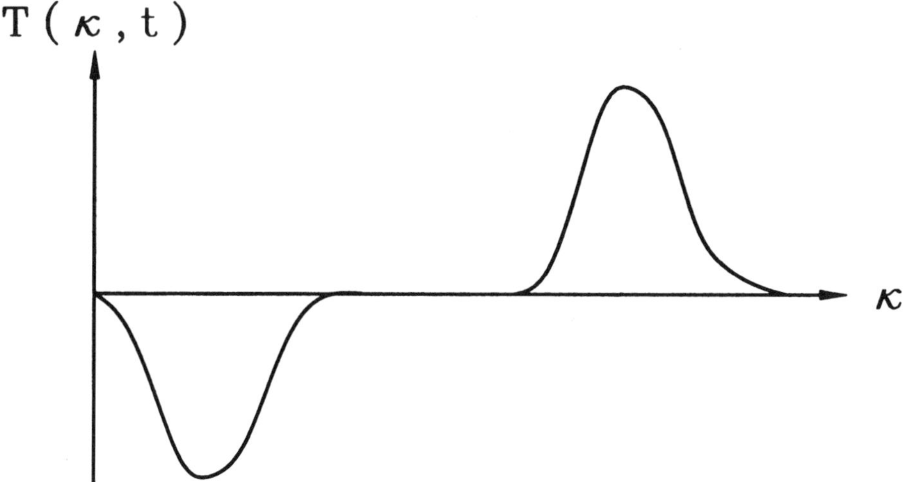

Figure 5.3.2 Spectral transfer function for high-turbulence Reynolds number.

$T(\kappa, t)$, the relatively long lifetimes of the energy-containing fluctuations, and the relatively short lifetimes of the fluctuations responsible for dissipating energy, these three features lead naturally to the energy cascade process. This observation provides the theoretical basis for associating the mean viscous dissipation with $\langle \varepsilon \rangle \propto k^{3/2}/L_l$ when the turbulence Reynolds number is large. Here L_l is associated with the large turbulence scales. In the global estimates for the mean viscous dissipation discussed earlier, we replace $k^{1/2}$ either by a mean velocity or by a difference in mean velocities and L_l by a global length, e.g., by L_δ; in such replacements both the numerator and denominator involve factors less than unity. Typically, $k^{1/2} \approx 0.1\ U_1$ and $L_l \approx 0.5\ L_\delta$. As a consequence, the estimate for $\langle \varepsilon \rangle$ based on $k^{3/2}/L_l$ is less than our earlier global estimates. However, the differences in the estimated Kolmogorov length are small since, as noted earlier, $L_K \propto \langle \varepsilon \rangle^{-1/4}$.

5.4 MODELS FOR SPECTRAL TRANSFER

Although Eq. (5.2.11) is restricted to the highly idealized case of isotropic, homogeneous turbulence, it is of fundamental importance to the mechanics of turbulence. Hence considerable attention has been devoted to its closure by means of models for the spectral transfer term $T(\kappa, t)$ of widely varying complexity. The direct interaction model of Kraichnan (cf., e.g., Kraichnan, 1966) and the quasi-normality approximation of Millionshtchikov (1941) (cf. also Proudman and Reid, 1954; Chandrasekhar, 1956) are important examples. Hinze (1975) provides a useful review of various assumed forms for the spectral transfer function.

A proper model of the physical processes of energy transfer in wave number space must describe the gain or loss of energy from or to wave numbers smaller than a given κ as well as the loss or gain to or from larger wave numbers. Representative of such models is one due to von Karman (1948),

$$T(\kappa, t) = 2\,\alpha \left\{ E^{\phi}\,\kappa^{\psi} \int_0^{\kappa} d\kappa'\ [E(\kappa', t)]^{(3/2)-\phi}\ \kappa'^{(1/2)\psi} - E^{(3/2)-\phi}\ \kappa^{(1/2)\psi} \int_{\kappa}^{\infty} d\kappa' [E(\kappa', t)]^{\phi}\ \kappa'^{\psi} \right\} \tag{5.4.1}$$

where α is a dimensionless constant and ϕ and ψ are parameters that are specific to each model. As $\kappa \to 0$ the second integral dominates and $T(\kappa, t) < 0$; that is, at small wave numbers there is a net energy loss to fluctuations at higher wave numbers. By contrast, as $\kappa \to \infty$ the first integral dominates and there is a net gain in energy from fluctuations associated with smaller wave numbers. In some intermediate range of wave numbers the two integrals are nearly equal and opposite and $T(\kappa, t)$ is suitably "small" in some sense. This behavior is in accord with the ideas set forth in the previous section and indeed motivates the form of Eq. (5.4.1).

Somewhat simpler models for $T(\kappa, t)$, restricted to the *equilibrium range of wave numbers* and based on gradient transport models for the transport of energy in wave number space, have been proposed by Heisenberg (1948) and Kovasznay (1948). Here a new function $S(\kappa, t)$ replaces $T(\kappa, t)$ as the quantity to be modeled, where

$$T(\kappa, t) = -\frac{\partial S}{\partial \kappa}$$

Integration yields

$$S(\kappa, t) \equiv -\int_0^{\kappa} d\kappa' T(\kappa', t) \tag{5.4.2}$$

so that $S(\kappa, t)$ represents the total energy transferred out of the range from $\kappa = 0$ to a given κ. As depicted in Fig. 5.3.2, this implies that $S > 0$ for small κ and that $S \to 0$ with increasing κ.

Kovasznay assumes that

$$S(\kappa, t) = \alpha\, \kappa^{5/2} E(\kappa, t)^{3/2} \tag{5.4.3}$$

where α is an unknown coefficient. Restriction to the equilibrium range is reflected in the following approximation. We integrate Eq. (5.2.11) with respect to κ from 0 to a value of κ sufficiently large so that

$$\int_0^{\kappa} d\kappa' E(\kappa') \approx \int_0^{\infty} d\kappa E(\kappa)$$

where we again drop the time dependence to reflect the quasi-steady nature of the energy spectrum function in the equilibrium range. The implication of this approximation is that the smallest value of the upper limit κ is sufficiently large that negligible energy is contained in $E(\kappa, t)$ for even larger wave numbers. In this case we have from Eq. (5.2.11) integrated from 0 to κ and Eqs. (5.2.12) and (5.4.3),

$$\langle \varepsilon \rangle = \alpha\, \kappa^{5/2} E(\kappa)^{3/2} + 2\nu \int_0^{\kappa} d\kappa' \kappa'^2 E(\kappa') \tag{5.4.4}$$

which is an equation for determining $E(\kappa)$.

If we assume a functional form for $E(\kappa)$, namely,

$$E(\kappa) \propto \frac{(1 - \bar{\kappa}^{4/3})^2}{\bar{\kappa}^{5/3}}$$

where $\bar{\kappa} \equiv \kappa L_K$ is a dimensionless wave number based on the Kolmogorov scale L_K [cf. Eq. (5.3.15)], substitution into Eq. (5.4.4) leads after some algebra to

$$\frac{E(\bar{\kappa})\, L_K}{\nu^2} = 2\, \frac{(1 - \bar{\kappa}^{4/3})^2}{\bar{\kappa}^{5/3}} \tag{5.4.5}$$

where the form of the solution is chosen to satisfy Eq. (5.3.16). This solution applies for $\bar{\kappa} \leq 1$, while for $\bar{\kappa} > 1$ we take $E(\bar{\kappa}) = 0$, representing a spectral cutoff at the wave number corresponding to the reciprocal of the Kolmogorov length. Equation (5.4.5) holds provided the coefficient in Eq. (5.4.3) is chosen to be $\alpha = 2^{-3/2}$. The combination of Eqs. (5.4.2), (5.4.3), and (5.4.5) gives

$$T(\bar{\kappa}) = 4\left(\frac{\nu}{L_K}\right)^3 \bar{\kappa}^{1/3}(1 - \bar{\kappa}^{4/3})^2 \tag{5.4.6}$$

As $\bar{\kappa} \to 0$, i.e., as $\kappa \to \kappa_e$, we see from Eq. (5.4.5) and (5.4.6) that

$$\frac{E(\bar{\kappa} \to 0)\, L_K}{\nu^2} = \frac{2}{\bar{\kappa}^{5/3}} \qquad T(\bar{\kappa} \to 0) = \lim_{\bar{\kappa} \to 0} 4\left(\frac{\nu}{L_K}\right)^3 \bar{\kappa}^{1/3} = 0 \tag{5.4.7}$$

i.e., we obtain the $-5/3$ decay of the energy spectrum function, the vanishing of the spectral transfer function as $\kappa \to \kappa_e$ attributed earlier to the inertial subrange [cf. Eqs. (3.5.18)], now with a value of $\alpha = 2$ compared with previously cited values in the range 1.4–1.5. Of course, it is the need to replicate this behavior which dictates the form for $S(\kappa, t)$ in Eq. (5.4.3). Note that if we eliminate L_K by use of its definition from the first of Eqs. (5.3.15), the limit $E(\bar{\kappa} \to 0)$ is independent of the kinematic viscosity, as discussed earlier in connection with Eq. (5.3.18). Furthermore, our earlier considerations connected with Eq. (5.3.15) imply that κ_d corresponds to $\bar{\kappa} \approx 0.1-0.2$. Equations (5.4.5) and (5.4.6) also give

$$\frac{E(\bar{\kappa} \to 1)\, L_K}{\nu^2} = \frac{32}{9}(1 - \bar{\kappa})^2 \qquad T(\bar{\kappa} \to 1) = \frac{64}{9}\left(\frac{\nu}{L_K}\right)^3 (1 - \bar{\kappa})^2 \tag{5.4.8}$$

In Fig. 5.4.1 we show the three-dimensional spectrum function and the spectral transfer function along with their asymptotic behavior as $\bar{\kappa} \to 0$. We shall refer to this figure in Section 5.5. Other models for either $T(\kappa, t)$ or its equivalent $S(\kappa, \mathrm{t})$ in the equilibrium range of wave numbers give results similar to those of Fig. 5.4.1.

The discussion in this and the previous section constitutes the theory of local similarity of the small turbulence scales due to Kolmogorov (1941). Because of the conceptual importance of these ideas, Eq. (5.2.11) has been assessed by means of several experiments utilizing grid flow to simulate isotropic, homogeneous turbulence. In an early experiment, Uberoi (1963) measured the term on the left side of this equation and the dissipation term on the right side and then *deduced* the spectral transfer function. Somewhat later, van Atta and Chen (1969) reported measurements of all three terms in grid turbulence and thereby tested the validity of Eq. (5.2.11). They concluded: "For large enough wave numbers the spectral transfer of turbulent energy in grid turbulence is adequately described (within experimental uncertainty) by the spectral balance equation for isotropic turbulence." However, because of the small turbulence Reynolds num-

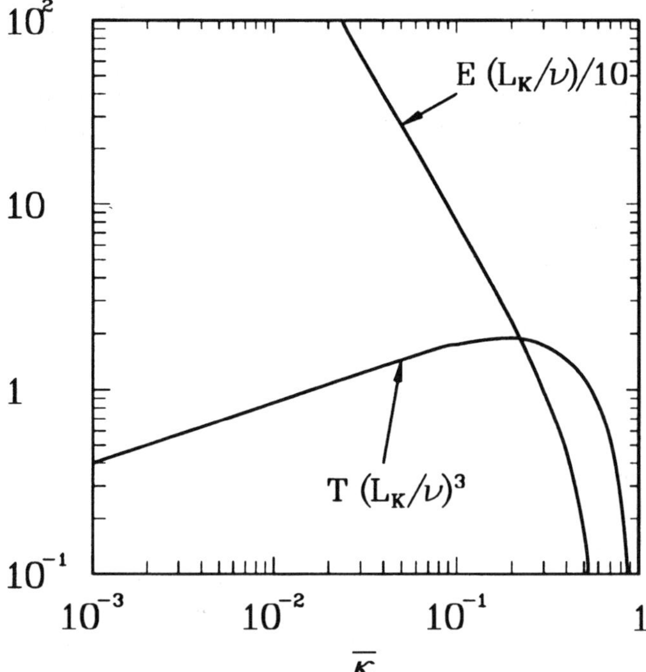

Figure 5.4.1 Three-dimensional energy function and spectral transfer function in the equilibrium range as given by the Kovasznay model.

bers associated with their flow, they found no range of wave numbers for which the spectral transfer function is essentially zero. In a related series of experiments involving higher turbulence Reynolds numbers, Helland and van Atta et al. (1977) showed that there is *a tendency* for $T(\kappa)$ to involve a region of $T \approx 0$. In the next section we discuss experiments in *turbulent shear flows* aimed at assessing the applicability of the theory discussed in this section outside the realm of grid turbulence.

5.5 IMPLICATIONS OF THE KOLMOGOROV THEORY

From the discussion in Sections 5.2–5.4 we surmise that homogeneous isotropic turbulence at large Reynolds number is characterized by fluctuations of large scale or small wave number that are effectively uncoupled from fluctuations of small scale or large wave number. We also learn that between these two spectral ranges, i.e., between the low-wave number, energy-containing range and the high-wave number dissipative range, there is an inertial subrange within which the energy spectrum function $E(\kappa, t)$ varies as $\kappa^{-5/3}$. Finally, we also surmise that large-scale fluctuations possess a relatively long lifetime, while small-scale fluctuations are short-lived. Also, we have seen that in grid turbulence which closely approximates isotropic homogeneous conditions, many of these ideas are well supported by experiment. In this section we address the implications of these findings for turbulent flows of applied interest, those with mean shear, such as jets, wakes, boundary layers, etc. We also discuss briefly the implications for flows with buoyancy.

The discussion in Section 5.1 indicates that the large-scale fluctuations in turbulent shear flows generally involve different length scales representing different correlation lengths in orthogonal directions and hence are anisotropic. However, if the Reynolds number is sufficiently large, we may inquire whether the same uncoupling of the large and small scales found to prevail in isotropic, homogeneous turbulence leads to isotropy of the small scales in flows with mean shear despite the characteristic anisotropy of the large scales.

We saw in Section 4.7 and shall see again in Section 6.3 that pressure fluctuations redistribute the energy associated with velocity fluctuations in the various spatial directions. Thus fluctuations in one direction may be created by a mean shear but transferred to another direction by means of pressure fluctuations, implying that the pressure fluctuations tend to make turbulence isotropic. Since we know that the velocity fluctuations of small scale have a short lifetime, they might be expected to be highly responsive to this redistributive effect. Thus, if the range of scales is suitably large—i.e., if the Reynolds number is suitably high—the small scales may be isotropic while the long-lived, relatively nonresponsive large scales are anisotropic. This reasoning is generally accepted as being applicable to turbulent flows of applied interest and has significant implications for turbulence modeling in that most closure models for dissipation *assume* local isotropy of the small scales.

We now pursue this reasoning further. Because of its importance to the fundamentals of turbulence as well as to modeling and computation, the assumption of the isotropy of the small scales or of *local isotropy* in large-Reynolds-number shear flows has received extensive experimental examination according to various criteria, e.g., by Champagne (1978), van Atta (1991), and Saddoughi and Veeravalli (1994). One criterion for isotropy is based on Eqs. (4.7.8), which implies for isotropic flow the following relations among the rates of strain:

$$\begin{aligned}
&\overline{\left(\frac{\partial u_1}{\partial x_1}\right)^2} = \frac{1}{2}\overline{\left(\frac{\partial u_2}{\partial x_1}\right)^2} = \frac{1}{2}\overline{\left(\frac{\partial u_3}{\partial x_1}\right)^2} \\
&\overline{\left(\frac{\partial u_1}{\partial x_2}\right)^2} = \overline{\left(\frac{\partial u_1}{\partial x_3}\right)^2} = \overline{\left(\frac{\partial u_2}{\partial x_1}\right)^2} = \overline{\left(\frac{\partial u_2}{\partial x_3}\right)^2} \\
&\qquad = \overline{\left(\frac{\partial u_3}{\partial x_1}\right)^2} = \overline{\left(\frac{\partial u_3}{\partial x_2}\right)^2} = 2\,\overline{\left(\frac{\partial u_1}{\partial x_1}\right)^2}
\end{aligned} \tag{5.5.1}$$

Corresponding relations involving a scalar such as temperature are

$$\overline{\left(\frac{\partial\theta}{\partial x_1}\right)^2} = \overline{\left(\frac{\partial\theta}{\partial x_2}\right)^2} = \overline{\left(\frac{\partial\theta}{\partial x_3}\right)^2}$$
$$\overline{\left(\frac{\partial\theta}{\partial x_1}\right)^3} = \overline{\left(\frac{\partial\theta}{\partial x_2}\right)^3} = \overline{\left(\frac{\partial\theta}{\partial x_3}\right)^3} \tag{5.5.2}$$

Since the correlations involved, being gradient, tend to be dominated by small-scale fluctuations, satisfaction of these equations in turbulent shear flows supports the notion of local isotropy of the small turbulent scales. In examining experimentally the extent to which these conditions are satisfied, the gradients in the streamwise direction can be determined from a single sensor by invoking the Taylor hypothesis but cross-stream gradients require multiple sensors. Champagne (1978) shows that Eqs. (5.5.1) are approximately satisfied in a variety of different turbulent flows.

Other experiments have been carried out at various Reynolds numbers and various mean rates of shear and involving other criteria. Spectral criteria have the advantage of allowing for the isotropy of the smallest scales, i.e., of avoiding contamination of the averages in Eqs. (5.5.1) and (5.5.2) by lower-frequency fluctuations which may not be isotropic (cf. van Atta, 1991). van Atta reviews a number of experimental investigations into the isotropy of small turbulent scales, including the wake of a cylinder considered by Townsend (1956) and the large mixing layer and the return flow in a large wind tunnel considered by Praskovsky et al. (1991), both of which appear to raise doubts as to the validity of the local isotropy assumption. The most careful experiment at the highest Reynolds number in a turbulent boundary layer appears to be that of Saddoughi (1993), who used hot wires to measure two velocity components in the middle of a thick turbulent boundary layer on the ceiling of a large wind tunnel. Spectral criteria establish that at the large Reynolds numbers and mean shear involved in this experiment, the smallest turbulence scales are essentially isotropic.

In addition to the experimental research cited here, there are a number of theoretical studies setting forth other criteria for the existence of local isotropy in high-Reynolds-number shear flows. Durbin and Speziale (1991) considered a transport equation for one component of the mean viscous dissipation ε_{ij} [cf. Eq. (4.5.2)] and argued that for local isotropy to be a valid approximation so that $\varepsilon_{ij} = \frac{2}{3}\varepsilon\delta_{ij}$, a condition almost universally employed in turbulence modeling (cf. Section 10.1),[†]

[†]Here, for the first time, we follow standard practice in the analysis of turbulence by moment methods such as those discussed in Sections 6.3–6.8 and Sections 10.1–10.8, of denoting the mean viscous dissipation by ε, that is, of dropping the averaging symbol.

$$\frac{S^*k}{\varepsilon} << 1 \tag{5.5.3}$$

must be satisfied, where S^* is a measure of the mean rate of strain and k is the turbulent kinetic energy. For example, in a simple shear flow (cf. Fig. 4.2.1 and Sections 6.2–6.5), $S^* = dU_1/dx_2$, where x_1 is in the streamwise direction and x_2 is in the normal direction. The criterion of Eq. (5.5.3) is analogous to, but differs from, those provided in a series of early studies by Uberoi (1957) and Corrsin (1958). For example, Uberoi argued that local isotropy requires that

$$\frac{\partial U_1/\partial x_2}{[\overline{(\partial u_1/\partial x_2)^2}]^{1/2}} << 1 \tag{5.5.4}$$

i.e., that the mean velocity gradient must be much smaller than the corresponding root mean square of the fluctuating gradient. Corrsin found that turbulence of wave number *greater than* κ_1 is isotropic, provided

$$\kappa_1 \left[\frac{\varepsilon}{(\partial U_1/\partial x_2)^3}\right]^{1/2} = \kappa_1 L_K \left[\frac{\nu^3}{L_K^6(\partial U_1/\partial x_2)^3}\right]^{1/2} \approx 10 \tag{5.5.5}$$

Thus if the wave number of the fluctuations greater than a specified fraction of L_K^{-1} are to be isotropic, the mean shear must be sufficiently small to satisfy Eq. (5.5.5). We are forced to conclude that in the experiments of Saddoughi and Veeravalli (1994), the mean shear is sufficiently small and the Reynolds number sufficiently high for the restrictions of Eqs. (5.5.3)–(5.5.5) to be satisfied within experimental accuracy. In certain of the other experiments cited earlier, either the mean shear is excessive for the Reynolds number involved and/or the criteria for the existence of local isotropy are inappropriate and/or the experimental techniques are flawed, with the consequence that local isotropy is called into question.

In sum, these considerations, both experimental and theoretical, point to certain circumstances under which isotropy of the small turbulent scales prevails in flows with mean shear. In most turbulence modeling and LES, local isotropy is implicitly assumed to hold. The extent of the inaccuracies which result from this assumption is unclear, while the whole matter of local isotropy continues to be studied (cf., e.g., Antonia and Kim, 1994).

The existence of local isotropy in turbulent flows *with buoyancy* is the subject of current research. However, the experiments of van Atta (1991) and of Thoroddsen and van Atta (1992) establish that stable stratification which is dynamically significant has a strong influence on even the smallest turbulence scales, an influence leading to their being anisotropic. They conclude "that small scale local isotropy is extremely vulnerable to anisotropic forcing at the large scales."

We next take up the *spectral implications* of the Kolmogorov theory. If the notion of an energy cascade applies to turbulent shear flow at high Reynolds numbers with the transfer of energy from large to small scales, then the power spectral density [cf. Eq. (3.5.3)] or the one-dimensional velocity spectrum tensor $\phi_{11}(\kappa_1; \mathbf{x})$ [cf. Eq. (3.5.14)] of fluctuations of the streamwise velocity, the component most frequently measured, may satisfy the Kolmogorov hypothesis of dependence only on the viscous dissipation in the inertial subrange. From Eqs. (3.5.15) and (3.5.16),

$$\phi_{11}(\kappa_1; \mathbf{x}) = \frac{2}{\pi} \int_0^\infty d\tau \, \overline{u_1(t; \mathbf{x}) u_1(t + \tau; \mathbf{x})} \cos \omega\tau \tag{5.5.6}$$

so that simple dimensional analysis with ϕ_{11} regarded as a function of κ_1 and ε leads to

$$\phi_{11}(\kappa_1; \mathbf{x}) = C\kappa_1^{-5/3}\varepsilon^{2/3} = C(\kappa_1 L_K)^{-5/3} \frac{\nu^2}{L_K} \tag{5.5.7}$$

where the second form suggests an appropriate scaling of experimental data.

The applicability of Eq. (5.5.7), which assumes that the Kolmogorov hypothesis for the inertial subrange applies to turbulent shear flows, is supported by a wide variety of data provided the Reynolds number is suitably high. As a consequence, experimentalists routinely display the extent of the $-5/3$ decay regime of their *one-dimensional spectra* in order to establish the adequacy of their experimental techniques and of the Reynolds number involved. Figure 5.5.1, which is taken from Saddoughi (1992), shows the one-dimensional spectra of the streamwise velocity from a large number of flows, including grid turbulence, wakes, jets, boundary layers, etc. Here the ordinate and abscissa are the same as those suggested by Eq. (5.5.7) provided η is replaced by L_K and $(\varepsilon\nu^5)^{1/4}$ by ν^2/L_K. The spectra are normalized so that they possess universality at the high wave numbers but depart from the $-5/3$ law at a wave number depending on the Reynolds number and the particular flow. The most extensive inertial subrange occurs in the data from the tidal channel flow of Grant et al. (1962) and in the boundary layer of Saddoughi and Veeravalli (1994).[†] From data such as those in Fig. 5.5.1, it should not be inferred that the small turbulent scales in these various flows are necessarily isotropic, although in some flows they may be within experimental accuracy—i.e., it is possible for velocity spectra to be correlated in this fashion even though Eqs. (5.5.1) are not satisfied.

[†]In a refinement of his 1941 theory, Kolmogorov (1962) introduced the notion of the nonuniformity in space of the dissipation, a notion which leads to a slight modification of the $-5/3$ law. Champagne (1978) noted that the modification leads to "a difficult if not impossible change to detect in most realizable high Reynolds number flows." The extensive results in Fig. 5.5.1 confirm this observation.

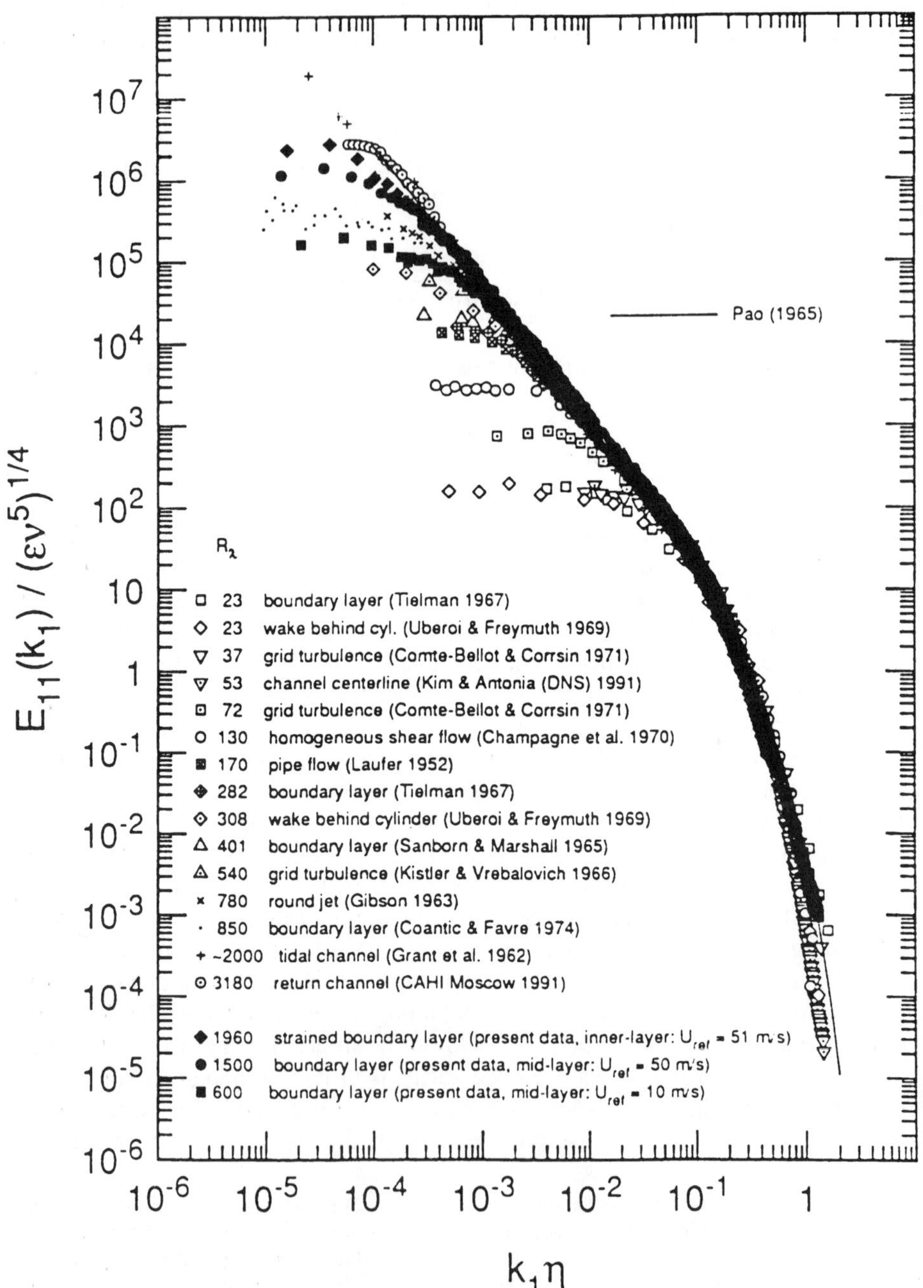

Figure 5.5.1 One-dimensional spectra normalized to high wave numbers from a wide variety of experiments. (From Saddoughi, 1992, with permission.)

We thus see that the Kolmogorov theory for isotropic homogeneous turbulence retains significance for more complex turbulent flows of applied interest, such as wakes, jets, boundary layers, etc.

5.6 REVIEW OF VELOCITY SCALES

In this chapter we identify the various length and time scales arising in turbulent flows. In particular, we relate the integral scale L_I obtained from either two-point, one-time correlations or time autocorrelation results supplemented by the Taylor hypothesis to the largest turbulence scales, those that essentially fill a shear flow, and to the energy-containing scales, those which in isotropic turbulence we associate with the wave number κ_e^{-1}.

The range of these lengths can be estimated in many flows according to the following considerations. We start with an estimate of the intensity of the turbulent kinetic energy and its integral scale L_I, recalling that in many flows $k^{1/2}/U_1$ is of the order of 10^{-1} and that L_I is somewhat less than but of the order of the global scale of the flow. We can then identify a turbulence Reynolds number $k^{1/2}\,L_I/\nu$ and a mean viscous dissipation with $k^{3/2}/L_I$. Then from Eqs. (5.2.8) and (5.2.12),

$$\frac{L_\lambda}{L_I} = \left[\frac{\nu}{k^{1/2}\,L_I}\right]^{1/2} \qquad \frac{L_K}{L_I} = \left[\frac{\nu}{k^{1/2}\,L_I}\right]^{3/4} \tag{5.6.1}$$

Thus, from our original estimates we can make related estimates for the two lengths L_λ and L_K. We see again in a different form that the range of length scale increases as the turbulence Reynolds number increases. The close connection of the various turbulence scales rationalizes the casualness with which Reynolds numbers are frequently referred to in discussions of turbulence—i.e., referring to "high Reynolds numbers" without specificity as to the velocity and length scale involved; implied is that *all* the Reynolds numbers are "large," including those associated with the Taylor microscale and the integral scale.

5.7 TEMPERATURE SCALES

Considerations similar to those in Sections 5.1–5.6 apply to the scales associated with advected turbulent scalar fields, e.g., when velocity and temperature fluctuations coexist. Space and time correlations of temperature fluctuations similar to those discussed in Section 5.1 can be obtained in grid flows with a mesh consisting of electric heating elements so that velocity *and* temperature inhomogeneities are introduced into the downstream flow. Turbulent shear flows can also involve velocity and temperature fluctuations, e.g., when the turbulent boundary layer grows on a surface with a temperature different from that of the free stream. In these circumstances the sets of anemometers required for multipoint, multitime velocity measurements must be supplemented by fast-response thermometers so that local instantaneous temperatures can be recorded and subsequently used to determine a variety of space–time correlations of temperature. If coordinated with anemometer data, correlations of temperature and velocity such as $\overline{u_i\theta}$. Furthermore, the notion of homogeneous isotropic turbulence and the interpretation of correlations and spectra obtained from two thermometers fixed in the flow at separation r extend to temperature fields. However, the role of the Prandtl number as an additional parameter produces spectra of increased complexity. In this regard it should be recalled that a wide range of Prandtl numbers occurs in liquids, with values both large and small compared with unity, while for most gases N_σ is close to unity.

Here we consider briefly the thermal counterpart of isotropic, homogeneous turbulence dealt with in Sections 5.2–5.5 (cf. Hinze, 1975). Since we have repeatedly related two-point, one-time measurements at fixed locations in grid flow to ensemble-averaged correlations in a frame of reference moving with the mean velocity, we immediately introduce the corresponding two-point, one-time ensemble correlation of temperature in such a frame, namely, $R_\theta(r, t) \equiv \langle\theta(t)\theta(t; r)\rangle$, the correlation of the temperature at two spatial locations r apart obtained from many realizations at a fixed time $t = x_1/U_1$ after passage through a heated grid. Comparison with the scalar correlation function $R(r, t)$ defined by Eq. (5.2.9) reveals the relative simplicity of obtaining $R_\theta(r, t)$ experimentally. The associated correlation coefficient is

$$f_\theta(r, t) \equiv \frac{\langle\theta(t)\ \theta(t; r)\rangle}{\langle\theta^2(t)\rangle} \tag{5.7.1}$$

As a parallel to the considerations of Section 5.2, we can employ $f_\theta(r, t)$ to define the Taylor microscale for temperature fluctuations,

$$L_{\lambda\theta} = \frac{2^{1/2}}{[-(\partial^2 f_\theta/\partial r^2)_{r=0}]^{1/2}} \tag{5.7.2}$$

and an integral scale,

$$L_{I\theta}(t) = \int_0^\infty dr\ f_\theta(r, t) \tag{5.7.3}$$

We consider next the Fourier transform pair

$$R_\theta(r, t) = \int_0^\infty d\kappa\, E_\theta(\kappa, t)\, \frac{\sin \kappa r}{\kappa r}$$
$$E_\theta(\kappa, t) = \frac{2}{\pi} \int_0^\infty dr\, R_\theta(r, t)\, \kappa r \sin \kappa r \tag{5.7.4}$$

From the first of these equations with $r = 0$ we have

$$\langle \theta^2(t) \rangle = \int_0^\infty d\kappa\, E_\theta(\kappa, t) \tag{5.7.5}$$

and thus, as expected from our earlier encounter with $E(\kappa, t)$, we see that the integral of $E_\theta(\kappa, t)$, the scalar energy spectrum function, over all wave numbers defines the intensity of the temperature fluctuations. The implication is that $E_\theta(\kappa, t)$ represents the contribution to the temperature variance in the wave number band from κ to $\kappa + d\kappa$. Note that the units of E_θ are temperature squared times L, the temperature squared per wave number.

Now, just as the spectral transfer equation for $E(\kappa, t)$ is found by taking the Fourier transform of the transport equations for momentum, the corresponding equation for $E_\theta(\kappa, t)$ obtained from Eq. (2.3.1), that for thermal energy transport, is

$$\frac{\partial}{\partial t} E_\theta(\kappa, t) = T_\theta(\kappa, t) - 2\, \frac{\nu}{N_\sigma}\, \kappa^2 E_\theta(\kappa, t) \tag{5.7.6}$$

Although the presence of the Prandtl number in this equation may appear benign, it has a profound effect on the solutions. Once again the first term on the right side arises from convection and represents the closure problem for the scalar field in this formulation. It describes spectral transfer and possesses the property encountered earlier in $T(\kappa, t)$, namely, the vanishing of its integral over all wave numbers. Thus from Eqs. (5.7.5) and (5.7.6) we obtain

$$\frac{d}{dt} \int_0^\infty d\kappa\, E_\theta(\kappa, t) = -2\, \frac{\nu}{N_\sigma} \int_0^\infty d\kappa\, \kappa^2 E_\theta(\kappa, t)$$
$$= \frac{d}{dt} \langle \theta^2(t) \rangle = -\langle \varepsilon_\theta \rangle \tag{5.7.7}$$

where the mean scalar dissipation is defined as the rate of change of the intensity of the temperature fluctuations and is related to the integral of $\kappa^2 E_\theta(\kappa, t)$. Equation (5.7.6) closely resembles Eq. (5.2.11). Note that the units of $E_\theta(\kappa, t)$ are temperature squared times length and those of $\langle \varepsilon_\theta \rangle$ are temperature squared per time.

An analysis of Eq. (5.7.6) closely following that for Eq. (5.2.11) in Sections 5.2 and 5.3 can be carried out so that the statistics of temperature fluctuations in low- and high-Reynolds-number isotropic homogeneous turbulence can be treated. Of particular interest is the existence for high Reynolds numbers of an equilibrium spectral range in which the two terms of the right side of Eq. (5.7.6) dominate. There also exists the counterpart of the inertial subrange within which $T_\theta(\kappa, t)$ is zero. It is through this subrange that the spectrally resolved intensity of the large-scale temperature fluctuations is transfered to the small scales, where it is dissipated by molecular thermal diffusivity. While we shall see that the spectral extent of this subrange is influenced by the Prandtl number, a sufficiently large turbulence Reynolds number guarantees that such a range can always be identified.

These observations suggest that we may follow the analysis in Section 5.4 and modify the Kovasznay model for spectral transfer in order to develop ideas about temperature spectra. Thus we again restrict attention to the equilibrium range and make the approximation

$$\int_0^\kappa d\kappa\ E_\theta(\kappa) \approx \int_0^\infty d\kappa\ E_\theta(\kappa)$$

where, as before, the value of κ in the upper limit is understood to be sufficiently large, and where we again drop the time dependence to reflect the quasi-steady nature of the scalar energy spectrum function in the equilibrium range. In this case we have, from Eqs. (5.7.7),

$$\langle \varepsilon_\theta \rangle = S_\theta(\kappa) + 2\frac{\nu}{N_\sigma}\int_0^\kappa d\kappa' \kappa'^2 E_\theta(\kappa') \tag{5.7.8}$$

where we follow Eq. (5.4.2) and introduce a spectral transfer function S_θ such that

$$T_\theta(\kappa, t) = -\frac{\partial S_\theta}{\partial \kappa}$$

Guidance for the required modifications of the Kovasznay model for $S_\theta(\kappa)$ is provided by several observations. First, the Batchelor length L_B related to the dissipation of temperature fluctuations corresponds to the Kolmogorov length (cf. Batchelor, 1959), where

$$\frac{L_B}{L_K} = \frac{1}{N_\sigma^{1/2}} \tag{5.7.9}$$

Considering L_B from the perspective of the variation of temperature between two spatial locations at a given instant of time, we see from Eq. (5.7.9) that $L_B > L_K$ for $N_\sigma < 1$, implying that under these circumstances the small scales of the temperature fluctuations tend to be smoother than those of the velocity. In this case the distance over which meaningful interpolation between two spatial points can be carried out is greater for the temperature field than for the velocity field.

We also realize from Eq. (5.7.9) that if $N_\sigma >> 1$, then $L_B << L_K$ and thus significant temperature fluctuations can occur within a distance involving a smooth velocity variation. The relative magnitude of these two lengths is particularly relevant at the edges of a heated turbulent jet or wake or at the edge of a heated boundary layer, as discussed in Section 8.6. Furthermore, Corrsin (1951) showed that in the inertial subrange,

$$E_\theta = \beta \frac{\langle \varepsilon_\theta \rangle}{\langle \varepsilon \rangle} \kappa^{-5/3} \tag{5.7.10}$$

where β is a constant. Hence the ubiquitous $-5/3$ decay law for the inertial subrange prevails for temperature fluctuations.

These considerations supplemented by dimensional arguments suggest the following model for $S_\theta(\kappa)$:

$$S_\theta(\kappa) = \alpha_\theta \, \kappa^{5/2} \left(\frac{\langle \varepsilon \rangle}{\langle \varepsilon_\theta \rangle} \right)^{1/2} E_\theta^{3/2} \tag{5.7.11}$$

where α_θ is a parameter to be determined. We seek a solution to Eq. (5.7.8) in the form

$$E_\theta \propto L_K^{5/3} \frac{\langle \varepsilon_\theta \rangle}{\langle \varepsilon \rangle^{1/3}} \frac{[1 - (\bar{\kappa}/N_\sigma)^{4/3}]^2}{\bar{\kappa}^{5/3}}$$

where again $\bar{\kappa} \equiv \kappa L_K$ is a dimensionless wave number, the same variable used in Eq. (5.4.5). Substitution into Eq. (5.7.8) yields the analog of Eq. (5.4.5),

$$\frac{E_\theta(\bar{\kappa})\nu}{(\langle \varepsilon_\theta \rangle \nu)^{1/2} L_K} = 2\, N_\sigma \frac{[1 - (\bar{\kappa}/N_\sigma)^{4/3}]^2}{\bar{\kappa}^{5/3}} \tag{5.7.12}$$

and

$$\alpha_\theta = \frac{1}{2^{1/2}\, N_\sigma^{3/2}} \tag{5.7.13}$$

Figure 5.7.1 shows the corresponding temperature energy spectrum function in appropriate dimensionless form for three different values of N_σ. As N_σ increases, E_θ moves to the right and upward. In particular, we see that the cutoff wave number $\bar{\kappa}_c = N_\sigma^{1/2}$. The prevalence of small-scale, high-wave-number fluctuations of temperature when the Prandtl number is large compared with unity is a consequence of the small ratio of thermal conductivity to viscosity coefficient. Thus in this case the turbulence involves significant temperature fluctuations embedded in a relatively smooth velocity field. The opposite is the case when the Prandtl number is small compared with unity; i.e., velocity fluctuations occur in a relatively smooth temperature field.

The simple spectral transfer model of Eq. (5.7.11) leading to the results in Fig. 5.7.1 provides a prediction of only some of the influence of Prandtl number

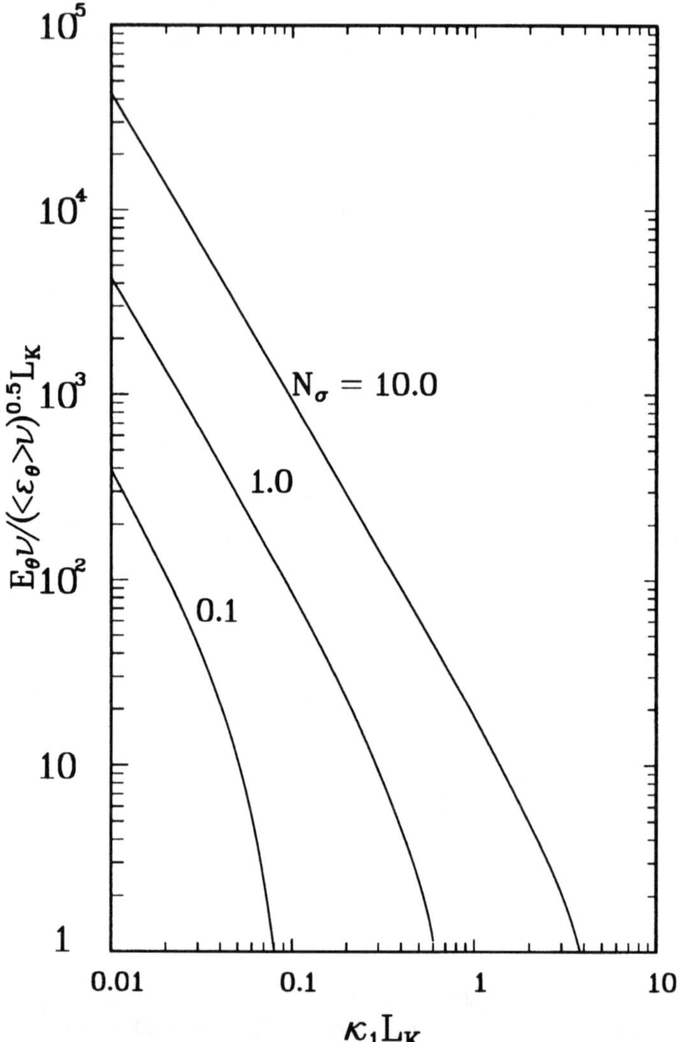

Figure 5.7.1 The dimensionless scalar energy spectrum function.

on the temperature spectra. In particular, when $N_\sigma >> 1$, more elaborate calculations and experiments show an intermediate spectral range between the inertial subrange with its $-5/3$ decay and cutoff with a nearly -1 decay (cf. Batchelor, 1959; Batchelor et al., 1959; Gibson and Kanellopoulos, 1988).

Another useful result from the analysis of temperature fluctuations paralleling that in Section 5.3 and Eq. (5.2.6) for the viscous dissipation in isotropic, homogeneous turbulence, is

$$\frac{d}{dt}(\overline{\theta^2}) = -\frac{12\,\nu\,\overline{\theta^2}}{N_\sigma L_{\lambda\theta}^2} \qquad (5.7.14)$$

where $L_{\lambda\theta}$ is the equivalent of the Taylor microscale for temperature fluctuations and provides a measure of the rate of decrease of the correlation $\overline{\theta(t;\,\mathbf{x})\theta(t;\,\mathbf{x}+\mathbf{h})}$ as $\mathbf{h} \to 0$. The right side of Eq. (5.7.2) is the scalar dissipation describing the influence of molecular diffusion on the intensity of temperature fluctuations.

We now discuss briefly the extension of this analysis of the temperature scales in isotropic homogeneous turbulence to turbulent shear flows. Significantly less attention has been paid to the circumstances under which local isotropy holds for temperature fluctuations compared to velocity fluctuations, but van Atta (1991) reviewed this matter and in addition analyzed temperature gradient spectra in a heated turbulent boundary layer; he found that under the conditions existing in this particular flow, the temperature fluctuations are locally isotropic. Spectra of temperature fluctuations in turbulent shear flows are frequently presented in the form suggested by Eq. (5.7.12).

5.8 COHERENT STRUCTURES

In concluding this chapter on the length and time scale of turbulence, it is appropriate to discuss the subject of coherent structures, which in recent years has generated much interest in the turbulence community. This interest relates to their origin, characteristics, and significance, and to the means for their incorporation into turbulence theory. Although not universally agreed upon, the adjective "coherent" is usually attributed to the large turbulent structures observed in many shear flows. Accepted features of coherent structures are their large size and thus their long lifetimes as discussed in Section 5.3, but details of their structure, the processes operative within their interior, their significance for the behavior of the flows in which they exist, and the means for their quantitative description are the subjects of current research.

The present interest in large turbulent structures is the result of a longstanding realization that many flows involve such structures and that the representation of turbulence in terms of unstructured fluctuations which is reflected as in the moment methods discussed in all of the subsequent chapters overlooks important phenomena. The discovery of intermittency by Corrsin (1943), the analysis of large eddies by Townsend (1956), and the identification of horseshoe vortices as important exchange structures in turbulent boundary layers by Theodorsen (1952) have contributed to this realization. These early developments and their later evolution have been reviewed by Cantwell (1981), while coherent structures in turbulent boundary layers have been surveyed by Robinson (1991). The genesis of the current interest in large turbulent structures can be traced to the photographs of Brown and Roshko (1974), one of which is reproduced here as Fig. 5.8.1. These photographs depict the two-dimensional turbulent mixing layer in terms of large spanwise vortices. These vortices pair as they proceed downstream and lead to the observed linear growth of the mean thickness of the layer. Their two-dimensionality has stimulated experimental research on their modification (cf. Ho and Huang, 1982), and to efforts to compute them by DNS (cf. Metcalf et al., 1987) and by random vortex methods (cf. Ashurst, 1979). However, coherent structures are not limited to these two-dimensional vortices. Indeed, in all other turbulent shear flows the large vortices are three-dimensional, with characteristic lengths in the direction of the principal mean velocity being much greater than in the two transverse directions (cf. Section 5.1). Furthermore, although they are more obvious in turbulence bounded by contiguous irrotational flows, relatively large volumes of fluid with distinctive turbulence characteristics surrounded by different turbulent fluid are observed in other circumstances, and these can also be considered coherent structures. Examples are the "puff" and "slugs" in transitional pipe flow of Wygnanski and Champagne (1973).

It is known from experiments employing conditional sampling that large structures account for a significant fraction of the turbulent kinetic energy and of the Reynolds stresses and fluxes in turbulent shear flows, a result consistent with the discussion in Sections 5.2 and 5.3. Thus they play an important role in the behavior of these flows. Unfortunately, conventional averaging and the resulting equations do not contain information concerning such structures.† In upstream regions of turbulent boundary layers, large structures are related to transition from laminar to turbulent flows; but far downstream a regenerating mechanism must be operative to allow upstream structures to spawn replicas which maintain their presence. An interaction between the outer and wall flows can provide such a mechanism, which is frequently studied by conditional sampling (cf. Smith and Abbott, 1978). In channel flows the interaction of the flows near the walls and in the center of the channel results in relatively large turbulent structures.

The experimental study of the *statistical features* of coherent structures involves conditional averaging (cf. Section 3.6) but is difficult not only because of uncertainity as to the appropriate trigger for conditional sampling but also because of the problem in distinguishing essential from inessential fluctuations (cf. Hussain, 1986; Hussain et al., 1987; Jeong and Hussain, 1995). The large turbulent structures found in some DNS can be subjected to analysis; although the turbulence Reynolds numbers in these computations are typically low, the possibility of studying accurately and repeatedly the statistical features of two- and three-dimensional structures makes their examination by DNS attractive (cf. Bradshaw, 1987).

†Analyses based on conditional averaging which incorporates at least partially the selective contributions of large turbulent structures in boundary layers, wakes, and jets provide some limited relief from this shortcoming (cf. Libby, 1976).

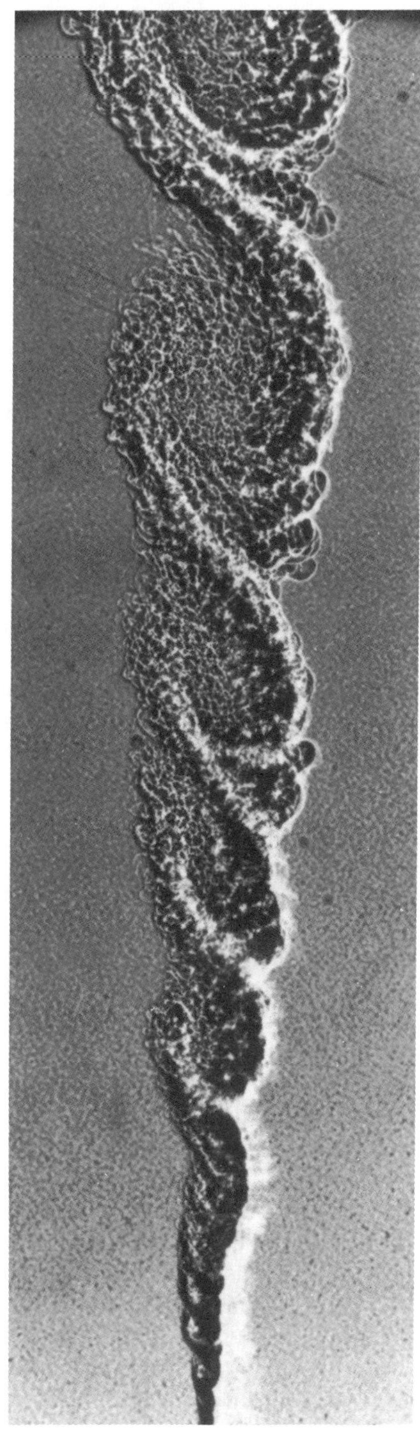

Figure 5.8.1 Large-scale two-dimensional structures in a turbulent mixing layer. (From Brown and Roshko, 1974, with permission.)

It seems clear that, because of their importance in determining the behavior of many turbulent flows, coherent structures will continue to attract the attention of experimentalists and theoreticians (cf. Sirovitch, 1987; Holmes, 1990).

A related perspective on the role of large turbulent structures arises from the examination of the possible connection between turbulence and the theory of dynamical systems (cf. Ruelle and Takens, 1971, and the references therein to the long history of this theory). A connection is sought by decomposing the velocity components of a flow evolving to a statistically stationary state according to a straightforward representation,

$$\tilde{u}_i(\mathbf{x}, t) = U_i(\mathbf{x}) + a_k(t)\, u_{ik}(\mathbf{x}) \qquad i = 1, 2, \ldots, N \tag{5.8.1}$$

where $u_{ij}(\mathbf{x})$ are a *known* set of orthonormal functions, i.e., functions satisfying the condition

$$\int_V dx_1\, dx_2\, dx_3\, u_{ik}\, u_{ij} = \delta_{ij} \tag{5.8.2}$$

where the integration is over the entire flow domain. Expansions of this nature are widely employed in physics and mechanics for linear problems such as providing the basis for Rayleigh-Ritz and Galerkin procedures, but they can also be applied to nonlinear systems such as the Navier-Stokes equations.

If Eqs. (5.8.1) are substituted into Eqs. (2.5.3) with body forces neglected and the orthogonality condition exploited, then the Galerkin method leads to a set of ordinary differential equations,

$$\frac{da_i}{dt} = b_{ik}\, a_k + c_{ikl}\, a_k\, a_l + d_{iklm}\, a_k\, a_l\, a_m \qquad i = 1, 2, \ldots, N \tag{5.8.3}$$

where we violate our usual rule and use the index m repeated to indicate summation because of the need for three summation indices. In this equation the b, c, d coefficients are known constants involving integrals of the $u_{ij}(\mathbf{x})$ functions, with the linear terms arising from the viscous stresses, the quadratic terms from convection, and the cubic terms from interactions of the velocities $U_i(\mathbf{x})$ and the $u_{ij}(\mathbf{x})$, again from convection. These differential equations are to be solved subject to appropriate initial conditions representing an initial state of the flow.

In principle, as $N \rightarrow \infty$ with the $u_{ij}(\mathbf{x})$ functions satsifying the boundary conditions of the flow in question, the solutions to Eqs. (5.8.3) result in an exact description of the evolution of the flow from its initial state. However, of interest in the context of coherent structures is the possibility of a description adequate for some purposes when N is a relatively small number, perhaps 3 or 4, a description which might account for most of the turbulent kinetic energy and Reynolds shear stresses provided the $u_{ij}(\mathbf{x})$ functions faithfully represent the large turbulent scales. For further study of this matter, which is the subject of current research, see Sirovitch (1987), Aubry et al. (1988), and Holmes (1990).

The notion of representing the energy-containing motions in turbulence by a few known, judiciously chosen functions is also used in fundamental turbulence studies, e.g., to determine the number of terms in a series such as that given by Eq. (5.8.1) to describe accurately turbulent correlations obtained either experimentally or computationally. Such determinations provide a connection between the large turbulent structures in many flows and the correlations related to the Reynolds stresses. A mathematical basis for these studies is provided by the proper orthogonal decomposition technique of Lumley (1967), a technique related to the Galerkin method. An illustrative application treating experimental data from the mixing associated with the injection of a turbulent jet normal to a mean flow is given by Winter et al. (1992). A recent application involving DNS results from channel flow has been made by Moin and Moser (1989) in order to determine those flow structures which account for the calculated Reynolds stresses. Similarly, Glauser and Gatski (1993) employ the data from a DNS of the near-wall region of channel flow to reconstruct by means of a limited number of known functions the turbulent kinetic energy, Reynolds shear stress, and mean viscous dissipation.

5.9 SUMMARY

This chapter has shown that the fluctuations in turbulent flows involve a wide range of scales. The large-scale fluctuations are long-lived and account for most of the Reynolds stresses and thus for most of the turbulent kinetic energy. The interaction of these stresses with the mean rates of strain results in the production of turbulence. The fluctuations of smallest scale possess short lifetimes, are dominated by viscous effects, and account for the dissipation of turbulence. The relative persistence of large turbulent scales leads to the notion of coherent structures, which arise in many turbulent shear flows and suggest special analytic approaches.

CHAPTER

SIX

SIMPLE TURBULENT FLOWS

The fundamental mechanics of turbulent flows are exposed by considering various simple situations. One is grid flow, which we know from previous chapters plays a central role in turbulence research. Accordingly, in Sections 6.1 and 6.2 we consider the downstream decay of velocity and temperature fluctuations introduced by a grid. Turbulent Couette flow, discussed in Section 6.3, represents a second situation; in this case the turbulence is in a state of local balance involving an equilibrium among production, redistribution, and dissipation. Both isothermal flows and those involving a gradient of mean temperature normal to the two surfaces establishing the flow are of interest. Indeed, in Section 6.4 we assume that the temperature inhomogeneities are sufficiently large that coupling between the velocity and temperature fields occurs; we thereby examine the influence of buoyancy on turbulence. Homogeneous shear flows provide a useful vehicle for assessing and comparing various turbulence theories based on moment methods under conditions that are difficult to realize in the laboratory but readily computed via direct numerical simulation. Thus, in Sections 6.5 and 6.6 we discuss these flows without and with rotation, respectively. With the exception of the turbulent Couette flow with an active temperature field, the flows dealt with in Sections 6.3–6.6 are dominated by mean shear stresses and are therefore relevant to a wide class of turbulent flows, jets, wakes, mixing layers, and boundary layers. Sections 6.7 and 6.8 are concerned with the evolution of turbulence under extensive and compressive rates of strain which occur in ducts and in external flows encountering obstacles and bodies. The circumstances considered in Section 6.8 are special and permit a distinct treatment which has broader applicability than to the flow in a contraction section we use to illustrate its features. Turbulent transport is negligible in all of these flows, so the closure problem arises primarily because of the influence of pressure fluctuations. This contrasts with the flows considered in the remaining chapters, in which turbulent transport transverse to the mean streamlines plays an essential role. Finally, we conclude this chapter by discussing two-dimensional turbulence in Section 6.9.

Our desire to deal at this stage in our considerations with these simple flows and to expose their main features leads to an expositional difficulty. Since these features are best treated by methods detailed in Chapter 10, we must anticipate later developments at several points in this chapter. We overcome this difficulty by specializing the complete equations developed later to the particular applications under consideration in this chapter and by discussing the origin of the various terms. Because of its ability to predict the anisotropy of the turbulence, an important feature of most of the flows considered in this chapter, we adopt the Reynolds stress theory rather than the simpler k–ε theory. For these simple flows the former theory is not excessively complicated, but it does involve a number of empirical coefficients to which we generally assign *standard* values. However, it should be kept in mind that there is not universal agreement on these values and, given the current development of the theory, future changes are to be expected. In addition, we shall find that some flows place restraints on these coefficients; in this case the considerations used for the determination of standard values are inapplicable. For our purposes the implied uncertainties are inessential.

6.1 ISOTHERMAL GRID FLOW

Earlier, we discussed the nature of grid flow. Briefly, recall that U_1 = constant, $U_2 \equiv U_3 \equiv 0$, and $\overline{u_i u_j} = \delta_{ij}\, \overline{u^2}(x_1)$. The first-moment equations for this idealized flow, Eqs. (4.1.1) and (4.2.1), are identically satisfied by the assumed uniformity of the mean velocity and the assumed equality of the intensities in the three coordinate directions. This last condition assures that the Reynolds shear stresses are zero. However, the transport equation for the turbulent kinetic energy, Eq. (4.7.1), becomes

$$U_1 \frac{dk}{dx_1} = -\varepsilon \tag{6.1.1}$$

As discussed in Section 5.3, the experimental determination of dissipation in grid flows and thus of the evolution in the streamwise direction of the Kolmogorov length is provided by Eq. (6.1.1). Required is the measurement of $k(x_1)$ at several downstream stations spaced so that the local derivative dk/dx_1 is obtained. Once again, we call attention here to the absence in Eq. (6.1.1) of a bar over ε, i.e., we follow standard practice in the turbulence literature and denote the *mean viscous dissipation* by ε.

We know from the discussion in Section 5.3 that the decay of the turbulent kinetic energy in grid turbulence involves two stages. In the initial stage there is vigorous turbulence, i.e., a relatively large turbulence Reynolds number, so the mean viscous dissipation depends on the turbulent kinetic energy and the integral scale of the turbulence. In this case the spectra of the velocity fluctuations involves an inertial subrange with a five-thirds rate of decay in frequency or wave number. Farther downstream, in the final period of decay, the turbulence is weak and the mean viscous dissipation depends on the kinematic viscosity. In this section we are concerned with the initial period.

To proceed further with Eq. (6.1.1), consider $\varepsilon = Ak^{3/2}/L_I$, where L_I is an integral scale taken to be roughly equal to the scale of the energy-containing fluctuations and corresponding to κ_e in Section 5.3, and where A is an empirical coefficient of the order of unity. We know from the previous chapter that the more rapid decay, the shorter lifetimes, of the high-wave-number fluctuations results in a *growth* with x_1 of the integral scale and that the turbulent kinetic energy *decays* with downstream distance. To proceed, however, we seek more quantitative estimates by assuming that $k \propto x_1^{-\alpha_1}$ and $L_I \propto x_1^{\alpha_2}$, where α_2 and α_2 are exponents to be determined. Then the combination of Eq. (6.1.1) and our model for the dissipation implies that

$$\frac{1}{2}\alpha_1 + \alpha_2 = 1 \tag{6.1.2}$$

A second relation requires more subtle considerations. In the initial decay downstream of the turbulence grid, various time scales can be identified: the transit time from the grid, x_1/U_1; the spectral transfer or turnover time [cf. the discussion connected with Eq. (5.3.3)], $L_I/k^{1/2}$; and the lifetime of the large-scale fluctuations, L_I^2/ν [cf. the discussion connected with Eq. (5.3.5)]. These three times can be reduced to two dimensionless ratios by dividing the first and third by the second so that there exists a relation

$$\frac{L_I^2/\nu}{L_I/k^{1/2}} = \frac{k^{1/2}\,L_I}{\nu} = f\left(\frac{x_1}{L_I}\,\frac{k^{1/2}}{U_1}\right) \tag{6.1.3}$$

One solution to this equation, one found to agree roughly with experiment, involves a constant value for the Reynolds number $k^{1/2}L_I/\nu$. In this case

$$-\frac{1}{2}\alpha_1 + \alpha_2 = 0 \tag{6.1.4}$$

Equations (6.1.2) and (6.1.4) then yield

$$\alpha_1 = 1 \qquad \alpha_2 = \frac{1}{2} \tag{6.1.5}$$

The assumed constancy of this Reynolds number leads to the simple result [cf. Eqs. (5.6.1)] that all three lengths, the integral scale L_I, the Taylor microscale L_λ, and the Kolmogorov scale L_K, all grow as $x_1^{1/2}$.

Clearly these power laws cannot apply as $x_1 \to 0$, since they call for an indefinite increase of the turbulent kinetic energy and an indefinite decrease in the turbulent length scales in that limit. We are thus led to the introduction of a virtual origin for grid flow; the notion of such an origin arises frequently in fluid mechanics. Suppose that experimental data on $\overline{u_1^2}$ at various stations x_1 measured from the grid are available and that we wish to determine the point where the power laws indicate the flow originates. If we assume that the data correspond to the initial period of decay and that α_1 is unity, then the reciprocal of the measured intensities is plotted versus x_1 and a least-square straight line is passed through the data points. The intercept with the x_1 axis determines the virtual origin.

Uncertainties are obviously associated with this procedure; other exponents for the decay in the intensities can be chosen, and the far-downstream data may correspond to sufficiently low values of the turbulence Reynolds number so as not to be applicable to the initial period of decay. Inclusion of these data can compromise the location of the virtual origin. Judgment is required in addressing these uncertainties. Mohamed and LaRue (1990) carefully analyze the results of a large number, 30 or more, of grid experiments by various research workers and conclude, *inter alia,* that $k \propto x_1^{-1.3}$. Given the large set of data involved and the care with which they are analyzed, this exponent must be considered definitive; but it is always possible to argue that data for higher values of the Reynolds number $U_1 M/\nu$ may lead to a different exponent. We thus see that despite the analytic appeal of $k \propto x_1^{-1}$ and of the reasoning leading thereto, real turbulent flows may behave somewhat differently. In this regard it is worth noting that as a consequence of its relative simplicity a large number of theoretical analyses of grid turbulence have been carried out, analyses based on various similarity considerations. As a consequence, a range of exponents has been predicted, from unity, as we have suggested here on the basis of simple dimensional considerations, to 6/5 [cf. Mohamed and LaRue (1990) for a listing of this literature and George (1992) for a current entry thereto]. It is interesting to note that early estimates of the behavior of the turbulent kinetic energy in grid flow fail to take into account the growth of the integral scales of the turbulence in the downstream direction and replace L_I in our dimensional considerations with the *fixed* mesh length L_M. As a consequence, a decay $k \propto x_1^{-1/2}$ is predicted and then found to be in agreement with early experiments (Goldstein, 1965)! We now know that both the theory and experiments are flawed.

We now consider a modified perspective for grid flow. In many flows of applied interest it is difficult to identify the length L_l that is appropriate for modeling ε, with the consequence that an alternative approach is required. To indicate this alternative we replace our algebraic model for the mean viscous dissipation ε with a differential model of the form

$$U_1 \frac{d\varepsilon}{dx} = -\left(C \frac{\varepsilon}{k}\right) \varepsilon = -C \frac{\varepsilon^2}{k} \tag{6.1.6}$$

where C is an empirical coefficient. The left side of this equation describes the rate of change of mean viscous dissipation experienced by a fluid element, while the right side accounts for the processes determining that rate. The principal argument in support of this right side is a dimensional one, namely, that we must multiply ε by a quantity with the dimensions of U/L, for example, by ε/k. If we again seek a power law solution to Eqs. (6.1.1) and (6.1.6) while retaining our earlier result $\alpha_1 = 1$, we find that

$$\varepsilon \propto x_1^{-2}$$

This behavior is consistent with our earlier algebraic model for ε, namely $\varepsilon \propto k^{3/2}/L_l$ if $k \propto x_1^{-1}$ and $L_l \propto x_1^{1/2}$. In Sections 10.1–10.10 we introduce a transport equation for the mean viscous dissipation in order to obtain length scale information.

6.2 THE TEMPERATURE FIELD DOWNSTREAM OF A HEATED GRID

Considerations similar to those of the previous section apply to the case of a turbulence grid consisting of heating elements so that both temperature and velocity fluctuations are added to the flow. The mean temperature Θ corresponds to the mean velocity U_1 and is constant throughout the flow downstream of the grid, while the temperature intensity $\overline{\theta^2}$ is high at the grid but decays in the downstream direction. The special form of Eq. (4.6.4) that is appropriate for this flow is

$$U_1 \frac{d}{dx_1} \overline{\theta^2} = -2\varepsilon_\theta \tag{6.2.1}$$

This equation is the counterpart of the equation for the turbulent kinetic energy [cf. Eq. (6.1.1)] and shows that, in the absence of one or more production mechanisms, the intensity of the temperature fluctuations decreases under the influence of molecular conduction.

A frequently adopted model for the mean scalar dissipation ε_θ under these circumstances is†

†Our earlier comments concerning the absence of an overbar on ε apply to ε_θ.

$$\varepsilon_\theta = \frac{\overline{\theta^2}\varepsilon}{k} = \hat{C}\,\frac{\overline{\theta^2}k^{1/2}}{L_I} \qquad (6.2.2)$$

where $\hat{C}$ is an empirical coefficient and L_I is an integral scale associated with the velocity field. An analysis closely following that for the decay of the turbulent kinetic energy applies to the decay of the temperature intensity $\overline{\theta^2}$ to yield

$$\overline{\theta^2} \propto x^{-1} \qquad (6.2.3)$$

where x_1 is measured from a suitably virtual origin, perhaps but not necessarily the same location as that found to prevail for the decay of the velocity fluctuations.

A considerably more complex situation prevails when the heat is added to the flow at other than the grid generating the velocity fluctuations, e.g., at a series of wires downstream of the turbulence grid. In this case the lengths associated with the dissipation of the velocity and temperature fluctuations are different and Eq. (6.2.2) does not provide an appropriate description of the scalar dissipation. At the present time there is some uncertainty about the behavior of the temperature field under these circumstances (cf. Sirivat and Warhaft, 1983).

Generalizations of turbulent flows created by a heated grid are of interest. For example, if the temperature added by the horizontal rods is not uniform but distributed so as to establish a uniform gradient of mean temperature in the x_2 direction, we deal with a scalar in a velocity field with a slowly decaying intensity and with homogeneity only in x_2–x_3 planes. In this case the statistical quantities characterizing the temperature field evolve in the streamwise direction. Indeed, if the temperature inhomogeneities are large enough, buoyancy effects can be significant so that the velocity and temperature fields are coupled. A variant of this flow involves a thermal mixing layer in grid turbulence. Suppose that the horizontal rods in the upper half of the wind tunnel are heated uniformly so as to establish a mean temperature distribution immediately downstream of the grid with two different values. There results a thermal mixing layer between streams of different mean temperatures in decaying grid turbulence. These flows are useful for fundamental studies of the behavior of passive scalars in turbulent flows (cf. Libby, 1975; LaRue and Libby, 1981).

6.3 TURBULENT COUETTE FLOW OF CONSTANT TEMPERATURE

In this and the following sections we deal with turbulent Couette flow. Consider two infinite plates in planes x_2 = constant, one fixed, the other moving in its plane at a fixed rate. With an appropriate orientation of a Cartesian coordinate system $U_2 = U_3 \equiv 0$ while in the central portion of the passage $U_1 \propto x_2$ with the proportionality constant depending on the velocity of the moving plate and the plate spacing. The mean momentum equation [Eq. (4.2.3)] in the x_1 direction reduces in this case to a statement that the mean shear $\bar{\tau}_{12}$, the sum of the stress due to viscosity and to the turbulent shear stress $\rho\overline{u_1u_2}$, is constant. Accordingly, in the immediate vicinity of each plate there is a viscous sublayer within which viscous stresses are significant, but under the high-Reynolds-number conditions of interest in this discussion the central portion of the space between the plates closely approximates the simple shear flow shown schematically in Fig. 4.2.1 with the Reynolds shear stress $\rho\overline{u_1u_2}$ constant.[†] Here we consider the isothermal case in which the two plates are at the same temperature, but in the following section we take the two temperatures to be different so that the flow involves uniform gradients in the x_2 direction of both the mean velocity U_1 and the mean temperature.

Statistical homogeneity in all three coordinate directions is assumed to prevail. Implied is that the various processes of convection, pressure redistribution (cf. Section 4.7), and dissipation are in local equilibrium. If the turbulence Reynolds number is high enough, the influence of viscous stresses is confined to dissipative effects. The Reynolds stress equations, Eqs. (4.4.1), with body forces neglected and with the dissipation given by Eqs. (4.5.2) and (4.5.3), become in this case

$$
\begin{aligned}
\overline{u_1u_2}\,\frac{dU_1}{dx_2} &= \overline{\frac{p}{\rho}\frac{\partial u_1}{\partial x_1}} - \varepsilon_{11} \\
0 &= \overline{\frac{p}{\rho}\frac{\partial u_2}{\partial x_2}} - \varepsilon_{22} \\
0 &= \overline{\frac{p}{\rho}\frac{\partial u_3}{\partial x_3}} - \varepsilon_{33} \\
\overline{u_2^2}\,\frac{dU_1}{dx_2} &= \overline{\frac{p}{\rho}\left(\frac{\partial u_1}{\partial x_2} + \frac{\partial u_2}{\partial x_1}\right)} - \varepsilon_{12}
\end{aligned}
\tag{6.3.1}
$$

The first three of these equations describe the balance for the three components of the turbulent kinetic energy, while the fourth does likewise for the Reynolds shear stress $\rho\overline{u_1u_2}$. The vanishing of the convective terms in this case implies that the Reynolds stresses are constant. Furthermore, the work done by the shear stresses on the moving plate is dissipated within the fluid across the passage. This balance is described by a dissipation equation which is not needed in the present example.

The only mechanism for maintaining turbulence in this flow is given by the left side of the $\overline{u_1^2}$ equation, i.e., of the first of Eqs. (6.3.1). Clearly, there must

[†]In Section 9.4 we provide a simple analysis of the plate velocity and spacing required to achieve constant gradients of the mean velocity and mean temperature.

be a mechanism for transferring energy created in the u_1 fluctuations into the u_2 and u_3 fluctuations if these fluctuations are to exist. This mechanism is discussed in Section 4.7: the redistributive influence of pressure fluctuations reflected in the first term on the right side of each of the first three equations. We can see this redistribution in the present context if these three equations are added to obtain

$$\overline{u_1 u_2} \frac{dU_1}{dx_2} = -\varepsilon \tag{6.3.2}$$

an equation balancing production and dissipation. The vanishing of the pressure–rate-of-strain terms by summation over the three coordinate directions implies that they are responsible for the transfer of energy among the various contributions to the turbulent kinetic energy. Thus, in the flow under consideration, energy is produced in the u_1 fluctuations and distributed by pressure–rate-of-strain effects into fluctuations of the other two velocity components. In contrast with grid flow treated earlier, a flow devoid of a production mechanism, turbulent Couette flow involves a local balance among the processes of production, transfer or redistribution, and dissipation.

A comment concerning Eq. (6.3.2) is appropriate here. In its present form this equation represents a balance between production on the left side and mean viscous dissipation on the right. This situation applies at least approximately in significant portions of other turbulent shear flows. This and derivative approximations supplemented by a model for ε frequently lead to useful, although approximate, results. An elaboration of this perspective is discussed and exploited in Section 10.12.

The discussion in connection with Eq. (6.3.2) suggests, as indeed is found to be the case experimentally, that in turbulent shear flows resembling Couette flow the three contributions to the turbulent kinetic energy are not equal. Frequently $\overline{u_1^2} \approx k$, $\overline{u_2^2} \approx \overline{u_3^2} \approx \frac{1}{2}k$, that is, the turbulence is anisotropic at the scales that determine the intensities of the velocity fluctuations. However, as discussed in Sections 5.3 and 5.4, even in these circumstances the turbulence at small scales may remain isotropic or nearly so, provided the turbulence Reynolds number is suitably high and the mean shear is suitably small.

If we examine the last of Eqs. (6.3.1), we see that the Reynolds shear stress is produced by the interaction term $\overline{u_2^2}\, dU_1/dx_2$, altered by the pressure–rate-of-strain term and dissipated.

We can develop a more quantitative understanding of turbulent Couette flow if we apply the Reynolds stress theory of Section 10.10 to its description.[†] When used to close Eqs. (6.3.1), we obtain the following algebraic equations:

$$\overline{u_1 u_2}\, S = -c_1 \frac{\varepsilon}{k}\left(\overline{u_1^2} - \frac{2}{3}k\right) + \left(\frac{4}{3}c_2 - 1\right)\overline{u_1 u_2}\, S - \frac{2}{3}\varepsilon$$

[†]Throughout this chapter we use the simplified form of Eqs. (10.10.5) with the C_{ij} and C terms on the right side dropped, since they are inessential for our purposes.

$$0 = -c_1 \frac{\varepsilon}{k}\left(\overline{u_2^2} - \frac{2}{3}k\right) - \frac{2}{3} c_2 \overline{u_1 u_2}\, S - \frac{2}{3}\varepsilon$$

$$0 = -c_1 \frac{\varepsilon}{k}\left(\overline{u_3^2} - \frac{2}{3}k\right) - \frac{2}{3} c_2 \overline{u_1 u_2}\, S - \frac{2}{3}\varepsilon \qquad (6.3.3)$$

$$\overline{u_2^2}\, S = -c_1 \frac{\varepsilon}{k} \overline{u_1 u_2} + c_2 \overline{u_2^2} S$$

Here $S \equiv dU_1/dx_2$ is the gradient of the mean velocity, the rate of strain parameter for this flow with dimensions of an inverse time. The terms on the right side of these equations involving the empirical coefficients c_1 and c_2 are models for the pressure–rate-of-strain terms in Eqs. (6.3.1). As required by the processes they represent, these terms in the first three equations sum to zero. The last terms relate to dissipation, which is clearly seen to be treated as isotropic (cf. Section 5.5 for a discussion of the validity of this treatment). The second and third of these equations imply that $\overline{u_2^2} \equiv \overline{u_3^2}$ and thus that the turbulent kinetic energy is

$$k = \frac{1}{2}\overline{u_1^2} + \overline{u_2^2} \qquad (6.3.4)$$

Accordingly, we can disregard the third of these equations. The empirical coefficients c_1 and c_2 have standard values of 1.5 and 0.6.

To nondimensionalize Eqs. (6.3.3) we use the turbulent kinetic energy k and the mean rate of strain S. In terms of these quantities, we let

$$G_{11} \equiv \frac{\overline{u_1^2}}{k} \qquad G_{22} \equiv \frac{\overline{u_2^2}}{k} \qquad G_{12} \equiv \frac{\overline{u_1 u_2}}{k} \qquad E \equiv \frac{\varepsilon}{kS} \qquad (6.3.5)$$

Thus we have the supplemental relation from Eq. (6.3.4)

$$1 = \frac{1}{2} G_{11} + G_{22} \qquad (6.3.6)$$

Note that since ε and k are positive definite, so too is the product of SE; thus if, for example, $S < 0$, then $E < 0$. The inverse of E, namely, kS/ε, is a dimensionless measure of the shear rate of strain which can be either positive or negative.

When Eqs. (6.3.5) are substituted into Eqs. (6.3.3), we obtain

$$G_{12} = -c_1 E\left(G_{11} - \frac{2}{3}\right) + \left(\frac{4}{3}c_2 - 1\right) G_{12} - \frac{2}{3}E$$

$$0 = -c_1 E\left(G_{22} - \frac{2}{3}\right) - \frac{2}{3} c_2 G_{12} - \frac{2}{3}E \qquad (6.3.7)$$

$$G_{22} = -c_1 E\, G_{12} + c_2\, G_{22}$$

Equations (6.3.6) and (6.3.7) are four nonlinear, algebraic equations; an approximate solution for standard values for c_1 and c_2 is $G_{11} = 1$, $G_{22} = \frac{1}{2}$, $G_{12} = -E$, and $E = \pm 0.36$. A more accurate solution is obtained if $E = \pm 0.354$ and $c_2 = 0.625$. As noted earlier, here the plus sign on E relates to $S > 0$, the negative to $S < 0$.

Not surprising in view of the local balance of the processes operative in this flow, turbulence production equals dissipation; to see this consider Eq. (6.3.2), which can be written as

$$\frac{P}{\varepsilon} = \frac{-\overline{u_1 u_2}\, S}{\varepsilon} = \frac{-G_{12}}{E} = 1$$

where the definition of P representing turbulence production is clear. The implication of this solution is that for a specified value of the turbulent kinetic energy the three intensities of the velocity fluctuations are known, and for specified values of both k and S the turbulent shear stress and mean viscous dissipation are known. In Sections 6.5 and 6.6 the ratio P/ε again arises.

Several comments are appropriate. The result relative to G_{12} implies that for $S > 0$,

$$\frac{\overline{u_1 u_2}}{k} = -0.36 \tag{6.3.8}$$

This compares with the value 0.30 adopted for this ratio and assumed to apply *throughout* a turbulent boundary layer in the theory of Bradshaw et al. (1967). The anisotropy of the turbulence mentioned earlier is now found here theoretically, namely, $\overline{u_1^2} = k$, $\overline{u_2^2} = \overline{u_3^2} = \frac{1}{2} k$. In Section 10.10 an equation for the mean viscous dissipation is developed; the present solution is consistent with that equation *provided that* two empirical coefficients appearing there, $c_{\varepsilon 1}$ and $c_{\varepsilon 2}$, are equal; their standard values, selected to achieve agreement between theory and experiment for a variety of shear flows, are 1.44 and 1.92, respectively. The implication is that, given the present state of turbulence theory, we must be prepared to alter standard values of empirical coefficients in special circumstances. Finally, to obtain the values of k and S for a particular plate velocity and plate spacing, the *entire flow* from one surface to the other must be analyzed; in Section 9.4 we discuss a relatively simple treatment of this flow.

6.4 TURBULENT COUETTE FLOW WITH A UNIFORM TEMPERATURE GRADIENT

A useful elaboration of the flow discussed in the previous section arises when we add to the gradient of the mean velocity U_1 a uniform temperature gradient $d\Theta/dx_2$ with x_2 vertical. Here we consider temperature gradients sufficiently large so that a body force arises from temperature inhomogeneities and the temperature is an *active scalar*. To achieve this situation the two infinite plates in the planes x_2 = constant in relative motion are considered to be at different temperatures and to be *horizontal*. In this case the equation for the mean temperature [Eq. (4.3.1)] implies that there is a uniform temperature flux from the hotter of the two surfaces. As a consequence, there are severe temperature gradients in the two viscous sublayers adjacent to the plates, and for the high turbulence Reynolds numbers of interest here a nearly uniform, smaller gradient in the central portion between the two surfaces, the portion again of interest here. This temperature gradient can be either positive or negative depending on which of the two plates is hotter. In the Boussinesq approximation there is a gravitational term in the time-dependent x_2 momentum equation, namely, $-g\beta\tilde{\theta}\,\delta_{i2}$, where $\beta \equiv \rho^{-1}\,\partial\rho/\partial\tilde{\theta}$ is the coefficient of thermal expansion with the dimensions of an inverse temperature. Generally, $\beta < 0$, and we shall assume this to be the case. Thus we treat a constant-shear-stress, constant-heat-flux layer.

When the gravitational term is carried through in the averaging procedure leading to Eqs. (6.3.1), the balance of the $\overline{u_2^2}$ intensity is influenced. Accordingly, we rewrite the second of Eqs. (6.3.1) as

$$0 = \overline{\frac{p}{\rho}\frac{\partial u_2}{\partial x_2}} + g(-\beta)\,\overline{u_2\theta} - \varepsilon_{22} \tag{6.4.1}$$

Inclusion of this buoyancy term introduces a second production mechanism supplementing that due to mean shear. Again the absence of convective terms implies a local balance among the processes altering the normal Reynolds stress $\overline{u_2^2}$. Homogeneity justifies neglect of a streamwise flux term $\overline{u_1\theta}$ in the fourth of Eqs. (6.3.1), compared with the production term on the left side.

In Section 4.3 we argued that if $d\Theta/dx_2 > 0$, then $\overline{u_2\theta} < 0$. These inequalities characterize the case of a stable temperature gradient in which, as seen in Eq. (6.4.1), the body force results in the *destruction* of the $\overline{u_2^2}$ intensity, leading to even greater degrees of anisotropy than in isothermal Couette flows. Indeed, if the damping effect of this body force is suitably large, it is possible that $\overline{u_2^2} << \overline{u_1^2}, \overline{u_3^2}$ and thus that the turbulence is nearly two-dimensional, i.e., the fluctuations are confined to x_1–x_3 planes. If this occurs, the production term in the $\overline{u_1u_2}$ equation, the fourth of Eqs. (6.3.1), is severely diminished. The alternative case of $d\Theta/dx_2 < 0$ corresponds to an unstable temperature distribution, yields according to our earlier argument $\overline{u_2\theta} > 0$, and results in a *production term* in Eq. (6.4.1) and to *increases* in the $\overline{u_2^2}$ intensity. Again, significant anisotropy can be anticipated with greater values of $\overline{u_2^2}$ than of the other two intensities. In the limit, $\overline{u_2^2} >> \overline{u_1^2}, \overline{u_3^2}$ and a one-dimensional state of turbulence is approached. In both of these cases, with a fixed velocity of the moving plate, a fixed plate spacing and a fixed temperature difference between them, the gradients dU_1/dx_2 and $d\Theta/dx_2$ in the central region of the flow are significantly altered from their values which prevail when the temperature inhomogeneities are so low as to make the temperature a passive scalar. To determine these alterations, the coupled velocity and temperature fields from one plate to the other must be treated.

The relative importance of production due to mean shear and to temperature gradient is reflected in the flux Richardson number,

$$N_{Rf} \equiv -\frac{g\,\beta\,\overline{u_2\theta}}{\overline{u_1u_2}(dU_1/dx_2)} = -\frac{g\,(-\beta)\,\overline{u_2\theta}}{-\overline{u_1u_2}\,(dU_1/dx_2)} \qquad (6.4.2)$$

With $\overline{u_1u_2}$ and dU_1/dx_2 of opposite sign and $-\beta > 0$, the sign of N_{Rf} is determined solely by $\overline{u_2\theta}$. If the flow is stable, i.e., if $d\Theta/dx_2 > 0$, then $\overline{u_2\theta} < 0$ and $N_{Rf} > 0$. On the other hand, if the flow is unstable, i.e., if $d\Theta/dx_2 < 0$, then $\overline{u_2\theta} > 0$ and $N_{Rf} < 0$. In fact, observations in a stable atmospheric boundary layer suggest that if N_{Rf} is greater than a critical value of the order of 0.5, the intensity of the vertical velocity fluctuations goes to zero (cf. Launder, 1989) and the turbulence is two-dimensional in horizontal planes. On the other hand, if the atmospheric layer is unstable, that intensity is enhanced at the expense of the intensities of the fluctuations of the horizontal velocity components and intense mixing ensues. If the absolute value of the flux Richardson number is suitably small, buoyancy effects are not significant and the velocity and temperature characteristics of the flow are uncoupled.

The experimental determination of the flux Richardson number involves measurement of the correlations $\overline{u_2\theta}$ and $\overline{u_1u_2}$. Under the assumption that gradient transport relations apply (cf. Sections 7.1 and 7.2), these correlations can be replaced by gradients of related mean quantities, which are frequently more readily determined by experimental means. Therefore, consider

$$\overline{u_1u_2} = -\nu_T\frac{dU_1}{dx_2} \qquad \overline{u_2\theta} = -\frac{\nu_T}{\sigma_T}\frac{d\Theta}{dx_2} \qquad (6.4.3)$$

ν_T and σ_T are a turbulent exchange coefficient and a turbulent Prandtl number, respectively. Then, from Eq. (6.4.2) we have

$$N_{Rf} \equiv \frac{g(-\beta)\,d\Theta/dx_2}{\sigma_T(dU_1/dx_2)^2} \qquad (6.4.4)$$

Now *define* the gradient Richardson number as

$$N_{Rg} \equiv \frac{g(-\beta)\,d\Theta/dx_2}{(dU_1/dx_2)^2} \qquad (6.4.5)$$

so that

$$\sigma_T = \frac{N_{Rg}}{N_{Rf}}$$

i.e., the turbulent Prandtl number in buoyant flows is the quotient of the two Richardson numbers.

Our earlier discussion of the anisotropy associated with stable ($N_{Rg} > 0$) and unstable ($N_{Rg} < 0$) temperature distributions applies to the gradient Richardson number. Thus, if the absolute value of N_{Rg} is suitably small, buoyancy effects may be neglected. Equation (6.4.5) indicates that such neglect is appropriate when the flow involves large velocity gradients, i.e., large velocity differences over small distances, and when the temperature gradients are small. On the other hand, small velocity gradients and large temperature gradients lead to significant buoyancy effects and altered turbulent exchange. Large-scale fires provide dramatic examples of buoyancy-dominated flows, since they involve relatively small velocity gradients and large temperature gradients. Similar very important but less dramatic examples arise in the atmospheric boundary layer and in the oceans.[†]

We now extend the analysis of the previous section to include bouyancy effects. Equation (6.4.1) precludes the equality of $\overline{u_2^2}$ and $\overline{u_3^2}$. However, if we supplement the definitions given by Eq. (6.3.5) and rewrite Eq. (6.4.2), we have

$$G_{33} \equiv \frac{\overline{u_3^2}}{k} \qquad N_{Rf} = \frac{g(-\beta)\,\overline{u_2\theta}}{S\,k\,G_{12}} \tag{6.4.6}$$

and we obtain the following algebraic equations, extensions of Eqs. (6.3.6):

$$\begin{aligned}
G_{12} &= -c_1\,E\left(G_{11} - \frac{2}{3}\right) + \left(\frac{4}{3}c_2 - 1\right)G_{12} - \frac{2}{3}E - \frac{1}{3}c_2\,N_{Rf}\,G_{12}\\
0 &= -c_1\,E\left(G_{22} - \frac{2}{3}\right) - \frac{2}{3}c_2\,G_{12} - \frac{2}{3}E + N_{Rf}\,G_{12}\left(\frac{2}{3}c_2 + 1\right)\\
0 &= -c_1\,E\left(G_{33} - \frac{2}{3}\right) - \frac{2}{3}c_2\,G_{12} - \frac{2}{3}E - \frac{1}{3}c_2\,N_{Rf}\,G_{12}\\
G_{22} &= -c_1\,E\,G_{12} + c_2\,G_{22}\\
1 &= \frac{1}{2}(G_{11} + G_{22} + G_{33})
\end{aligned} \tag{6.4.7}$$

Equations (6.4.7) determine G_{11}, G_{22}, G_{33}, G_{12}, and E with N_{Rf} as a parameter and c_1 and c_2 as empirical coefficients assigned the values cited earlier. The $c_2\,N_{Rf}\,G_{12}$ terms in these equations arise from the modeling of the influence of body forces on the pressure [cf. Eq. (2.4.1)] and thus on the pressure–rate-of-strain terms in Eqs. (6.3.1). As required by their redistributive influence, they sum to zero [cf. Launder (1989a) and Craft et al. (1994) for a discussion of this

[†]See Coantic (1978) for a complete discussion of geophysical turbulence and the Boussinesq approximation, Bradshaw and Wood in Bradshaw (1978) for an additional discussion of such turbulence, and Linden and Turner (1975) for an entry to the research literature on this subject. See also Lumley and Panofsky (1964) and Monin and Yaglom (1971).

modeling]. There are solutions to Eqs. (6.4.7) for both positive and negative values of E. However, we need be concerned only with $E \geq 0$, since the values of G_{11}, G_{22}, and G_{33} and the absolute value of G_{12} are unchanged if E changes sign.

If the first three of Eqs. (6.4.7) are added, we readily find that

$$-\frac{G_{12}}{E} = \frac{1}{1 - \frac{1}{2}N_{Rf}} \tag{6.4.8}$$

As a consequence, these three equations can be solved explicitly for the three dimensionless intensities to obtain

$$\begin{aligned} G_{11} &= \frac{2}{3}\left(1 - \frac{1}{c_1}\right) + \frac{2 - \frac{1}{3}c_2(4 - N_{Rf})}{c_1(1 - \frac{1}{2}N_{Rf})} \\ G_{22} &= \frac{2}{3}\left(1 - \frac{1}{c_1}\right) + \frac{\frac{2}{3}c_2(1 - N_{Rf}) - N_{Rf}}{c_1(1 - \frac{1}{2}N_{Rf})} \\ G_{33} &= \frac{2}{3}\left(1 - \frac{1}{c_1}\right) + \frac{\frac{1}{3}c_2(2 + N_{Rf})}{c_1(1 - \frac{1}{2}N_{Rf})} \end{aligned} \tag{6.4.9}$$

Furthermore, the fourth of Eqs. (6.4.7) gives

$$E^2 = \frac{G_{22}}{c_1}(1 - c_2)\left(1 - \frac{1}{2}N_{Rf}\right) \tag{6.4.10}$$

This equation used in conjunction with the second of Eqs. (6.4.9) determines E; G_{12} is then obtained from Eq. (6.4.8) and the entire solution is known with N_{Rf} as the sole parameter. These results all reduce to those given by Eqs. (6.3.9) when $N_{Rf} = 0$.

It is of interest to calculate from the second of Eqs. (6.4.9) the value of N_{Rf}, denoted N_{Rfcr}, resulting in $G_{22} = 0$. We find

$$N_{Rfcr} = 2\,\frac{c_1 + c_2 - 1}{c_1 + 2c_2 + 2} \tag{6.4.11}$$

When standard values for c_1 and c_2 are inserted in this equation, we find $N_{Rfcr} = 0.47$. Although there may be special solutions to Eqs. (6.4.8)–(6.4.10) for $N_{Rf} > N_{Rfcr}$, their physical significance is questionable and we therefore do not consider them further. We thus see that if the flux Richardson number exceeds a critical value, the intensity of the fluctuations in the vertical direction vanish and the turbulence is two dimensional.

Figure 6.4.1 shows the distributions of the various intensities as reflected in G_{11}, G_{22}, and G_{33} and of the Reynolds shear stress as reflected in G_{12}. As suggested earlier, for stable temperature distributions G_{22} is significantly reduced

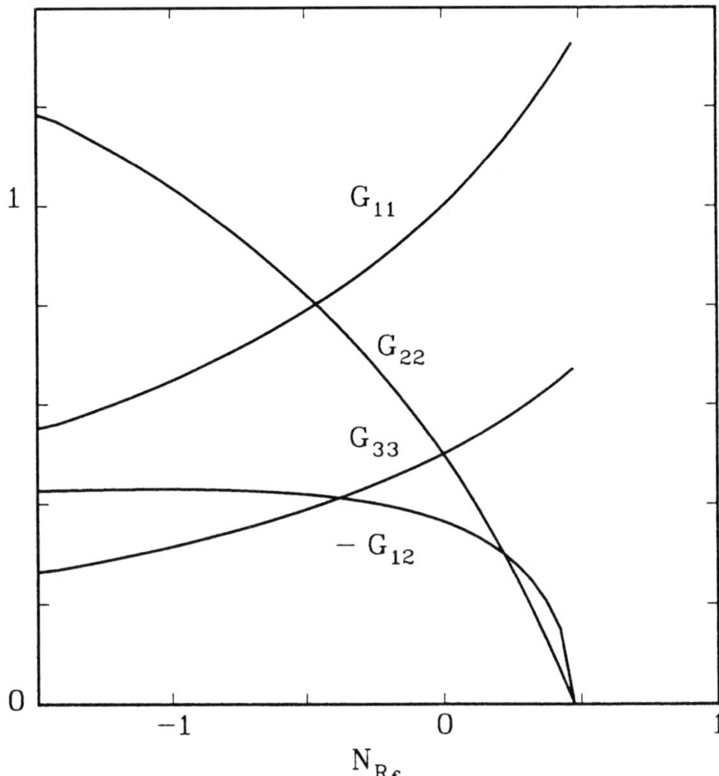

Figure 6.4.1 The dimensionless Reynolds stresses in Couette flow subject to buoyancy.

such that at $N_{Rf} = 0.47$ $G_{22} = G_{12} = 0$. On the other hand, for unstable temperature distributions both G_{11} and G_{33} decrease continuously, and G_{22} increases continuously as $-N_{Rf}$ increases such that at suitably large degrees of instability the turbulence is dominated by fluctuations in the x_2 direction. If bouyancy effects are negligible so that $|N_{Rf}| << 1$, we see from Fig. 6.4.1 that we recover our earlier solutions to Eqs. (6.3.6).

These considerations indicate in a simple fashion the profound influence of buoyancy on turbulence. Note that in the calculation in this section we have not determined the mean temperature gradient $d\Theta/dx_2$, which, combined with a rate of strain dU_1/dx_2, results in a particular flux Richardson number. Rather, we have *specified* N_{Rf} and determined the nature of the resulting turbulence. The analysis of a Couette flow with given values of dU_1/dx_2 and $d\Theta/dx_2$ is more complex, since transport equations for $\overline{u_2\theta}$ and $\overline{\theta^2}$ must be incorporated in the formulation. A consequence of this shortcoming is that the gradient Richardson number and the turbulent Prandtl number *are not given* by the present analysis. In Section 9.4 a more complete treatment of Couette flow with mean velocity and suitably small temperature gradients is given.

6.5 HOMOGENEOUS SHEAR FLOWS

In Sections 6.3–6.4 we discussed turbulence in simple, homogeneous shear flows in dynamic equilibrium involving a local balance among the various operative processes of production, redistribution, and dissipation. In this section we consider a simple shear flow which evolves in *time* from an arbitrary initial state. We again *assume* that the turbulence is homogeneous in all three coordinate directions. Although we discuss later the difficulties of establishing such a flow in the laboratory, for present purposes it is sufficient to imagine a turbulence-generating grid with a nonuniform spacing of the horizontal rods such that a linear distribution in the x_2 direction of the mean x_1 velocity with $U_2 \equiv U_3 \equiv 0$ is established downstream. The rate-of-strain parameter $S \equiv dU_1/dx_2$ in the previous section is again taken to be constant and, without loss of generality, positive. We describe the flow in terms of a coordinate system moving with the mean velocity along an arbitrary streamline, the perspective used in Sections 5.2–5.4 for grid turbulence.[†] In this system all of the turbulence quantities, i.e., the intensities, the Reynolds shear stress $\rho\overline{u_1u_2}$, and the mean viscous dissipation, vary with time.

Homogeneous shear flows are of fundamental interest in the theory of solids, fluids, and rheological materials. Accordingly, it is not surprising that we consider them in several sections; we discuss simple homogeneous shear in Section 4.2 to identify the correlations $\overline{u_iu_j}$ with stresses and in Section 7.4 to treat mixing-length theory. Here we consider the evolution of turbulence in homogeneous shear flows [cf. Gence (1983) for a review of homogeneous turbulence with and without mean shear].

According to a Reynolds stress theory, the time evolution of the turbulence is described by Eqs. (6.3.3) with convective terms included. Thus rearrangement of these equations into a convenient form and incorporation of this convection leads to the following equations:

$$\frac{d}{dt}\overline{u_1^2} = -c_1\frac{\varepsilon}{k}\left(\overline{u_1^2} - \frac{2}{3}k\right) + \left(\frac{4}{3}c_2 - 2\right)\overline{u_1u_2}\,S - \frac{2}{3}\varepsilon$$

$$\frac{d}{dt}\overline{u_2^2} = -c_1\frac{\varepsilon}{k}\left(\overline{u_2^2} - \frac{2}{3}k\right) - \frac{2}{3}c_2\,\overline{u_1u_2}\,S - \frac{2}{3}\varepsilon$$

[†]Inconsistencies with respect to the assumption of homogeneity arise when a Eulerian description of this flow is adopted (cf. Tavoularis, 1985).

$$\frac{d}{dt}\overline{u_3^2} = -c_1\frac{\varepsilon}{k}\left(\overline{u_3^2} - \frac{2}{3}k\right) - \frac{2}{3}c_2\,\overline{u_1u_2}\,S - \frac{2}{3}\varepsilon$$

$$\frac{d}{dt}\overline{u_1u_2} = -c_1\frac{\varepsilon}{k}\overline{u_1u_2} + (c_2 - 1)\,\overline{u_2^2}\,S \tag{6.5.1}$$

The previous identification of the source of the various terms in Eq. (6.3.3) clearly applies here, since only the convective terms on the left side are new. We must supplement these equations with an evolution equation for the mean viscous dissipation ε; from Section 10.10 we have for this flow

$$\frac{d}{dt}\varepsilon = \frac{\varepsilon}{k}(-c_{\varepsilon 1}\,\overline{u_1u_2}\,S - c_{\varepsilon 2}\,\varepsilon) \tag{6.5.2}$$

where $c_{\varepsilon 1}$ and $c_{\varepsilon 2}$ are two empirical coefficients with, as noted earlier, standard values of 1.44 and 1.92, respectively.[†] In Eq. (6.5.2) the two terms on the right side represent the production and dissipation of the mean viscous dissipation.

To nondimensionalize Eqs. (6.5.1) and (6.5.2), we modify Eqs. (6.3.5) by introducing the turbulent kinetic energy at the initial time k_0; accordingly, we define

$$G_{11} \equiv \frac{\overline{u_1^2}}{k_0} \qquad G_{22} \equiv \frac{\overline{u_2^2}}{k_0} \qquad G_{33} \equiv \frac{\overline{u_3^2}}{k_0}$$

$$G_{12} \equiv \frac{\overline{u_1u_2}}{k_0} \qquad E \equiv \frac{\varepsilon}{k_0S} \qquad \hat{t} = S\,t \tag{6.5.3}$$

In terms of these variables, Eqs. (6.5.1) become

$$\frac{dG_{11}}{d\hat{t}} = -c_1\frac{E}{K}\left(G_{11} - \frac{2}{3}K\right) + \left(\frac{4}{3}c_2 - 2\right)G_{12} - \frac{2}{3}E$$

$$\frac{dG_{22}}{d\hat{t}} = -c_1\frac{E}{K}\left(G_{22} - \frac{2}{3}K\right) - \frac{2}{3}c_2\,G_{12} - \frac{2}{3}E$$

$$\frac{dG_{33}}{d\hat{t}} = -c_1\frac{E}{K}\left(G_{33} - \frac{2}{3}K\right) - \frac{2}{3}c_2\,G_{12} - \frac{2}{3}E \tag{6.5.4}$$

$$\frac{dG_{12}}{d\hat{t}} = -c_1\frac{E}{K}G_{12} + (c_2 - 1)\,G_{22}$$

$$\frac{dE}{d\hat{t}} = \frac{E}{K}(-c_{\varepsilon 1}\,G_{12} - c_{\varepsilon 2}\,E)$$

[†]Equation (6.5.2) provides the justification for our earlier requirement that in turbulent Couette flow these two coefficients be equal since there $-\overline{u_1u_2}\,S = \varepsilon$ and $d\varepsilon/dt = 0$.

Here

$$K \equiv \frac{k}{k_0} = \frac{1}{2}(G_{11} + G_{22} + G_{33}) \tag{6.5.5}$$

is a dimensionless turbulent kinetic energy normalized with k_0; clearly $K(0) = 1$ and thus $G_{11}(0) + G_{22}(0) + G_{33}(0) = 2$. If we add the first three of these equations, we obtain the evolution equation for $K(\hat{t})$,

$$\frac{dK}{d\hat{t}} = -G_{12} - E \tag{6.5.6}$$

Comparison of this equation with Eqs. (4.7.3) and (6.3.2) makes clear that the present flow involves the more general case of turbulent kinetic energy changing with time under the influence of shear and dissipation.

Equations (6.5.4) are to be solved subject to initial conditions at $\hat{t} = 0$. However, it is more interesting to reformulate these equations in terms of the anisotropy tensor b_{ij} defined by Eq. (4.7.9),

$$b_{ij} \equiv \frac{\overline{u_i u_j}}{2\,k} - \frac{1}{3}\delta_{ij}$$

which from Eqs. (6.5.3) and (6.5.5) yields

$$\begin{aligned} b_{11} &= \frac{1}{2}\frac{G_{11}}{K} - \frac{1}{3} \qquad & b_{22} &= \frac{1}{2}\frac{G_{22}}{K} - \frac{1}{3} \\ b_{33} &= \frac{1}{2}\frac{G_{33}}{K} - \frac{1}{3} \qquad & b_{12} &= \frac{1}{2}\frac{G_{12}}{K} \end{aligned} \tag{6.5.7}$$

Now, if we write the first four of Eqs. (6.5.4) in terms of these b_{ij} quantities, we find

$$\begin{aligned} \frac{db_{11}}{d\hat{t}} &= \left(b_{11} + \frac{1}{3}\right)(2\,b_{12} + \beta) - c_1\,\beta\,b_{11} + \left(\frac{4}{3}c_2 - 2\right)b_{12} - \frac{1}{3}\beta \\ \frac{db_{22}}{d\hat{t}} &= \left(b_{22} + \frac{1}{3}\right)(2\,b_{12} + \beta) - c_1\,\beta\,b_{22} - \frac{2}{3}c_2 b_{12} - \frac{1}{3}\beta \\ \frac{db_{33}}{d\hat{t}} &= \left(b_{33} + \frac{1}{3}\right)(2\,b_{12} + \beta) - c_1\,\beta\,b_{33} - \frac{2}{3}c_2 b_{12} - \frac{1}{3}\beta \\ \frac{db_{12}}{d\hat{t}} &= b_{12}(2\,b_{12} + \beta) - c_1\,\beta\,b_{12} + (c_2 - 1)\left(b_{22} + \frac{1}{3}\right) \end{aligned} \tag{6.5.8}$$

where we introduce $\beta \equiv E/K > 0$. Using Eqs. (6.5.2) and (6.5.6), we supplement Eqs. (6.5.8) with the following equation for $\beta(\hat{t})$:

$$\frac{d\beta}{d\hat{t}} = \beta[(2\,b_{12} + \beta) - (2\,c_{\varepsilon 1}b_{12} + c_{\varepsilon 2}\beta)] \tag{6.5.9}$$

Equations (6.5.8) and (6.5.9) are alternatives to Eqs. (6.5.4) and are again to be solved subject to initial conditions which in this case are restricted by the requirement that $b_{11}(0) + b_{22}(0) + b_{33}(0) = 0$. However, we now examine the existence of a special state, an equilibrium state in some sense defined by values of b_{11}, b_{22}, b_{33}, b_{12}, and β which make the right side of each of these equations zero. We identify these values with a subscript ∞. Notice that the second and third *algebraic* equations for $b_{22\infty}$ and $b_{33\infty}$ are identical, so $b_{22\infty} \equiv b_{33\infty}$. Thus we must consider only four equations, the first, second, and fourth of Eqs. (6.5.8) and Eq. (6.5.9).

An important intermediate result is obtained from setting the right side of the last of Eqs. (6.5.8) equal to zero, namely

$$\frac{-2\, b_{12\infty}}{\beta_\infty} = \frac{1 - c_{\varepsilon 2}}{1 - c_{\varepsilon 1}} \tag{6.5.10}$$

which if we use Eqs. (6.5.3) and (6.5.7) leads to

$$-\frac{\overline{u_1 u_2}_\infty S}{\varepsilon_\infty} = \left(\frac{P}{\varepsilon}\right)_\infty = \frac{1 - c_{\varepsilon 2}}{1 - c_{\varepsilon 1}} \tag{6.5.11}$$

We thus see that the ratio of production to mean viscous dissipation in the flow associated with the equilibrium point is a constant with a value that depends on the two empirical coefficients, $c_{\varepsilon 1}$ and $c_{\varepsilon 2}$. This finding is consistent with our earlier observation that for a simple shear flow in equilibrium $P/\varepsilon = 1$ and in this case these two coefficients must be equal. With standard values assigned $c_{\varepsilon 1}$ and $c_{\varepsilon 2}$, we have $(P/\varepsilon)_\infty = 2.09$, i.e., production *exceeds* mean viscous dissipation. Accordingly, in the flow associated with the equilibrium point the components of the anisotropy tensor are constant but the Reynolds stresses, turbulent kinetic energy, and mean viscous dissipation all grow since production exceeds dissipation. Indeed, these turbulence quantities grow exponentially fast, as may be seen as follows. Using Eq. (6.5.5) and the definition of β, we rewrite Eq. (6.5.6) as

$$\frac{dK}{d\hat{t}} = -(2\, b_{12} + \beta)\, K \tag{6.5.12}$$

When the right side is evaluated at the equilibrium point, we obtain

$$\frac{dK}{d\hat{t}} = -\lambda K$$

where $\lambda \equiv 2b_{12\infty} + \beta_\infty$. The solution to this equation grows as $\exp(\lambda \hat{t})$ so that the constancy of the components of the anisotropy tensor and of $\beta \equiv E/K$ at the equilibrium point implies that all of the dimensionless Reynolds stresses and the mean viscous dissipation grow at the same rate. A further consequence of Eqs. (6.5.10) and (6.5.11) is that we are interested only in negative values of $b_{12\infty}$, since the mean viscous dissipation is nonnegative and $S > 0$.

With the empirical coefficients assigned the values $c_1 = 1.5$, $c_2 = 0.6$, $c_{\varepsilon 1} = 1.44$, and $c_{\varepsilon 2} = 1.92$, Eqs. (6.5.10) can be solved numerically to yield†

$$\begin{aligned} b_{11\infty} &= 0.213 \qquad b_{22\infty} = b_{33\infty} = -0.108 \\ b_{12\infty} &= -0.190 \qquad \beta_\infty = 0.182 \qquad \lambda = 0.199 \end{aligned} \tag{6.5.13}$$

These values appear to be unique, i.e., there is only one physically acceptable equilibrium point. Moreover, Speziale et al. (1990b) show that the solutions to Eqs. (6.5.8) and (6.5.9) approach the equilibrium point in a stable fashion. Such a point is an *attractor* in the parlance of dynamic systems theory. The implication is that solutions to Eqs. (6.5.8) subject to *arbitrary initial conditions* approach this equilibrium point as $\hat{t} \to \infty$. The values given by Eq. (6.5.13) are in good agreement with the LRR solution by Speziale et al. (1990b) provided differences in the empirical coefficients are taken into account. In this regard it is worth noting the variability in the literature of the *standard* values of these coefficients; Speziale et al. set $c_1 = 1.8$, $c_2 = 0.6$, $c_{\varepsilon 1} = 1.45$, and $c_{\varepsilon 2} = 1.90$.

Figures 6.5.1 and 6.5.2 show the evolution of homogeneous turbulence which is initially isotropic but subject to a constant shear rate of strain; thus $b_{11}(0) = b_{22}(0) = b_{33}(0) = b_{12}(0) = 0$, $G_{11}(0) = G_{22}(0) = G_{33}(0) = \frac{2}{3}$, and $G_{12}(0) = 0$. In this case $E(0)$ is the sole additional parameter characterizing the initial state.‡ We arbitrarily assign $E(0) = 1$, a representative value. We see in Fig. 6.5.1 that as time evolves, the components of the anisotropy tensor approach the equilibrium point; while in Fig. 6.5.2 the dimensionless Reynolds stresses and mean viscous dissipation increase indefinitely, indeed exponentially fast. In more complex, inhomogeneous shear flows, countervailing influences, e.g., inhomogeneities, result in turbulent diffusion and inhibit this growth.

We now examine some of the implications of the turbulence at the equilibrium point. If we define a length scale $l \equiv k^{3/2}/\varepsilon$ as being characteristic of the large turbulent fluctuations, we realize that the constancy of β and the exponential growth of both k and ε result in l *growing* exponentially with time. Consider next the gradient transport assumption connecting the Reynolds stresses with the mean rate of strain (cf. Section 7.1), a connection which is *not* used in the Reynolds stress theory but which in the present case would call for $\rho\overline{u_1 u_2} = -\mu_T S$, where μ_T is a turbulent exchange coefficient. One model for this coefficient (cf. Section 10.1) sets $\mu_T = \rho c_\mu k^2/\varepsilon$ with c_μ an empirical coefficient. Now,

†An inverse numerical procedure is effective; if a trial value for $b_{12\infty}$ is chosen, then β_∞ is obtained from Eqs. (6.5.10) and the remaining three equations are linear and can be solved for $b_{11\infty}$, $b_{22\infty}$, *and* $b_{12\infty}$. Comparison of the trial and computed values of $b_{12\infty}$ leads to a one-dimensional search for the correct solution which must respect the requirements that $b_{11\infty}$, $b_{22\infty}$ be greater than $-\frac{1}{3}$ and less than $\frac{2}{3}$.

‡It is recalled that $E(0)$ is the ratio of two lengths at $\hat{t} = 0$, one associated with the rate of strain $k_0^{1/2}/S$ and a second with the large turbulent scales $k_0^{3/2}/\varepsilon(0)$. That the behavior of the turbulence depends on this length–scale ratio is to be expected.

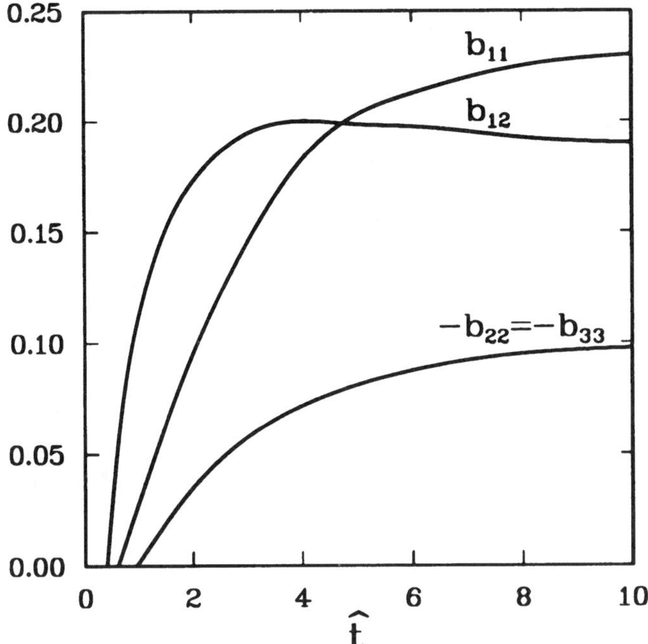

Figure 6.5.1 The evolution of the components of the anisotropy tensor in homogeneous shear.

in the flow at the equilibrium point, all of the Reynolds stresses, including that associated with shear, grow exponentially with time while S remains constant; we see that gradient transport is consistent with this behavior because of the exponential growth of the transport coefficient.

Because of their fundamental importance, simple shear flows have been the subject of a variety of experimental investigations starting with Champagne et al. (1970) and continuing to Rohr et al. (1988), Gibson and Kanellopoulos (1988), and Tavoularis and Karnik (1989). The initial experiments involved a turbulence grid with nonuniform spacing of the horizontal rods, but later experiments employed multiple slots fed by individual blowers at volumetric flow rates, resulting in a linear variation in the transverse direction of the streamwise mean velocity. By providing heaters in each slot, both mean velocity and mean temperature gradients can be established independently. In early experiments the evolution of the turbulence was not followed to a sufficient downstream distance, with the consequence that no growth of the turbulence was observed. With greater downstream distances a linear growth was obtained, but with even greater downstream distances available in the most recent experiments an exponential growth rate as suggested by the present analysis is found [cf. Rohr et al. (1988) for a history of experiments on homogeneous shear flows].

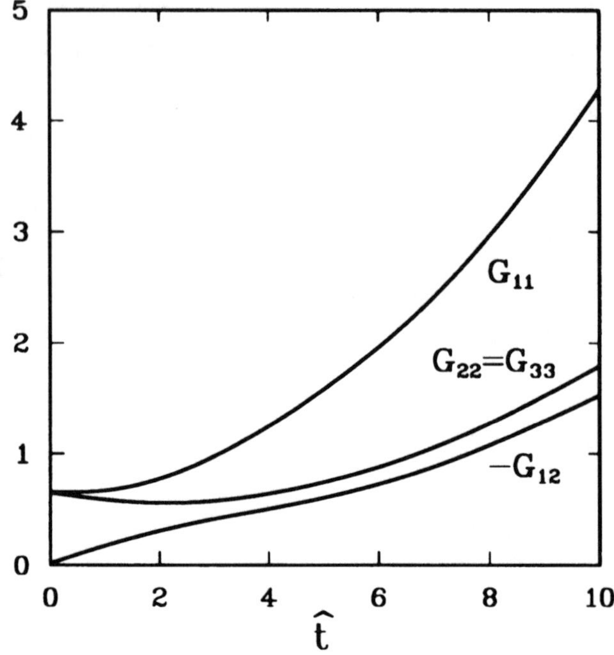

Figure 6.5.2 The evolution of the dimensionless Reynolds stresses in homogeneous shear.

It is difficult to replicate in these experiments conditions corresponding to theory; for example, the initial turbulence level is influenced by either the turbulence grid or slot flow and is not homogeneous. Downstream this effect decreases, but then the finite length of any facility and the growth of boundary layers on the tunnel walls raise questions as to the adequacy of the available streamwise length. It is also clear that any growth with time of the large turbulent scales must in due course be inhibited by tunnel walls. Indeed, in a suitably long tunnel the flow far downstream must be a fully developed, three-dimensional flow with a constant streamwise gradient of the mean pressure that is independent of the initial flow. Thus it seems clear that, as confirmed by the historical record, comparison of theoretical predictions and experiment in this area of homogeneous shear is difficult and prone to uncertainties.

The results of direct numerical simulation applied to simple shear flows can be compared with the predictions of various moment methods. In this case an arbitrary initial distribution of turbulence in a box at time zero is subject to a simple steady shear and its evolution with time calculated by solution of the full Navier-Stokes equations (cf., e.g., Rogers et al., 1987). It is found that the ratio P/ε and the components of the anisotropy tensor obtained by averaging over the entire computational domain at each time step settle down to constant values while the individual Reynolds stresses grow exponentially with time until the calculation must be terminated due to the growth of the large turbulent scales. One such calculation (cf. Rogers et al., 1987) gives

$$\left(\frac{P}{\varepsilon}\right)_\infty = 1.8 \qquad b_{11} = 0.2 \qquad b_{12} = b_{22} = -0.15 \qquad b_{33} = -0.05 \tag{6.5.14}$$

The differences between this asymptotic state and that given by Eqs. (6.5.13) warrant comment. If P/ε is to be 1.8 rather than 2.09, either $c_{\varepsilon 1}$ or $c_{\varepsilon 2}$ or both must be modified from their standard values, e.g., $c_{\varepsilon 1} = 1.44$ and $c_{\varepsilon 2} = 1.79$ yield $P/\varepsilon = 1.8$. However, the inequality of b_{22} and b_{33} obtained from DNS is in essential disagreement with the results of the Reynolds stress theory employed here, so changes in this theory may be called for.

Despite the difficulties in comparing solutions to either Eqs. (6.5.4) or (6.5.10) with experiment, homogeneous shear flows are of value in comparing the predictions of various theories in a simple setting which removes the uncertainties associated with numerical complexity. In addition, such flows are useful vehicles for examining the influence of various complications that are difficult to examine in the laboratory. For example, Speziale and MacGiolla Mhuiris (1989) studied the influence of rotation about the x_3 axis on a homogeneous shear flow, an influence that is discussed in the next section.

6.6 THE INFLUENCE OF ROTATION ON TURBULENCE

The influence of rotation on turbulent fluid is significant and is the subject of considerable study. Thus, for example, turbulent jets involving swirl have been treated by Fu et al. (1987); experimental results on grid turbulence subject to rotation about a streamwise axis have been presented by Wigeland and Nagib (1978); and turbulence in a fully developed two-dimensional channel rotating about an axis parallel to the long side of the channel was studied experimentally by Johnston (1973). Moreover, a variety of computational and theoretical investigations of the influence of rotation have been carried out; thus DNS of homogeneous turbulence subject to rotation has been reported by Bardina et al. (1985), while Speziale et al. (1990b) assessed various moment theories within the context of homogeneous shear flows subject to rotation. Cambon (1994) provides a review of a Euromech meeting concerned with the alteration of turbulence by distortion and rotation, while Hossain (1994) describes the anisotropy of turbulence under strong rotation.

The starting point for consideration of a flow subject to additional accelerations, translational and/or rotational, is a set of conservation equations in a noninertial coordinate system. Simple examples calling for this perspective are the fuel tanks in a space vehicle during launch or the flow in a turbine viewed in a coordinate system fixed on a blade. In this connection consider Fig. 6.6.1,

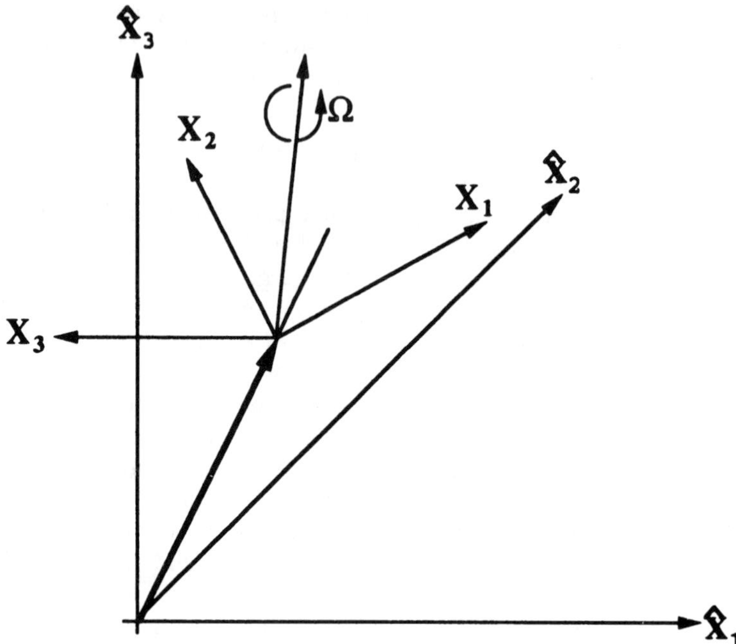

Figure 6.6.1 Inertial and noninertial coordinate systems extension in the axial direction and a compression in the radial direction.

which shows two coordinate systems; one is fixed, i.e., inertial, with coordinates $\hat{\mathbf{x}}$ and velocity components $\hat{\mathbf{v}}$. A second, with coordinates $\mathbf{x}$ and velocities $\mathbf{v}$ is moving with a velocity, in general varying with time, $\mathbf{V}_0$ and rotating about an instantaneous axis with an instantaneous magnitude, both indicated by the vector $\boldsymbol{\Omega}$. To describe the flow in terms of the second set of coordinates and velocities we must transform the variables of the conservation equations of Chapter 2 written in terms of $\hat{\mathbf{x}}$ and $\hat{\mathbf{v}}$ into those of $\mathbf{x}$ and $\mathbf{v}$. In addition to the usual Eulerian acceleration in the latter variables, this transformation introduces a translational acceleration and centrifugal and Coriolus accelerations associated with the motion of the $\mathbf{x}$-coordinate system relative to an inertial frame of reference. When these new equations are applied to turbulent flow, i.e., averaged, these additional terms introduce new correlations indicating the manner in which the mean velocities and the Reynolds stresses and fluxes are altered by accelerations of the coordinate system (cf. Speziale, 1989).

As an example of the equations of motion in a noninertial frame of reference, consider the case of $\mathbf{V}_0 \equiv 0$ but of a *steady* rotation $\boldsymbol{\Omega}$. In this case the conservation equations become

$$\frac{\partial \tilde{u}_k}{\partial x_k} = 0$$
$$\frac{\partial \tilde{u}_i}{\partial t} + \frac{\partial}{\partial x_k}\tilde{u}_k\tilde{u}_i = -\frac{1}{\rho}\frac{\partial \tilde{p}}{\partial x_i} + \nu\frac{\partial^2 \tilde{u}_i}{\partial x_k\,\partial x_k} - 2e_{ilk}\Omega_l\tilde{u}_k \tag{6.6.1}$$

Thus, for example, if $i = 1$, we obtain a pair of contributions to conservation of momentum in the x_1 direction, $-2\ \Omega_2\ \tilde{u}_3$ and $2\ \Omega_3\ \tilde{u}_2$. If the only nonzero component of $\boldsymbol{\Omega}$ is Ω_3, i.e., if we consider a constant rate of rotation about the x_3 axis, then to the three instantaneous momentum equations corresponding to $i = 1, 2, 3$ in the second of Eqs. (6.6.1) we have the additional terms $2\ \Omega_3\ \tilde{u}_2$, $-2\ \Omega_3\ \tilde{u}_1$, and zero, respectively. The closure of Eqs. (6.6.1) must incorporate the influence of rotation on the pressure (cf. Section 2.4), with the consequence that the modeling of the pressure–rate-of-strain correlation is altered (cf. Speziale, 1989).

To illustrate the effect of rotation on turbulence, we revisit the homogeneous shear flows dealt with in Section 6.5 but include a steady rotation about the x_3 axis of a magnitude Ω, as shown schematically in Fig. 6.6.2. Thus the flow involves centrifugal and Coriolis forces which alter the dynamic balance of the turbulence. We follow our earlier perspective and view the flow from a frame of reference moving with the mean velocity along an arbitrary mean streamline from an initial time. Accordingly, we obtain variants of Eqs. (6.5.1) and (6.5.2) which include not only the influence of rotation seen in Eqs. (6.6.1) but alterations of the modeling as a consequence of rotation.

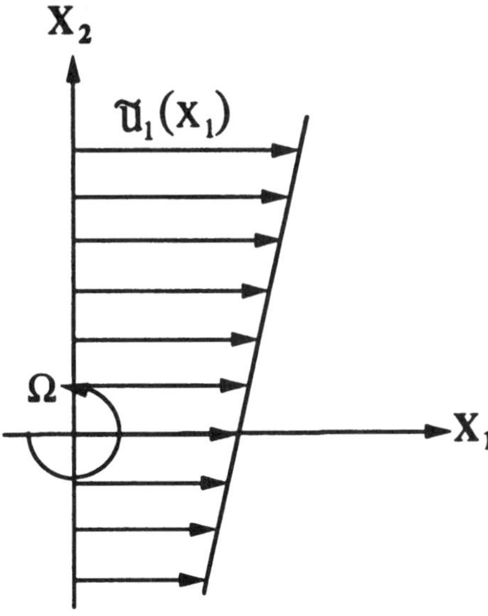

Figure 6.6.2 A simple shear flow subject to rotation about the x_3 axis.

From Speziale et al. (1990b) we find that the Reynolds stress theory of Section 10.10 when modified to include rotation for this flow yields the following equations:

$$\frac{d\overline{u_1^2}}{dt} = -c_1\frac{\varepsilon}{k}\left(\overline{u_1^2} - \frac{2}{3}k\right) + \left(\frac{4}{3}c_2 - 2\right)\overline{u_1u_2}\,S - 2\,(c_2 - 2)\,\overline{u_1u_2}\,\Omega - \frac{2}{3}\varepsilon$$

$$\frac{d\overline{u_2^2}}{dt} = -c_1\frac{\varepsilon}{k}\left(\overline{u_2^2} - \frac{2}{3}k\right) + \frac{2}{3}c_2\,\overline{u_1u_2}\,S + 2\,(c_2 - 2)\,\overline{u_1u_2}\,\Omega - \frac{2}{3}\varepsilon$$

$$\frac{d\overline{u_3^2}}{dt} = -c_1\frac{\varepsilon}{k}\left(\overline{u_3^2} - \frac{2}{3}k\right) - \frac{2}{3}c_2\,\overline{u_1u_2}\,S - \frac{2}{3}\varepsilon \tag{6.6.2}$$

$$\frac{d\overline{u_1u_2}}{dt} = -c_1\,\frac{\varepsilon}{k}\,\overline{u_1u_2} + (c_2 - 1)\,\overline{u_2^2}\,S + (c_2 - 2)(\overline{u_1^2} - \overline{u_2^2})\,\Omega$$

$$\frac{d\varepsilon}{dt} = \frac{\varepsilon}{k}(-c_{\varepsilon1}\,\overline{u_1u_2}\,S - c_{\varepsilon2}\,\varepsilon)$$

If $\Omega = 0$, these equations are identical with Eqs. (6.5.1), but here the terms involving Ω consist of those free of, and those multiplied by, c_2; the former arise from the Ω_i term in Eqs. (6.6.1), while the latter arise from the influence of rotation on the pressure (cf. Section 2.4) reflected in the pressure–rate-of-strain modeling as mentioned earlier. Equations (6.6.2) imply that, in homogeneous shear flows, rotation results in a redistribution of the energy in the x_1 and x_2 coordinate directions; e.g., if $\Omega > 0$, rotation increases the $\overline{u_1^2}$ contribution and decreases the $\overline{u_2^2}$ contribution to the turbulent kinetic energy.

We could introduce into Eqs. (6.6.2) the dimensionless Reynolds stresses defined by Eqs. (6.5.3), but we obtain more interesting results by pursuing the development of the previous section, i.e., by rewriting Eqs. (6.6.2) in terms of the components of the anisotropy tensor. We thus obtain

$$\frac{db_{11}}{d\hat{t}} = \left(b_{11} + \frac{1}{3}\right)(2\,b_{12} + \beta) - c_1\,\beta\,b_{11} + \left(\frac{4}{3}c_2 - 2\right)b_{12} - 2\,(c_2 - 2)\,b_{12}\,\frac{\Omega}{S} -$$

$$\frac{db_{22}}{d\hat{t}} = \left(b_{22} + \frac{1}{3}\right)(2\,b_{12} + \beta) - c_1\,\beta\,b_{22} - \frac{2}{3}c_2\,b_{12} + 2\,(c_2 - 2)\,b_{12}\,\frac{\Omega}{S} - \frac{1}{3}\beta$$

$$\frac{db_{33}}{d\hat{t}} = \left(b_{33} + \frac{1}{3}\right)(2\,b_{12} + \beta) - c_1\,\beta\,b_{33} - \frac{2}{3}c_2\,b_{12} - \frac{1}{3}\beta \tag{6.6.3}$$

$$\frac{db_{12}}{d\hat{t}} = b_{12}(2\,b_{12} + \beta) - c_1\,\beta\,b_{12} + (c_2 - 1)\left(b_{22} + \frac{1}{3}\right) + (c_2 - 2)(b_{11} - b_{22})\,\frac{\Omega}{S}$$

$$\frac{d\beta}{d\hat{t}} = \beta[(2\,b_{12} + \beta) - (2\,c_{\varepsilon1}b_{12} + c_{\varepsilon2}\,\beta)]$$

Here again,

$$K \equiv \frac{k}{k_0} = \frac{1}{2}(G_{11} + G_{22} + G_{33})$$

is a dimensionless turbulent kinetic energy normalized by an initial value k_0. The equation for $\beta(\hat{t})$ is identical with the last of Eqs. (6.5.9) and is obtained from the last of Eqs. (6.6.2), from adding the first three of those equations to obtain an equation for $dK/d\hat{t}$ identical with Eq. (6.6.5), and finally from noting that $\beta \equiv E/K$. The absence of an explicit effect of rotation on the evolution of the turbulent kinetic energy implies that its influence is reflected in *both* the mean rate of production and the dissipation.

Equations (6.6.3) are identical with Eqs. (6.5.8) if rotation is zero. We thus see that the new parameter characterizing the influence of rotation is the dimensionless quotient Ω/S, the ratio of two times: one associated with rotation Ω^{-1} and a second with the rate of strain S^{-1}. The implication is that the influence of rotation on turbulence in a simple shear flow depends on the magnitude of the shearing rate of strain.

With initial values for each of the dependent variables and with a fixed value for Ω/S, Eqs. (6.6.3) determine the evolution of the components of the anisotropy tensor and of β with time. Furthermore, if we calculate the history of $K(\hat{t})$ from Eq. (6.5.12) by integration, the corresponding evolution of the dimensionless Reynolds stresses can be determined. However, it is instructive to defer consideration of the evolution of the turbulence and to examine the dependence of the equilibrium point on the rotation parameter Ω/S; accordingly, we set the right sides of these equations to zero and denote the resulting values of the components of the anisotropy tensor and β with the subscript infinity. Solutions to the algebraic equations determining the equilibrium point are shown in Fig. 6.6.3 for a range of values of Ω/S. The most significant finding shown there is that $b_{12\infty}$ is negative, i.e., that turbulence production and thus the mean viscous dissipation is positive, only for a limited range of the rotation parameter. With the values for the empirical coefficients used here, $-0.075 < \Omega/S < 0.36$. The implication is that for Ω/S increasing from zero, rotation is destabilizing and increases the production of turbulence; however, further increases beyond $\Omega/S \approx 0.15$ and decreases from $\Omega/S = 0$ are stabilizing and reduce such production. We also see from Fig. 6.6.3 that rotation has little influence on b_{33}, i.e., on the contribution of the fluctuations of the velocity in the x_3 coordinate direction but, as suggested earlier, switches the relative contributions to the turbulent kinetic energy from the fluctuations of the velocity in the x_1 and x_2 coordinate directions.

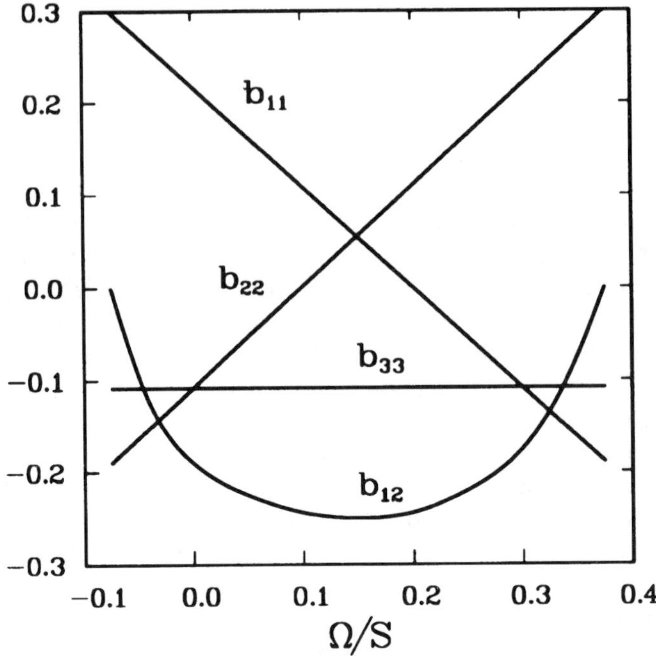

Figure 6.6.3 Variation of the components of the anisotropy tensor with rotation at the equilibrium.

It is now of interest to consider solutions to Eqs. (6.6.3) subject to arbitrary initial conditions. There are a wide variety of such conditions; we select as indicative a set corresponding to isotropic turbulence on which a constant shearing rate of strain and rotation are imposed. Thus we take $b_{11}(0) = b_{22}(0) = b_{33}(0) = b_{12}(0) = 0$. Here for $E(0)$ we select the value 0.296 used by Bardina et al. (1985) in their LES study of the influence of rotation. Because of the restriction we have imposed on b_{12}, we present the results in terms of the evolution of this component of the anisotropy tensor in Fig. 6.6.4 for a range of values of Ω/S. For both $\Omega/S = 0$ and 0.2, values within the range that yield equilibrium solutions, we see by referring to Fig. 6.6.3 that the evolving flows approach those corresponding to the equilibrium point as time increases. This is also the case for the other components of the anisotropy tensor. Furthermore, we see in Fig. 6.6.4 that for values of Ω/S outside the range yielding equilibrium points, turbulence production is initially positive but afterward becomes negative and the solutions become pathological.

The analysis in this section indicates that the influence of rotation on turbulence can be profound and that this influence arises from both explicit dependence of momentum transport on rotation but also from the alteration of the turbulence by rotation. A related influence, that of surface curvature on these shear layers, is discussed in Section 9.14.

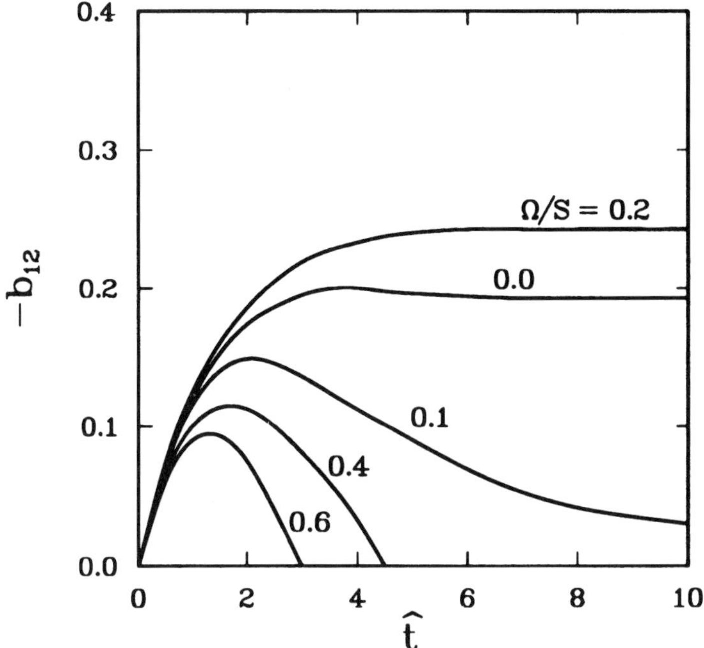

Figure 6.6.4 The evolution of $-b_{12}$ under the influence of rotation.

6.7 TURBULENT FLOWS SUBJECT TO PLANE RATES OF STRAIN

Understanding the characteristics and evolution of turbulence in the shear flows treated in Sections 6.3–6.6 is important for application to jets, wakes, mixing layers, and boundary layers. In this and the next section we discuss the response of turbulence to normal stresses, i.e., to compression and extension, of interest in connection with turbulent duct flows and turbulent external flows encountering bodies. Thus in Section 6.8 we discuss the behavior of turbulence in an axisymmetric contraction section which imposes an extension in the axial direction and a compression in the radial direction. The particular circumstance under which this rate of strain field is assumed to occur permits a special analysis of more general applicability; however, of importance to the present discussion is the influence of that field on the turbulence, an influence which results in an overall reduction of turbulent intensity with an increase in anisotropy.

In this section we deal with another rate-of-strain field, one associated with the flow in a *constant-area duct* with varying shape such as that shown in Fig. 6.7.1. In practical applications there arise circumstances in which the cross section of a duct changes shape, e.g., from a circle to a square, without a change in area, with a consequent distortion of the mean flow and alteration of the turbulence. The flow considered here is an idealization of that circumstance.

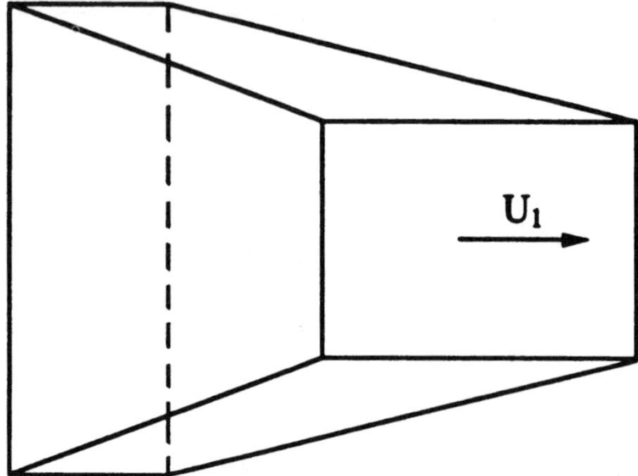

Figure 6.7.1 Schematic representation of a constant-area duct with its cross section rotated through 90° in the streamwise direction.

Because in the absence of area change there is no rate of strain in the streamwise direction, these flows involve plane rates of strain. Turbulence subject to such rates of strain has been studied both experimentally and computationally; experiments have been carried out by Tucker and Reynolds (1968), Gence and Mathieu (1980), and Choi and Lumley (1984). DNS has been applied to these flows by Lee and Reynolds (1985).

If we confine attention to the neighborhood of the axis of the duct shown in Fig. 6.7.1, we take U_1 to be constant and $U_2 = (\partial U_2/\partial x_2)x_2 \equiv S(x_1)\, x_2$ and $U_3 = (\partial U_3/\partial x_3)x_3 = -S(x_1)\, x_3$, where $S(x_1)$ is a rate-of-strain parameter and where we have used the continuity equation to relate the two velocity gradients. In this case the turbulence is unstrained in the flow direction and with $S > 0$ extended in the x_2 coordinate direction and compressed in the third direction. Since the rates of strain are confined to x_2–x_3 planes, this flow involves plane rates of strain. For the duct shown in Fig. 6.7.1, $S > 0$; without loss of generality we confine attention to this case, since with a rotation of the duct with the coordinate system fixed the alternative case of $S < 0$ arises but adds nothing new to the present considerations. As a final preliminary, we note that because we assume double symmetry about the x_1 coordinate, the two Reynolds shear stresses $\overline{u_1u_2}$ and $\overline{u_1u_3}$ are zero with linear dependencies on x_2 and x_3, respectively. The intensities of the velocity fluctuations in the three coordinate directions upstream of the distorting section and the distribution of the rate-of-strain parameter $S(x_1)$ are assumed known. We are interested in the evolution of the turbulence with x_1.

Several comments regarding $S(x_1)$ are appropriate. The assumption that $S(x_1)$ is given is based on the notion that appropriate contours for the walls of the duct can impose any desired rate of strain along the x_1 axis even if the turbulence interacts with the mean flow. This same assumption is implicit in all previous sections in this chapter; however, in the next section we consider such an interaction more carefully. Here we employ two different descriptions of $S(x_1)$; in fundamental studies the rate-of-strain parameter is taken to be constant and in the first of our cases we do likewise. An alternative which is more realistic is to assume that $S(x_1)$ is distributed to reflect application to a duct flow. Thus, as a second application we consider the distribution

$$\hat{S}(\hat{x}_1) \equiv \frac{S(\hat{x}_1)}{S(0)} = \exp(-\hat{x}_1^2) \tag{6.7.1}$$

where

$$\hat{x}_1 \equiv \frac{S(0)\, x_1}{U_1}$$

is a dimensionless streamwise coordinate. In the case of a constant rate of strain, $\hat{S} \equiv 1$.

The Reynolds stress theory for this flow [cf. Eq. (10.10.5)] in dimensional variables becomes

$$\begin{aligned} U_1 \frac{d\overline{u_1^2}}{dx_1} &= -c_1 \frac{\varepsilon}{k}\left(\overline{u_1^2} - \frac{2}{3}k\right) + \frac{2}{3} c_2 \left(-\overline{u_2^2} + \overline{u_3^2}\right) S - \frac{2}{3}\varepsilon \\ U_1 \frac{d\overline{u_2^2}}{dx_1} &= -2\,\overline{u_2^2}\, S - c_1 \frac{\varepsilon}{k}\left(\overline{u_2^2} - \frac{2}{3}k\right) + \frac{2}{3} c_2 \left(2\overline{u_2^2} + \overline{u_3^2}\right) S - \frac{2}{3}\varepsilon \\ U_1 \frac{d\overline{u_3^2}}{dx_1} &= 2\,\overline{u_3^2}\, S - c_1 \frac{\varepsilon}{k}\left(\overline{u_3^2} - \frac{2}{3}k\right) - \frac{2}{3} c_2 \left(\overline{u_2^2} + 2\overline{u_3^2}\right) S - \frac{2}{3}\varepsilon \end{aligned} \tag{6.7.2}$$

Again we must supplement these equations with a transport equation for the mean viscous dissipation, which from Eq. (10.10.7) becomes

$$U_1 \frac{d\varepsilon}{dx_1} = \frac{\varepsilon}{k}\left[c_{\varepsilon 1}\left(-\overline{u_1^2} + \overline{u_3^2}\right) S - c_{\varepsilon 2}\, \varepsilon\right] \tag{6.7.3}$$

The turbulent kinetic energy is

$$k = \frac{1}{2}\left(\overline{u_1^2} + \overline{u_2^2} + \overline{u_3^2}\right) \tag{6.7.4}$$

With the exception of the first terms on the right side of the second and third of Eqs. (6.7.2), additional production terms, each of the remaining terms in these equations and in Eq. (6.7.3) have their counterparts in Eqs. (6.3.3) and (6.5.1) but, as expected given the different rate-of-strain field, the production terms involving S are altered significantly. In particular, these terms are proportional to the *difference* in the two intensities, $\overline{u_2^2}$ and $\overline{u_3^2}$; thus any quantity which influences these intensities can be expected to have a large effect on the turbulence. Turbulent diffusion in the streamwise direction is again neglected.

To nondimensionalize these equations we use the turbulent kinetic energy far upstream of the distortion section $x_1 \to -\infty$, that is, k_0, and the rate of strain at the origin $S(0)$. Thus we let

$$G_{11} \equiv \frac{\overline{u_1^2}}{k_0} \qquad G_{22} \equiv \frac{\overline{u_2^2}}{k_0}$$
$$G_{33} \equiv \frac{\overline{u_3^2}}{k_0} \qquad E \equiv \frac{\varepsilon}{k_0\, S(0)} \qquad K \equiv \frac{k}{k_0} = \frac{1}{2}\,(G_{11} + G_{22} + G_{33}) \tag{6.7.5}$$

In terms of the variables of Eqs. (6.7.1) and (6.7.5), Eqs. (6.7.2) and (6.7.3) become

$$\frac{dG_{11}}{d\hat{x}_1} = -c_1\,\frac{E}{K}\left(G_{11} - \frac{2}{3}\,K\right) + \frac{2}{3}\,c_2\,(-G_{22} + G_{33})\,\hat{S} - \frac{2}{3}\,E$$
$$\frac{dG_{22}}{d\hat{x}_1} = -2G_{22}\hat{S} - c_1\,\frac{E}{K}\left(G_{22} - \frac{2}{3}\,K\right) + \frac{2}{3}\,c_2\,(2G_{22} + G_{33})\,\hat{S} - \frac{2}{3}\,E \tag{6.7.6}$$
$$\frac{dG_{33}}{d\hat{x}_1} = 2G_{33}\hat{S} - c_1\,\frac{E}{K}\left(G_{33} - \frac{2}{3}\,K\right) - \frac{2}{3}\,c_2\,(G_{22} + 2G_{33})\,\hat{S} - \frac{2}{3}\,E$$
$$\frac{dE}{d\hat{x}_1} = \frac{E}{K}\,[c_{\varepsilon 1}(-G_{22} + G_{33})\,\hat{S} - c_{\varepsilon 2}\,E]$$

If we assume that the turbulence is isotropic far upstream of the distortion, these equations are to be solved subject to conditions at $\hat{x}_1 \to -\infty$:

$$G_{11} = G_{22} = G_{33} = \frac{2}{3} \qquad E = E_0 \tag{6.7.7}$$

With standard values assumed to prevail for the four empirical coefficients, E_0 is the only parameter in these equations. Its physical significance can be understood as follows. Suppose that we introduce a length scale of the turbulence far upstream defined as $l_0 \equiv k_0^{3/2}/\varepsilon_0$, where ε_0 is the mean viscous dissipation there. Then we can write

$$E_0 = \frac{\varepsilon_0}{k_0 S(0)} = \frac{k_0^{1/2}}{U_1}\,\frac{U_1/S(0)}{l_0} \tag{6.7.8}$$

We see that E_0 is a product of the relative intensity of the turbulence far upstream and the ratio of a mean flow length $U_1/S(0)$ to a turbulence length l_0, i.e., the product of small and large quantities. In fundamental studies, Speziale et al. (1991b) set $E_0 = 2$ and we do likewise, taking this to be representative.

Since the turbulent kinetic energy is found to vary differently with the two representations of $\hat{S}(\hat{x}_1)$, it is instructive to present the results in terms of the components of the anisotropy tensor and of the normalized turbulent kinetic energy $K(\hat{x}_1)$. Thus Figs. 6.7.2 and 6.7.3 show the distributions of $b_{11}(\hat{x}_1)$, $b_{22}(\hat{x}_1)$, $b_{33}(\hat{x}_1)$, and $K(\hat{x}_1)$. In each case we start the calculation at $\hat{x}_1 = -2$, a value which is arbitrary if $\hat{S} \equiv 1$ and which corresponds to negligible rate of strain if $\hat{S}(\hat{x}_1)$. In both cases we assume that the turbulence is initially isotropic. When the rate of strain is constant, we see from Fig. 6.7.2 that, as expected, extension in the x_2 direction and compression in the x_3 direction lead to a decrease and an increase, respectively, in the relative contribution to the turbulent kinetic energy from $\overline{u_2^2}$ and $\overline{u_3^2}$. In addition, we see that the turbulent kinetic energy initially decreases and subsequently increases because of changes in the production term which depends on differences in G_{22} and G_{33}. Our results are in good qualitative agreement with those of Speziale et al. (1991b), although their Reynolds stress theory is somewhat different from that used here (cf. Section 10.13).

As seen in Fig. 6.7.3, the response of the turbulence when the rate of strain is distributed according to Eq. (6.7.1) is significantly different; the decrease in $\overline{u_2^2}$ and the increase in $\overline{u_3^2}$ seen in Fig. 6.7.2 is now restricted to the neighborhood of the origin, where the rate of strain is localized, but because of the relative

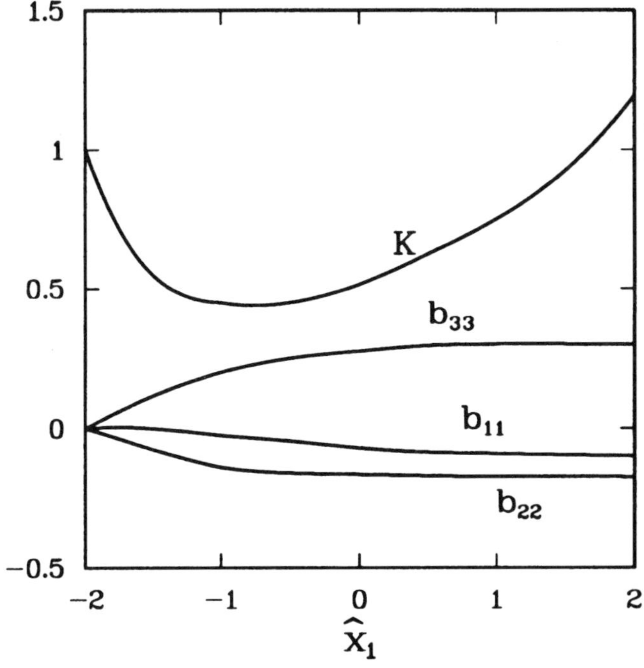

Figure 6.7.2 Distribution of turbulence quantities under a constant plane rate of strain.

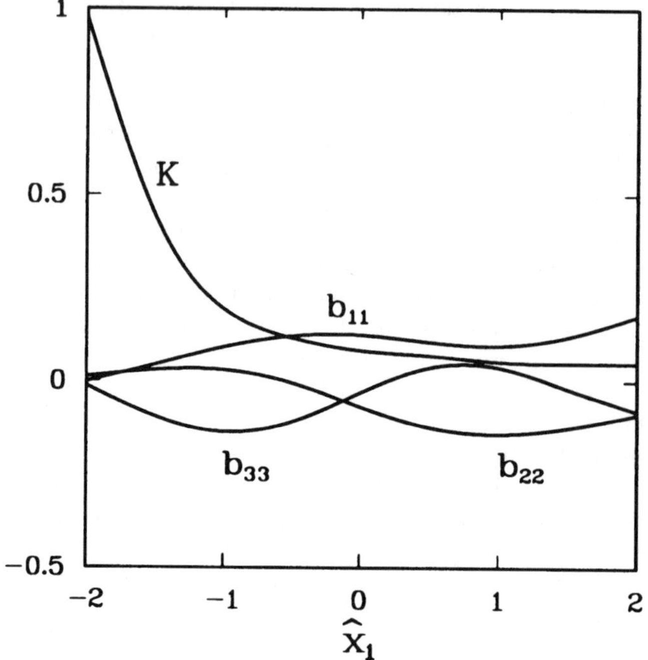

Figure 6.7.3 Distribution of turbulence quantities under a varying plane rate of strain.

magnitudes of G_{22} and G_{33} the turbulent kinetic energy decreases monotonically in this case. We thus see that, as suggested earlier, turbulence production and thus the response of the turbulence subject to plane rates of strain are sensitive to the distribution of those rates.

6.8 TURBULENT FLOWS SUBJECT TO RAPID DISTORTION

There exists a class of turbulent flows described by rapid distortion theory (RDT), e.g., the impingement of weak turbulence on an obstacle such as a building or underwater pier (cf. Bearman, 1972) or turbulence within the contraction section of a wind tunnel. In these flows the mean rate of strain in the main flow direction, compressive in the first example and extensive in the second, is so large and the turbulence so weak that the usual turbulent exchange mechanisms become inoperative. The result is that the describing equations can be linearized and a large body of mathematical machinery can be brought to bear on their solution. Since our previous discussion emphasizes the important role of *nonlinearity* in turbulent flows, it is clear that RDT pertains to a special class of flows. Following the initial work of Taylor (1935a), RDT has evolved in several directions, with contributions continuing at the present time. Accordingly, there is an extensive literature on RDT (cf. Savill, 1987).

A sense of the theory can be gained by adopting a current second-moment theory to the original application of RDT, namely, to the study of the influence of a contraction section on turbulence. This approach exposes in a direct and clear fashion the principal features of RDT. At the same time, the influence on turbulence of a simple extensive rate of strain in the streamwise direction complements our discussion of plane rate of strain treated in the previous section. Indeed, we shall find here various points in common with the analysis of that section.

Consider an axisymmetric contraction section as shown in Fig. 6.8.1 and confine attention to the neighborhood of the axis. Let k_0 and W_0 denote the turbulent kinetic energy and the mean axial velocity at the inlet to the section, and locate there the origin of the z coordinate, lying along the axis. There are two length scales in this flow; an obvious one is the length of the section denoted d. We assume that the walls of the contraction section are shaped such that in this length the mean axial velocity increases from W_0 to $W_1 = cW_0$, where $c >> 1$ is the contraction ratio, with typical values in the range from 5 to 15. A second length is associated with the large turbulence scales at the inlet section. We identify this length with the turbulent kinetic energy and the mean viscous dissipation at $z = 0$, that is, with $l_0 \equiv k_0^{3/2}/\varepsilon_0$ [cf. Eq. (6.7.8)]. The ratio l_0/d enters the analysis later.

In the neighborhood of the axis, only two turbulent intensities need be considered: $\overline{w^2}$, corresponding to fluctuations in the axial velocity, and $\overline{u^2}$, corresponding to the intensity of the radial velocity. The intensity of the azimuthal velocity is indistinguishable from that of the radial velocity, with the consequence that the turbulent kinetic energy is

$$k = \frac{1}{2}\,\overline{w^2} + \overline{u^2} \qquad (6.8.1)$$

Furthermore, in the neighborhood of the axis the various dependent variables vary with the radius r in ways dictated by physical considerations. When these dependencies are imposed, the transport equations reduce to ordinary differential form. We start by assuming that the mean radial velocity is given by

$$U\left(\frac{r}{d}, \hat{z}\right) = -\frac{r}{d}\, W_0\, F'(\hat{z}) \qquad (6.8.2)$$

where $\hat{z} \equiv z/d$ is a dimensionless axial coordinate and $F(\hat{z})$ is a dimensionless streamfunction. The minus sign is inessential but convenient; it reflects the radial *inflow* to the axis in a contracting section. Since the principal function of a contraction section is to provide a uniform axial flow at its discharge end, we require $F'(1) = 0$, but to do so we must accept the nonuniform distribution at the inlet section required by the analysis.

The mean continuity equation in cylindrical coordinates in the absence of a mean azimuthal velocity is

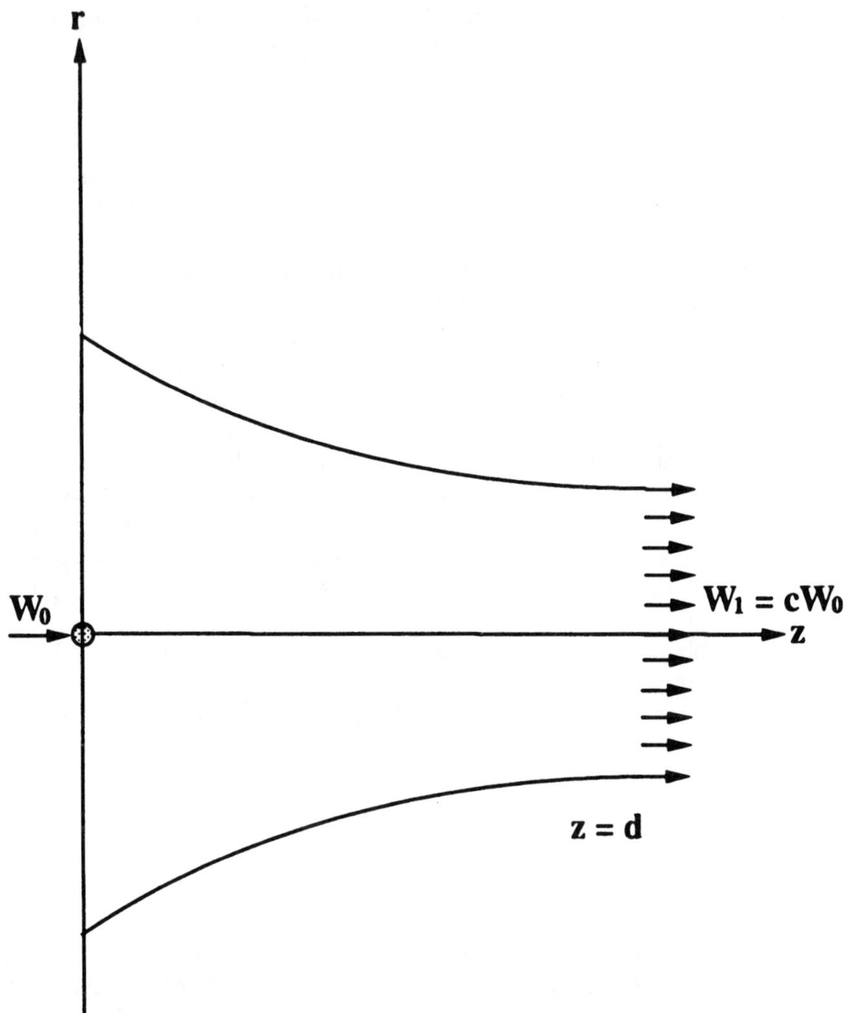

Figure 6.8.1 Schematic representation of an axisymmetric contraction section.

$$\frac{1}{r}\frac{\partial}{\partial r} r\, U + \frac{\partial W}{\partial z} = 0 \tag{6.8.3}$$

Substitution of Eq. (6.8.2) into (6.5.3) and integration leads to

$$W = 2\, W_0\, F(\hat{z}) \tag{6.8.4}$$

provided $F(0) = \frac{1}{2}$ as required by $W(0) = W_0$. In order that the mean axial velocity at the discharge end of the contraction section be $W_1 = c\, W_0$, we see that $F(1) = \frac{1}{2}\, c$.

A second assumed form relates to the distribution of the mean pressure in the neighborhood of the axis; consistency with the mean momentum equations in both the radial and axial directions is achieved if

$$P\left(\frac{r}{d}, \hat{z}\right) = \rho W_0^2 \left[P_0(\hat{z}) + \frac{1}{2}\left(\frac{r}{d}\right)^2 Q \right] \tag{6.8.5}$$

where $P_0(\hat{z})$ is obtained from the mean momentum equation in the axial direction *after* $F(\hat{z})$ is known. The distribution of $P_0(\hat{z})$ is not of interest here and is not pursued. The quantity Q is an unknown constant to be determined as part of the solution. The r^2 variation in Eq. (6.8.5) reflects the physically correct behavior of the mean pressure in the neighborhood of the axis, while Q determines the radial curvature of that pressure.

The final radial dependence relates to turbulence quantities. We introduce the two intensities and the Reynolds shear stress in the form

$$G_{uu} \equiv \frac{\overline{u^2}}{k_0} \qquad G_{ww} = \frac{\overline{w^2}}{k_0} \qquad G_{uw} \equiv \frac{r}{d}\frac{\overline{uw}}{k_0} \tag{6.8.6}$$

We make all turbulence quantities proportional to the intial turbulent kinetic energy k_0. The two intensities are taken to be independent of radial position, while the Reynolds shear stress is required to be zero on the axis and to vary linearly with the radial coordinate. The first two of these forms suggest introduction of a dimensionless turbulent kinetic energy, namely, from Eq. (6.8.1),

$$K(\hat{z}) \equiv \frac{k}{k_0} = \frac{1}{2} G_{ww}(\hat{z}) + G_{uu}(\hat{z}) \tag{6.8.7}$$

When Eqs. (6.8.2) and (6.5.4)–(6.5.6) are substituted into the mean radial momentum equation, we obtain

$$-\frac{1}{2}\delta^2 G'_{uw} + FF'' + \frac{1}{2}(Q - F'^2) = 0 \tag{6.8.8}$$

where $\delta^2 \equiv k_0/W_0^2$ is the relative intensity of the turbulence at the initial station. The FF'' and F'^2 terms in Eq. (6.8.8) arise from convection, the Q term from the radial gradient of the mean pressure, and the G_{uw} term from the Reynolds shear stress.

We now invoke the first of several approximations related to RDT: The turbulence is weak so that $\delta^2 << 1$. From this perspective we can consider that $F(\hat{z}; \delta^2)$, i.e., the mean velocity, depends on both $\hat{z}$ and δ^2, but a small magnitude for δ^2 suggests the solution be found in terms of a regular series expansion in that quantity. The first term in that series is obtained by setting δ^2 equal to zero in Eq. (6.8.8). Since we limit attention to this first term only, in the interest of simplicity we do not introduce either a sub- or superscript to denote this restriction, but it should be kept in mind. The reason for introducing the quantity Q is now evident: The resulting equation for $F(\hat{z})$ is of second order, but we seek to impose three boundary conditions and must therefore have another parameter, e.g., Q, at our disposal.

A quadratic function of $\hat{z}$ satisfies the altered Eq. (6.8.8), so we easily find that the solution respecting the boundary conditions is

$$F(\hat{z}) = \frac{1}{2}\,[1 + (c - 1)(2\,\hat{z} - \hat{z}^2)] \tag{6.8.9}$$

with $Q = c\,(c - 1)$. Since this solution results from the neglect of the δ^2 term in Eq. (6.8.8), we see from Eq. (6.8.9) that the approximation involved is consistent only if $c >> 1$, that is, if the contraction is large, otherwise *all* of the terms in Eq. (6.8.8) could be small. The implication is that for the present analysis to prevail, the turbulence must not only be weak but must be subject to a large rate of strain, a second approximation associated with RDT.

In this context it is useful to employ the solution given by Eq. (6.8.9) to calculate the mean rates of strain experienced by fluid elements in the neighborhood of the axis. We find

$$\begin{aligned} S_{rr} &= \frac{\partial U}{\partial r} = -\frac{W_0}{d}\,(c - 1)(1 - \hat{z}) \\ S_{zz} &= \frac{\partial W}{\partial z} = 2\,\frac{W_0}{d}\,(c - 1)(1 - \hat{z}) \end{aligned} \tag{6.8.10}$$

Thus we see that the rates of strain are dependent on the contraction ratio c with compression in the radial direction and extension in the axial direction. Because of the assumption of uniform axial velocity at the discharge end of the section, the rate of strain vanishes there.

Figure 6.8.2 shows the distribution of $F(\hat{z})$ which is closely associated with the mean axial velocity [cf. Eq. (6.8.4)] for three values of the contraction ratio c. We see the monotone increase in $F(\hat{z})$ from $F(0) = \frac{1}{2}$ to $F(1) = \frac{1}{2}\,c$.

With the mean velocity field known, we next consider the distributions of turbulence quantities. As in the previous sections in this chapter, we employ the Reynolds stress theory of Section 10.10 to obtain equations for $G_{uu}(\hat{z})$ and $G_{ww}(\hat{z})$. Although an equation for $G_{uw}(\hat{z})$ could be included, it is of little interest and is not discussed. However, as in several of the flows treated earlier, we shall need an equation for the mean viscous dissipation. Since the turbulent kinetic energy is independent of the radial coordinate and since we expect on physical grounds that the turbulent length scale is similarly independent of r, we conclude that the mean viscous dissipation depends only on $\hat{z}$. In addition, $\varepsilon(\hat{z})$ must be scaled so that in the balance equation for the turbulent kinetic energy the dissipation remains as $\delta^2 \to 0$. These considerations lead to the introduction of

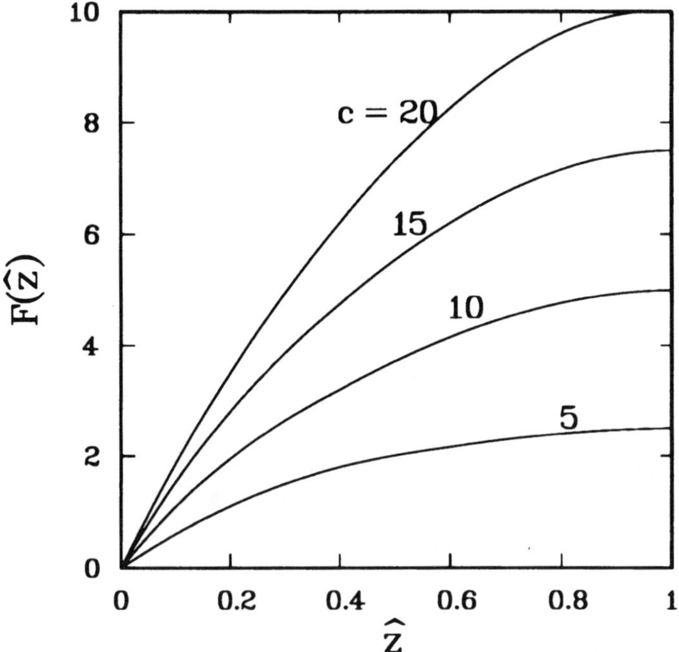

Figure 6.8.2 Distribution of mean axial velocity on the axis of an axisymmetric contraction section in terms of $F(\hat{z})$.

$$E(\hat{z}) \equiv \frac{\varepsilon\, d}{k_0 W_0} \tag{6.8.11}$$

where alternative denominators $k_0^{3/2}$ and W_0^3 do not lead to the desired results. If Eq. (6.8.11) is applied at $\hat{z} = 0$ and the definition of the turbulence length l_0 is used to eliminate $\varepsilon(0)$, we have

$$E(0) = \frac{k_0^{1/2}}{W_0} \frac{1}{l_0/d} = \frac{\delta}{l_0/d} \tag{6.8.12}$$

We thus see that if $E(0)$ is to be independent of the limit process $\delta \to 0$, then $l_0/d << 1$ and $E(\hat{z})$ is the quotient of two small quantities. We thus have another approximation related to RDT: The turbulence scales must be suitably small. The quantity $E(0)$ is a parameter adding to c in defining the turbulent flow in the contraction section under consideration.

With these as preliminaries we write down the transport equations for $G_{uu}(\hat{z})$, $G_{ww}(\hat{z})$, and $E(\hat{z})$ obtained for a turbulent flow which is axisymmetric in the mean and restricted to the neighborhood of the axis:

$$-2FG'_{uu} = -2(1 - c_2)F'G_{uu} - c_1 \frac{E}{K}\left(G_{uu} - \frac{2}{3}K\right) - \frac{4}{3}c_2F'(G_{uu} - G_{ww}) - \frac{2}{3}E$$

$$-2FG'_{ww} = 4(1 - c_2)F'G_{ww} - c_1 \frac{E}{K}\left(G_{ww} - \frac{2}{3}K\right) - \frac{4}{3}c_2F'(G_{uu} - G_{ww}) - \frac{2}{3}E \qquad (6.8.13)$$

$$-2FE' = -2c_{\varepsilon 1}F'\frac{E}{K}(G_{uu} - G_{ww}) - c_{\varepsilon 2}\frac{E^2}{K}$$

As in Sections 6.5 and 6.7, in each of these equations we have dropped a turbulent diffusion term; here we do so on the formal grounds that it is multiplied by δ^2. Thus we again consider only the first terms in expansions in δ^2 for the dependent variables $G_{uu}(\hat{z})$, $G_{ww}(\hat{z})$, and $E(\hat{z})$. Furthermore, as noted earlier, we can add to this system an equation for the Reynolds shear stress function $G_{uw}(\hat{z})$, but it is uncoupled from the other three and is of little interest. Note that $K(\hat{z})$ is given by Eq. (6.8.7) and that $F(\hat{z})$ and $F'(\hat{z})$ relate to the mean velocity components [cf. Eq. (6.8.9)]. Thus we determine via Eqs. (6.8.13) the characteristics of turbulence in a *known* mean velocity field. Finally, c_1, c_2, $c_{\varepsilon 1}$, and $c_{\varepsilon 2}$ are empirical coefficients with standard values cited earlier.

Since Eqs. (6.8.13) are first order, we specify initial values, i.e., of $G_{uu}(0)$, $G_{ww}(0)$, and $E(0)$. Although the first two of these quantities can be chosen arbitrarily subject to the restraint implied by Eq. (6.8.7), namely, $K(0) = 1$, for our purposes it is appropriate to assume that the turbulence is isotropic at the inlet section to the contraction cone. Thus we take $G_{uu}(0) = G_{ww}(0) = \frac{2}{3}$. The mean viscous dissipation function at the origin $E(0)$ is arbitrary; in other related studies it is found to have values of approximately unity, which we take as a representative value.

Figures 6.8.3 and 6.8.4 show the distributions along the axis of the contraction section of quantities of interest from the solution of Eqs. (6.8.13) for two values of the contraction ratio, $c = 5$ and $c = 20$, respectively. Consider the first of these figures. Isotropic turbulence becomes highly anisotropic under the influence of the rate-of-strain field in the contraction section, with the intensity of the radial velocity component decreasing and that of the axial component increasing significantly. As a consequence, the turbulent kinetic energy likewise increases. However, consider the *relative intensity* of the turbulence, that is, $k^{1/2}/W$, normalized by its initial value; we find that

$$\frac{k^{1/2}/W}{(k^{1/2}/W)_0} = \frac{K^{1/2}}{2F}$$

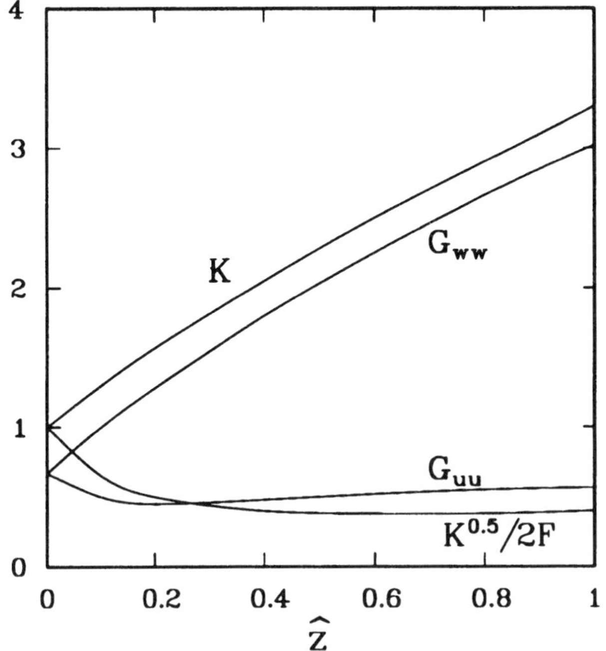

Figure 6.8.3 Distributions of dimensionless turbulence quantities in a contraction section: $c = 5$.

Accordingly, Fig. 6.8.3 shows the distribution of this quantity; we see that although the turbulent kinetic energy increases in the contraction section, the relative intensity decreases significantly, with this decrease largely confined to the initial portion of the section where the rates of strain are greatest [cf. Eqs. (6.8.10)]. We do not show the distribution of the mean viscous dissipation in terms of $E(\hat{z})$, since it is of interest only in examining the influence of contraction on the integral scale of turbulence, a scale proportional to $K^{3/2}/E$; both K and E increase with $\hat{z}$, but the length scale decreases slightly.

Consider now Fig. 6.8.4, which corresponds to a larger contraction ratio, $c = 20$. Qualitatively the results are in accord with those for the smaller contraction ratio, although the intensity of the radial velocity component, while considerably smaller than that of the axial component, actually increases somewhat with $\hat{z}$. Notice that the relative intensity in terms of $K^{1/2}/2F$ is reduced to a small fraction of its initial value with strong contraction, i.e., to 13%.[†]

[†]In a diffuser, i.e., a diverging axisymmetric section, the rates of strain are reversed and the intensities reflected in $G_{uu}(\hat{z})$ and $G_{ww}(\hat{z})$ *increase* and *decrease*, respectively. The present analysis applies to this case provided $c << 1$, that is, there is strong compression in the axial direction and extension in the radial direction. In this case we must be concerned with separation of the boundary layer on the diffuser wall.

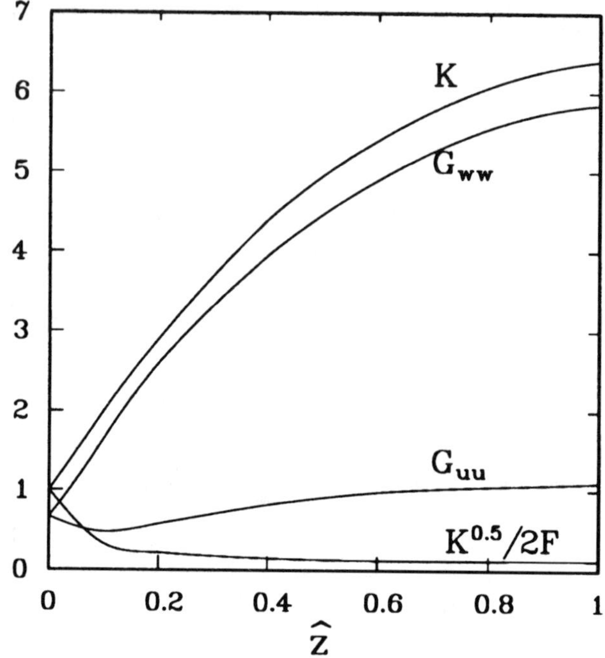

Figure 6.8.4 Distributions of dimensionless turbulence quantities in a contraction section: $c = 20$.

Several concluding comments regarding this calculation are indicated. The realistic rate of strain considered here differs from the usual constant rate of strain considered in studies of the influence of a contraction on turbulence (cf. Tucker and Reynolds, 1968). However, the results from the two calculations are qualitatively the same. A similar analysis involving a different rate-of-strain field can be applied to stagnating turbulence along the dividing streamline approaching a bluff two-dimensional body (cf. Bearman, 1972; Champion and Libby, 1991). Turbulent flow from an axisymmetric jet impinging on a nearby wall provides a further example of flows which can be analyzed in a similar fashion (cf. Champion and Libby, 1994). Finally, we see from this analysis that RDT pertains to weak turbulence subject to high rates of strain; under these circumstances a simplified calculation of the response of the turbulence to a known distribution of mean velocities is possible.

6.9 TWO-DIMENSIONAL TURBULENCE

In Section 2.5 our discussion of vorticity including its alteration by rotation and stretching in nonuniform flows emphasizes that three-dimensionality is an essential feature of turbulence. However, we know from Section 6.4 that stable stratification and from Section 6.8 that large extensive rates of strain result in nearly two-dimensional turbulence. Thus, in this section on simple flows we call attention to studies of such turbulence.

As an introductory comment on this subject, we cannot improve on the following statement by Kraichnan and Montgomery (1980): "Two dimensional turbulence has the special distinction that it is nowhere realised in nature or the laboratory but only in computer simulations. Its importance is two-fold: first, that it idealises geophysical phenomena in the atmosphere, oceans and magnetosphere and provides a starting point for modelling these phenomena; second, that it presents a bizarre and instructive statistical mechanics." This reference provides a comprehensive review of two-dimensional turbulence and its applications to geophysical and other flows of applied interest. For other reviews, see Rhines (1979) and Salmon (1982), who discuss two-dimensional turbulence within the context of geostrophic flows, which, as suggested earlier, provide an important motivation for the study of this special turbulence. Each of these publications includes an extensive list of references to the relevant literature.

In two-dimensional turbulence there are only two velocity components, $\tilde{u}_1$ and $\tilde{u}_2$ in the usual x_1–x_2 coordinate system with $\tilde{u}_3 \equiv 0$. Thus there is a clear distinction with the two-dimensional flows we discuss repeatedly, i.e., flows with mean velocity components U_1, U_2, and with *three intensities,* $\overline{u^2}$, $\overline{u_2^2}$, and $\overline{u_3^2}$. Because of the simplified velocity field there is only one vorticity component in two-dimensional turbulence, $\tilde{\omega}_3$, in the coordinates we have chosen, and since $\tilde{u}_3 \equiv 0$ there is no stretching of vorticity. Thus, if viscous effects are negligible, the vorticity of each fluid element is conserved; moreover, the integral of the vorticity squared is constant:

$$\Omega = \int_S dx_1 \, dx_2 \, \tilde{\omega}_3^2 = \text{constant} \qquad (6.9.1)$$

This simple behavior is not found in three-dimensional turbulence with its three components of vorticity leading to their complex interactions.

In addition to the constancy of the vorticity in the sense implied by Eq. (6.9.1), the total kinetic energy per unit mass is also constant if viscous effects are neglected.

$$E = \int_S dx_1 \, dx_2 \, (\tilde{u}_1^2 + \tilde{u}_2^2) = \text{constant} \qquad (6.9.2)$$

A consequence of the constancy of Ω and E in two-dimensional turbulence is that spectral transfer is drastically different from that in conventional turbulence. In the former, turbulence energy is shifted from a given wave number in the direction of both smaller and larger wave numbers, whereas in three-dimensional turbulence the transfer is generally to higher wave numbers and thus to dissipative scales.

Numerical solutions for two-dimensional turbulence are frequently carried out in square domains subject to cyclical boundary conditions (cf., e.g., Carnevale and Vallis, 1990). In this case the initial conditions correspond to arbitrary distributions over a square cell of the two velocity components constrained only by the requirement that the two-dimensional continuity equation be respected. These initial velocities correspond to an initial distribution of vorticity $\tilde{\omega}_3(x_1, x_2, t)$. As the flow evolves in time, well-defined vortex structures surrounded by a background of small-scale vorticity emerge. Close encounters of well-defined vortices of like sign lead to merging, i.e., to pairing, and larger vortices, while those of opposite sign remain separated. Again, this simple behavior of pairing and separation is not realized in three-dimensional turbulence. These calculations lose their validity and are terminated when the largest vortex structures approach the size of the computational domain.

If weak viscous effects exist, this behavior is slightly modified. For example, the vorticity of a fluid element decays as a result of dissipation at small scales, but the total kinetic energy, being transferred to large scales by vortex pairing, remains appproximately invariant.

6.10 SUMMARY

In this chapter we have discussed a series of turbulent flows with geometries sufficiently simple so that reasonably accurate descriptions of their features are possible. Grid flow, which plays a central role in turbulence research, involves no production mechanism, with the consequence that the fluctuations generated at the grid decay in the downstream direction. On the other hand, there is uniform shear in the central portion of a turbulent Couette flow, so an equilibrium between production and dissipation prevails. This is a special case of homogeneous shear flows, which are of significant fundamental interest for understanding the interaction of various turbulent processes and for the assessment of moment methods applicable to flows of applied interest. Moreover, homogeneous shear flows permit the systematic study of special effects, e.g., the influence of rotation and curvature on turbulence. A large percentage of the turbulent flows of applied interest are dominated by shear rates of strain, but turbulence in ducts and around obstacles involves compressional and extensional rates of strain; in Sections 6.7 and 6.8 we analyzed the response of turbulence to such rates. Finally, two-dimensional turbulence represents a special limiting situation, closely approximating some geophysical flows and possessing its own special features.

CHAPTER

SEVEN

CLOSURE BY CLASSICAL MEANS

In Chapter 4, averaging the fundamental conservation and transport equations of fluid mechanics was seen to introduce additional unknowns so that the resulting system of equations is not closed. Various methods have been used to overcome this difficulty, so a variety of systems of equations are available for the description of various turbulent flows. Chapter 10 discusses second-moment and related methods which lead to equations with greater fluid mechanical content and which are the object of current research. This chapter focuses on classical methods based on gradient transport notions which close the equations at the first level in the hierarchy of moment equations. Prior to the 1960s, when the availability of high-speed computers made second-moment methods practicable, turbulent flows of applied interest were treated exclusively by these classical methods; in fact, such methods continue to be preferred for many industrial problems involving turbulence, although the simplest of the second-moment methods is assuming increasing importance in this setting.

7.1 REYNOLDS STRESSES AND MEAN RATES OF STRAIN

The mean momentum equations, Eqs. (4.2.1), involve the mean rate-of-strain tensor

$$S_{ij} \equiv \frac{1}{2}\left(\frac{\partial U_i}{\partial x_j} + \frac{\partial U_j}{\partial x_i}\right) \tag{7.1.1}$$

and the Reynolds stresses $\rho\,\overline{u_i\,u_j}$, which are identified with shear if $i \neq j$ and with normal stresses if $i = j$. It will be useful for later discussion to recall that the mean viscous stresses in Eqs. (4.2.1) involve the viscosity coefficient μ according to

$$T_{ij} = \mu\left(\frac{\partial U_i}{\partial x_j} + \frac{\partial U_j}{\partial x_i}\right) = 2\mu S_{ij} \tag{7.1.2}$$

Classical closure of the momentum equations expresses the Reynolds stresses in terms of S_{ij} by partial analogy with Eq. (7.1.2). Formally,

$$-\rho\overline{u_iu_j} = \mu_T\left(\frac{\partial U_i}{\partial x_j} + \frac{\partial U_j}{\partial x_i}\right) - \frac{2}{3}\rho k\delta_{ij} \tag{7.1.3}$$

where, as earlier, $k \equiv \frac{1}{2}\overline{u_ku_k}$ is the turbulent kinetic energy and μ_T is a scalar turbulent viscosity coefficient, one of several turbulent exchange coefficients we shall introduce. Equation (7.1.3) satisfies the requirement that upon contraction, i.e., upon setting $i = j$ and summing over i, the definition of the turbulent kinetic energy is recovered. It is also Gallilean invariant provided μ_T is invariant; i.e., addition of a constant velocity vector alters none of the terms in this equation. Note that if Eq. (7.1.3) is divided by the density, the turbulent viscosity coefficient μ_T is replaced by a turbulent kinematic coefficient ν_T, which is generally a more convenient quantity for the treatment of constant density flows.

Several comments regarding Eq. (7.1.3) are appropriate. If the turbulent exchange coefficient ν_T, the mean velocity components $U_i(\mathbf{x})$, and the turbulent kinetic energy $k(\mathbf{x})$ are known, then *all* of the Reynolds stresses $\rho\overline{u_i\,u_j}(\mathbf{x})$ are known. Finally, Eq. (7.1.4) implies that the tensor $\overline{u_iu_j} - \frac{2}{3}\delta_{ij}k$ is colinear with the rate-of-strain tensor S_{ij}. To examine collinearity experimentally in a general three-dimensional turbulent flow is difficult, since it requires measurement of nine gradients of the mean velocities and of the six Reynolds stresses. However, we can compare Eq. (7.1.3) with the theoretical results obtained in several of the flows analyzed in Chapter 6 in terms of the anisotropy tensor. From Eqs. (4.7.9) and (7.1.3), it is readily found that

$$b_{ij} = -\nu_T\frac{S_{ij}}{k} \tag{7.1.4}$$

In turbulent Couette flow as treated in Section 6.3, for which the only component of the mean rate-of-strain tensor is S_{12}, we find that $b_{11} = \frac{1}{6}$ rather than zero and $b_{22} = b_{33} = -\frac{1}{12}$ rather than zero. Similar observations apply to the asymptotic state associated with homogeneous shear flow, i.e., small disagreements relative to the normal stresses. The most important applications of Eq. (7.1.3) are those related to turbulent shear flows, jets, wakes, and boundary layers, so the discrepancies suggested by these comparisons are generally of conceptual rather than practical importance. However, in Section 7.3 the *fundamental shortcomings* of gradient transport are discussed; in particular, when Eq. (7.1.3) is used to compute the individual contributions of the turbulent kinetic energy in these shear flows, an essential discrepancy is encountered.

7.2 REYNOLDS FLUXES AND MEAN TEMPERATURE GRADIENTS

A similar formal analogy applies to the mean turbulent flux of temperature arising in Eq. (4.3.1). Thus,

$$-\rho\overline{u_i\theta} = \frac{\mu_T}{\sigma_T}\frac{\partial\Theta}{\partial x_i} \tag{7.2.1}$$

where σ_T is a turbulent Prandtl number and the quotient $\mu_T/\rho\sigma_T$ is a turbulent thermal diffusivity, another turbulent exchange coefficient. An alternative form for Eq. (7.2.1) is obtained by replacing μ_T/σ_T with λ_T/c_p, where λ_T is a turbulent thermal conductivity and c_p is a coefficient of specific heat at constant pressure, a thermodynamic quantity. In this case a model for λ_T is needed to close Eq. (4.3.1).

In some, but not all, flows, it is justified to take the turbulent Prandtl number to be a constant, generally close to unity, so that the turbulent thermal diffusivity is simply related to the turbulent viscosity coefficient. The following perspective is called for: Both exchange coefficients μ_T and μ_T/σ_T have the dimensions ρUL, where U is a measure of the velocity fluctuations—for example, $k^{1/2}$—and L is a measure of the integral scales of the fluctuations. The notion of a constant turbulent Prandtl number is valid provided the two integral scales are essentially equal. This is the case, for example, in the wake of a heated cylinder where the heat is added and the momentum subtracted at the cylinder. However, consider a small heated wire extending across a turbulent grid flow; in this case the diffusion of heat is associated with scales which, at least initially, are considerably smaller than those of the large velocity fluctuations, so the notion of a turbulent Prandtl number must be abandoned. In this case separate models for μ_T and μ_T/σ_T are required. In this regard, note that the thermal wake of a wire in grid turbulence is treated in Section 8.4.

7.3 GENERAL COMMENTS ON GRADIENT MODELS

Gradient models such as those of Eqs. (7.1.3) and (7.2.1) have a long history, "possibly starting with de St. Venant (1843), certainly with Boussinesq (1877),"† and play a central role in the phenomenology of turbulent shear flows. We shall

†From Corrsin (1974).

see in Sections 10.10 and 10.11 that even in advanced methods for the treatment of turbulent flows, gradient models are used to eliminate *third-moment and other terms* and thereby achieve closure of the transport equations for second-moment quantities. Despite the widespread and pervasive use of gradient descriptions of turbulent transport, concerns as to their validity are well expressed by Corrsin (1974): "Simple gradient transport assumptions are virtually the only kind used for practical calculations in turbulent transport in 1973 in spite of the fact that it has long been realized that a gradient transport model requires (among other things) that the characteristic scale of the transporting mechanism must be small compared with the distance over which the mean gradient of the transported property changes appreciately. . . . There has been an astonishing lack of amazement at the partial success of these models." An alternative perspective is offered by Saffman (1978): "The continual preaching against the eddy diffusivity hypothesis . . . has not served any useful purpose. The effort would have been better spent trying to understand the reasons for the apparent success and the circumstances in which it must (not ought to) fail."

A further concern relative to Eq. (7.1.3) is exposed if we consider again its application to flows dominated by shear rates of strain, e.g., to one with only $\partial U_1 / \partial x_2$ not equal to zero; this is the case in many flows of applied interest, namely, wakes, jets, channel flows, and boundary layers, with the x_1 coordinate along the principal flow direction and x_2 normal thereto. In this case we see from Eq. (7.1.3) that

$$\overline{u_i^2} = \frac{2}{3} k \qquad -\rho\, \overline{u_1 u_2} = \mu_T \frac{\partial U_1}{\partial x_2} \tag{7.3.1}$$

The second of these equations is taken up later, but of immediate interest is the finding that in this important class of flows the gradient assumption predicts *equal* contributions to the turbulent kinetic energy from the velocity fluctuations in the three coordinate directions; i.e., the energy is equipartitioned. This is not in accord with either experiment or more advanced theories (cf. Section 10.10), and must be considered an essential shortcoming of the model.

In these circumstances it is not surprising that improved and alternative models to Eqs. (7.1.3) and (7.2.1) are suggested from time to time. In Section 10.9 one alternative arising from a general second-moment closure is discussed, the algebraic stress model of Rodi (1976), a model which results in implicit algebraic relations *among* the Reynolds stresses and the mean rates of strain. A conceptually useful exploitation of this model is due to Pope (1975) and Gatski and Speziale (1993), who show that these relations can be solved to yield *explicit* expressions for the Reynolds stresses in terms of the mean rates of strain, in the case of the latter work to three-dimensional turbulent flows involving body forces due to curvature and rotation. Such relations clearly represent a considerable extension of Eq. (7.1.3).

In the context of this discussion we call attention to the argument of Townsend (1956), which addresses squarely the uncertainty connected with gradient transport—turbulent diffusion due to large turbulent structures—for which the approximation of Eqs. (7.1.3) and (7.2.1) are most questionable. The suggestion is that such diffusion in this case is due to an effective bulk velocity in the x_2 direction and not to a gradient of the associated mean quantity. This notion was used by Bradshaw et al. (1967) in a theory of turbulent boundary layers to model the turbulent diffusion of turbulent kinetic energy. An interesting consequence of this nongradient description is that the partial differential equations for the mean velocity and mean Reynolds shear stress in this boundary-layer theory are hyperbolic rather than parabolic. In the same spirit, Keffer (1965) summed large eddy and gradient diffusion in an analysis of the turbulent wake and assigned equal weight to each contribution.

Having set forth the concerns for the validity of the gradient model, we now nevertheless take up its utilization. Suppose that we have a suitable representation for μ_T. Then Eq. (7.1.3) completes Eqs. (4.2.1) if and only if we know $k(\mathbf{x})$. However, Eq. (4.7.1), the equation for the turbulent kinetic energy, possesses its own closure problem, so as it stands Eq. (7.1.3) does not provide closure. However, consider once again the determination of the mean velocity and mean temperature in many flows of applied interest, those discussed earlier and involving a single shear stress $\rho\overline{u_1u_2}$ and a single thermal flux $\rho\overline{u_2\theta}$ in the usual notation. In this case the normal stresses $\overline{u_i^2}$, $i = 1, 2, 3$ while not distributed properly, do not appear in the formulation. Thus the turbulent kinetic energy is not needed and Eq. (7.1.3), rewritten as the second of Eqs. (7.3.1), provides closure if the exhange coefficient ν_T is described in some fashion. Used along with Eq. (7.2.1), we have a closed set of equations for U_i, $i = 1, 2, 3$, P, and Θ. More generally, the need to calculate the turbulent kinetic energy to exploit fully the gradient transport model leads naturally to the methods discussed in Sections 10.1–10.9.

The development of models for μ_T is based largely on intuition and dimensional arguments. When divided by the density, the coefficients μ_T and μ_T/σ_T have the units of UL, so the various models differ only in the "velocity" and "length" they introduce. Frequently, in the course of arguments leading to a model for μ_T, one unknown quantity is replaced by another unknown, presumably more readily dealt with, but the rationale for the replacement may not be obvious a priori. Moreover, even final forms generally involve replacement of μ_T with other quantities which must be assumed. Given this situation, it is appropriate to discuss in the next two sections several descriptions of the turbulent exchange coefficient.

7.4 MIXING LENGTH

The first model for the turbulent viscosity coefficient we consider is based on a further extension of the analogy with viscous transport by introducing a length over which fluid elements exchange momentum and/or thermal energy—i.e., the mixing length. Despite the limitations of this perspective suggested by the earlier comments of Corrsin, the resulting model for μ_T is widely applied in a variety of flows, as we shall see in Chapters 8–9.

Consider a simple shear flow with $U_1 = U_1(x_2)$ and $U_2 \equiv U_3 \equiv 0$ as in Fig. 4.2.1 and with the origin of the x_2 coordinate at an arbitrary reference plane. To determine $\rho\overline{u_1 u_2}$ we consider a suitably large number of realizations associated with a sequence of times t_n. Each realization corresponds to a fluid element which is located at $x_2 = 0$ at $t = t_n$ and has an x_2 velocity component u_{2n}. However, at some prior time $t = t_n - \tau_n$, this same fluid element was at $x_2 = x_{2n}$ and had an x_1 velocity close to $U_1(x_{2n})$. The locations x_{2n} can be either positive or negative, but if $x_{2n} > 0$, then $u_{2n} < 0$ and vice versa. We make the crucial assumption that during the transit from x_{2n} to $x_2 = 0$ the x_1 velocity component of the fluid element involved in each realization is nearly unaltered.

According to these considerations, the stress in question is obtained by ensemble averaging over many realizations (cf. Section 3.1) so that

$$\rho\overline{u_i u_2} = \rho \lim_{N\to\infty} \frac{1}{N}\left[\sum_{n=1}^{N} [U_1(x_{2n}) - U_1(0)]u_{2n}\right] \tag{7.4.1}$$

We now make a second crucial assumption, that the maximum absolute value of x_{2n} is sufficiently limited so that

$$U_1(x_{2n}) - U_1(0) \approx x_{2n}\frac{dU_1}{dx_2}(0) \tag{7.4.2}$$

Clearly, if the velocity gradient is large, the permitted range of x_{2n} is reduced. More to the point, however, if the fluid element arriving at the plane $x_2 = 0$ at $t = t_n$ is associated with a large turbulent structure and thus with a large value of x_{2n}, then $dU_1/dx_2(0)$ must be small. Nevertheless, with this approximation Eq. (7.4.1) yields

$$\rho\overline{u_1 u_2} \approx \rho\langle x_2 u_2\rangle \frac{dU_1}{dx_2} \tag{7.4.3}$$

where we drop as irrelevant the specialization to the plane $x_2 = 0$ and where the averaging in each realization involves the product $x_2 u_2$. Note that with $dU_1/dx_2 > 0$, on physical grounds we have $\langle x_2 u_2\rangle < 0$.

Equation (7.4.3) offers insight into the mechanism of turbulent exchange but possesses little practical advantage over the primative form of Eq. (7.3.1) since it involves an unknown quantity, the correlation $\langle x_2 u_2\rangle$, rather than the unknown μ_T. To proceed, we introduce further modeling and take

$$-\langle x_2 u_2\rangle \approx (\overline{u_2^2})^{1/2} l \tag{7.4.4}$$

where l is the mixing length. The notion here is that u_2 varies with the intensity of the fluctuations of the u_2 velocity component and that Eq. (7.4.4) *defines* l. Since in Eq. (7.4.4) we have substituted two unknowns for one, an apparently retrograde step, two tasks remain before a useful result is achieved from a combination of Eqs. (7.4.3) and (7.4.4). To eliminate the intensity of the velocity fluctuations, we refer to the first of Eqs. (6.3.5). If we assume that turbulence production there, represented by the left side, is balanced by dissipation on the right side and if dissipation is assumed proportional to $(\overline{u_2^2})^{3/2}/l$ as is appropriate for high-Reynolds-number turbulence, then some algebra yields

$$(\overline{u_2^2})^{1/2} \approx l \left| \frac{dU_1}{dx_2} \right| \tag{7.4.5}$$

where various multiplicative constants are swept into the mixing length l. We thus find from Eqs. (7.4.3) and (7.4.4) that

$$\rho \overline{u_1 u_2} = -\rho l^2 \left| \frac{\partial U_1}{\partial x_2} \right| \frac{\partial U_1}{\partial x_2} \tag{7.4.6}$$

where we introduce the absolute value to assure the proper sign of the Reynolds shear stress in flows with either positive or negative values of $\partial U_1/\partial x_2$ and the partial derivatives to reflect the applicability to turbulent shear flows with a slow evolution in the x_1 direction, e.g., in turbulent boundary layers, jets, and wakes.

Equation (7.4.6) is the statement of mixing-length theory. Comparison with Eq. (7.1.3) indicates that we have replaced the unknown exchange coefficient μ_T with the unknown mixing length l. However, fluid mechanical arguments supported by experimental evidence for the Reynolds shear stresses and the associated rates of strain are used to provide descriptions for $l(x_2)$ in a variety of flows and thus to complete the modeling. With such descriptions the single mean momentum equation needed for the shear flows considered here is closed, and predictions of the behavior of such flows can be carried out. As an example of such arguments, we note that as a solid wall is approached the length scale of the fluctuations in velocity and temperature is expected to decrease and thus the mixing length is likewise expected to decrease. Furthermore, remote from solid walls, the large turbulent scales are roughly the global dimension of the flow and the mixing length approaches a constant, a fraction of the integral scale of the turbulence. Thus, in many flows analytic approximations reflecting these characteristics lead to convenient and useful closure of the equations despite the absence of a fundamental basis for the model.

The underlying concern relative to the gradient transport model stated earlier by Corrsin resides specifically in the approximations attendant with Eq. (7.4.2), approximations that are generally not valid in turbulent shear flows of applied interest with their turbulent structures filling the flow. Nevertheless, the analyses in Chapters 8 and 9, although dependent significantly on Eq. (7.4.6) and on assumed distributions of the mixing length, result in satisfactory agreement with experiment. In the present context this approach is justified on the grounds of its physical appeal and simplicity and of the adequacy of its treatment of turbulence at the level of mean velocities, Reynolds shear stress, mean temperatures, and Reynolds flux, the level discussed in these chapters.

An important conceptual result follows from Eqs. (7.4.3) and (7.4.6). Suppose that we form the ratio of the turbulent to the viscous shear stresses for simple shear flow. We have

$$\frac{\rho\overline{u_1u_2}}{\mu(\partial U_1/\partial x_2)} = \frac{\rho(\overline{u_2^2})^{1/2}\, l(\partial U_1/\partial x_2)}{\mu(\partial U_1/\partial x_2)} = \frac{(\overline{u_2^2})^{1/2} l}{\nu} \approx \frac{k^{1/2} l}{\nu} \tag{7.4.7}$$

We thus see that if a turbulence Reynolds number based on the turbulent kinetic energy and the mixing length is large, then the viscous shear stress is negligible compared to the turbulent shear stress. Clearly, near walls, where both factors in the product $k^{1/2}l$ become small, the viscous stresses are dominant. Accordingly, turbulent flows involving walls are essentially different from free shear flows. Thus Chapters 8 and 9 deal with free shear and wall-bounded flows separately.

It is clear from this development that when the notion of a turbulent Prandtl number is applicable, we have for the turbulent heat flux of Eq. (7.2.1)

$$-\rho\overline{u_2\theta} = \frac{\rho l^2\, |\partial U_1/\partial x_2|}{\sigma_T}\frac{\partial \Theta}{\partial x_2} \tag{7.4.8}$$

Thus the calculation of the heat transfer involves the additional parameter σ_T. Many experimental results suggest that σ_T is roughly unity, varying from 0.7 to 1.2 depending on the circumstances. We note once again that Eq. (7.4.8) is not appropriate for flows in which the length scales of the temperature fluctuations are not the same as those of the velocity fluctuations.

The development presented here is essentially that for the mixing length after Prandtl (1925, 1942). Other related theories for the lengths characterizing turbulent exchange are due to Taylor (1935b) and von Karman (1930) but for our purposes need only be noted. To repeat an earlier remark, these theories result in different velocities and lengths yielding ν_T.

7.5 TURBULENT EXCHANGE IN FREE SHEAR FLOWS

In the classical approach to turbulence, the mixing-length theory with either simple or elaborate descriptions for $l(x_2)$ is generally used for flows involving walls. However, for free shear flows a simpler model may be applied, since in the absence of a restraining wall, turbulent exchange in these flows is dominated by large-scale turbulence, i.e., by scales on the order of the global dimensions of the flow (e.g., the thickness of the mixing layer). In particular, we discuss here the turbulent exchange coefficient for jets, wakes, and mixing layers.

We start with two-dimensional jets and wakes and orient the x_1 axis with the plane of symmetry and the x_2 axis normal. If we recall that ν_T has the dimensions of UL and that the mixing length is roughly constant remote from walls, it is reasonable to use for such flows the following model:

$$\nu_T(x_1) = \kappa_1 \, |U_{1\infty} - U_1(x_1, 0)| \, \delta(x_1) \tag{7.5.1}$$

where κ_1 is an empirical coefficient and $\delta(x_1)$ is the thickness of the shear layer defined in some fashion. The absolute-value signs permit Eq. (7.5.1) to apply to both jets which correspond to $U_1(x_1, 0) > U_{1\infty}$ and to wakes which correspond to $U_{1\infty} < U_1(x_1, 0)$. Without additional considerations we cannot determine from Eq. (7.5.1) the behavior of ν_T with streamwise distance. In two-dimensional jets and wakes the velocity difference decreases with x_1 but the thickness δ increases, and we must know the rates involved to determine the variation of $\nu_T(x_1)$ with streamwise distance. In this flow—indeed, in all the flows considered in this section—analytic estimates based on mass and momentum conservation can be made for the streamwise variation of the velocity difference and shear layer thickness (e.g., Schlicting, 1955). These estimates are well supported by experimental data.

Far downstream in planar jets in *moving streams* and in planar wakes, the velocity difference decreases as $x_1^{-1/2}$ while the width grows as $x_1^{1/2}$ so ν_T is constant, not only with respect to the transverse coordinate x_2, but with respect to x_1 as well. On the other hand, for an axisymmetric wake, e.g., the wake behind a sphere, the x_1 coordinate becomes the axial distance. In this case the velocity difference decreases as $x_1^{-2/3}$ and the radius increases as $x_1^{1/3}$. As a consequence, ν_T *decreases* with $x_1^{-1/3}$.

For a planar jet discharging into a *quiescent ambient,* the velocity difference again decreases as $x_1^{-1/2}$, but that the width grows as x_1. Thus in this case ν_T grows parabolically. For a circular jet with a quiescent ambient, the velocity difference between the axis and the surrounding ambient decreases with x_1 while the radius increases with x_1, so ν_T is constant.

Equation (7.5.2) with an obvious modification for the velocity difference also applies to the two-dimensional mixing layer. With a velocity U_{11} on the high-speed side and U_{12} on the low-speed side, we take

$$\nu_T(x_1) = \kappa_2(U_{11} - U_{12}) \, \delta(x_1) \tag{7.5.2}$$

where $\delta(x_1)$ is an appropriately defined thickness of the mixing layer. In this flow the linear growth of $\delta(x_1)$ results in a corresponding growth of the turbulent exchange coefficient; that is, $\nu_T \propto x_1$.

Both Eqs. (7.5.1) and (7.5.2) can be rewritten to show that a Reynolds number based on the velocity difference, the width of the jet or wake, and the *turbulent exchange coefficient* is equal to κ_1^{-1} or κ_2^{-1}. Typically these Reynolds numbers are assigned values in the range 40–80.

In nonisothermal free shear flows the temperature and momentum changes are generally from the same source, with the consequence that the notion of a turbulent Prandtl number is applicable. As noted earlier, typical values of roughly 0.7–1.2 lead to accord with experimental data.

7.6 TURBULENT EXCHANGE UNDER THE INFLUENCE OF BUOYANCY

We know from the discussion of Section 6.4 that in nonisothermal flows with relatively large temperature gradients in the vertical direction and with small velocity gradients $\partial U_1/\partial x_2$, where x_2 is taken to be vertical, buoyancy effects can influence turbulent exchange significantly, augmenting that exchange if the temperature gradient is destabilizing and diminishing it if stabilizing. As suggested in that section, because of its importance in the atmospheric boundary layer and in oceanic phenomena, there exists a considerable literature on this influence. However, the means for extending mixing-length and the other models discussed in this chapter to account for the influence of buoyancy are unclear. Accordingly, we defer this matter to Section 10.12, where the algebraic stress method describes this influence efficiently.

7.7 SUMMARY

In this chapter we have introduced gradient models which represent classical closure of the moment equations at the first level in the hierarchy of the moment equation. Involved is a turbulent exchange coefficient with the dimensions of a velocity multiplied by a length; the different models involve different velocities and different lengths. The shortcomings of gradient transport from a fundamental point of view are discussed. Models of such coefficients that are applicable to various flows are given.

CHAPTER
EIGHT

FREE SHEAR FLOWS

We now have at our disposal the tools for analyzing several classes of turbulent shear flows. The simplest and most amenable to theoretical treatment are flows which do not involve solid boundaries: wakes, jets, and mixing layers. With some idealization, these can be described in terms of similarity variables so that the two space variables in the original partial differential equations are reduced to a single independent variable. As a consequence, we deal with ordinary differential equations.

8.1 THE FAR WAKE OF A CYLINDER IN AN ISOTHERMAL FLOW

Figure 8.1.1 shows schematically the turbulent flow in the wake of a cylinder of arbitrary cross section orthogonal to a uniform steady stream with velocity U_∞ aligned with the x_1 axis. The wake is characterized by a momentum defect associated with the drag of the cylinder. We are not concerned with the details of the complex, near-wake region, but rather with the evolution of the momentum wake far from the cylinder, perhaps 10^2 diameters downstream. In this downstream region the streamlines external to the wake are nearly straight, so we can reasonably assume that the pressure is everywhere uniform. Moreover, we can assume that the mean quantities are two-dimensional, i.e., independent of x_3,

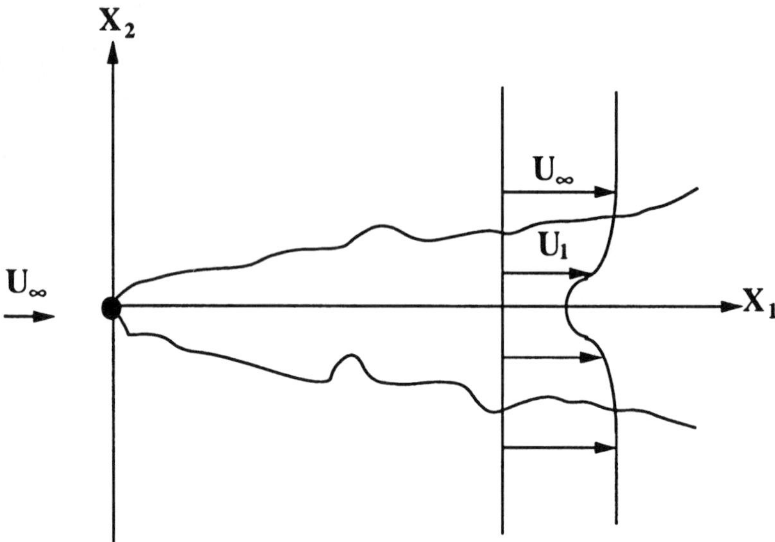

Figure 8.1.1 The turbulent wake of a cylinder.

that a thin flow approximation applies, and that with the turbulence Reynolds number sufficiently high viscosity may be neglected. This implies that although we treat the far wake, the *ultimate* downstream behavior which returns the stream to complete uniformity is not described, since when the turbulence becomes suitably weak, viscous stresses are no longer negligible.

The following equations obtained from Eqs. (4.1.1) and (4.2.1) apply:

$$\frac{\partial U_k}{\partial x_k} = 0$$
$$\frac{\partial}{\partial x_k}(U_k U_1) = -\frac{\partial}{\partial x_2}\overline{u_1 u_2} \tag{8.1.1}$$

These equations involve the three dependent variables U_1, U_2, and $\overline{u_1 u_2}$ and are closed later by an appropriate assumption for the Reynolds shear stress. Because we confine our attention to the far wake, we do not specify initial conditions but rather the boundary conditions on the plane of symmetry and in the external stream: $U_1(x_1, x_2 \rightarrow \infty) = U_\infty$, $U_2(x_1, 0) = \overline{u_1 u_2}\,(x_1, 0) = 0$.

Our analysis starts with the introduction of a similarity variable $\eta = \eta(x_1, x_2)$ and of a form for $U_1(x_1, x_2)$ leading to a self-consistent formulation. We noted in Section 7.5 that the classical turbulent exchange coefficient ν_T in a two-dimensional wake is a constant; in this case our equations become mathematically identical with those for laminar flow, which is described in terms of a similarity variable $\eta \propto x_2 x_1^{-1/2}$ and a velocity defect $U_\infty - U_1(x_1, 0)$ decaying as $x_1^{-1/2}$. However, transformation back to physical x_1, x_2 variables leads to considerably different rates of wake growth and velocity defect decay for the two flows.[†] Accordingly, we take

$$U_1(\tilde{x}_1, \eta) \approx U_\infty \left[1 - \frac{{}^{(0)}f(\eta)}{\tilde{x}_1^{1/2}} - \frac{{}^{(1)}f(\eta)}{\tilde{x}_1} + \ldots \right] \tag{8.1.2}$$

where

$$\eta \equiv \frac{x_2}{[(x_1 - x_0)d]^{1/2}}$$

$$\tilde{x}_1 \equiv \frac{x_1 - x_0}{d}$$

are two, dimensionless independent variables. Here x_0 is the location from the cylinder of the virtual origin of the variable $\tilde{x}_1$ and d is a characteristic dimension of the cylinder, e.g., the diameter if the cross section is circular. We defer discussion of the experimental determination of the location of the virtual origin. The sequence of $f(\eta)$ functions in Eq. (8.1.2) satisfying appropriate boundary conditions at $\eta = 0$ and $\eta \to \infty$ leads to the proper evolution of the momentum wake.

If Eq. (8.1.2) is substituted into the first of Eqs. (8.1.1), if the variables are changed from x_1, x_2 to $\tilde{x}_1$ and η, and if the resultant equation is integrated, we obtain

$$-U_2(\tilde{x}_1, \eta) \approx U_\infty \left[\frac{{}^{(0)}g(\eta)}{\tilde{x}_1} + \frac{{}^{(1)}g(\eta)}{\tilde{x}_1^{3/2}} + \ldots \right] \tag{8.1.3}$$

where

$${}^{(0)}g(\eta) = \frac{1}{2} \int_0^\eta d\eta \, (\eta\, {}^{(0)}f)' = \frac{1}{2} \eta\, {}^{(0)}f$$

$${}^{(1)}g(\eta) = \int_0^\eta d\eta \left({}^{(1)}f + \frac{1}{2} \eta\, {}^{(1)}f' \right)$$

and where we impose the boundary condition on U_2 at the plane of symmetry. Note that the leading term in $U_2(\tilde{x}_1, \eta)$ is proportional to $\tilde{x}_1^{-1}$; that is, that in the far wake the U_2 velocity is small compared with the U_1 velocity.

If, in accord with our earlier discussion, we let [cf. Eq. (7.3.4)]

$$-\overline{u_1 u_2} = \nu_{T0} \frac{\partial U_1}{\partial x_2} \tag{8.1.4}$$

[†]These differences are reflected in ν_{T0} and $f_0(0)$, respectively, parameters which appear later.

then the first two terms in $\tilde{x}_1$, that is, in $\tilde{x}_1^{-3/2}$ and $\tilde{x}_1^{-2}$, in the second of Eqs. (8.1.1) yield

$$\begin{aligned} -\frac{1}{2}(\eta^{(0)}f)' &= \left(\frac{\nu_{TO}}{U_\infty d}\right){}^{(0)}f'' \\ -\left({}^{(1)}f + \frac{1}{2}\eta^{(1)}f' - \frac{1}{2}{}^{(0)}f^2\right) &= \left(\frac{\nu_{TO}}{U_\infty d}\right){}^{(1)}f'' \end{aligned} \tag{8.1.5}$$

which are to be solved subject to the conditions

$${}^{(0)}f'(0) = {}^{(0)}f(\eta \to \infty) = 0 \qquad {}^{(1)}f'(0) = {}^{(1)}f(\eta \to \infty) = 0$$

Note that the dominant contribution to the left side of the second of Eqs. (8.1.2) arises from the $U_1\ \partial U_1/\partial x_1$ term; in contrast with boundary-layer theory, here the second term $U_2\ \partial U_1/\partial x_2$ is smaller than $U_1\ \partial U_1/\partial x_1$ and not of the same order. Thus the present analysis is based not on that theory but rather on an expansion in terms of the distance from the cylinder.

The first of Eqs. (8.1.5) can be solved immediately to obtain

$${}^{(0)}f(\eta) = {}^{(0)}f(0)\exp\left[-\frac{1}{4}\left(\frac{U_\infty d}{\nu_{TO}}\right)\eta^2\right] \tag{8.1.6}$$

where ${}^{(0)}f(0)$ is an arbitrary constant. We see from this solution that symmetry at $\eta = 0$ is automatically satisfied, with the consequence that an additional condition, the value of ${}^{(0)}f(0)$, can be imposed so as to achieve agreement with experiment.

It is useful to examine further the solution given by this first term in the series introduced by Eq. (8.1.2); to that end we write the mean velocity defect as

$$\frac{U_\infty - U_1}{U_\infty} = \frac{{}^{(0)}f(0)}{\tilde{x}_1^{1/2}}\exp\left[-\frac{1}{4}\left(\frac{U_\infty d}{\nu_{TO}}\right)\eta^2\right] \tag{8.1.7}$$

We see, as noted earlier, that the velocity defect decreases in the streamwise direction as $x_1^{1/2}$. The variation of the corresponding Reynolds shear stress becomes, from Eq. (8.1.4),

$$-\rho\frac{\overline{u_1u_2}}{U_1^2} = \frac{1}{2}\rho\frac{{}^{(0)}f(0)}{\tilde{x}_1}\eta\exp\left[-\frac{1}{4}\left(\frac{U_\infty d}{\nu_{TO}}\right)\eta^2\right] \tag{8.1.8}$$

We see from this equation that the Reynolds shear stress is zero on the plane of symmetry as required by symmetry considerations, reaches a maximum, and then decays as the free stream is approached.

A compact solution equivalent to Eq. (8.1.6) is obtained if we introduce a new independent variable:

$$ {}^{(0)}f(\xi) = {}^{(0)}f(0)\, \exp(-\xi^2) \tag{8.1.9} $$

where

$$ \xi \equiv \frac{1}{2}\left(\frac{U_\infty d}{\nu_{TO}}\right)^{1/2} \eta = \xi(x_1, x_2) $$

is a scaled similarity variable.

There are two empirical constants in the solutions given by Eqs. (8.1.6) and (8.1.9): the explicit ${}^{(0)}f(0)$ and the implicit Reynolds number $U_\infty d/\nu_{TO}$. The available experimental data are not unequivocal as to the appropriate values to be assigned these quantities, but the following ranges are indicated:

$$ 37 < \frac{U_\infty d}{\nu_{TO}} < 62 \qquad 1.12 < {}^{(0)}f(0) < 1.17 \tag{8.1.10} $$

Thus with some uncertainty we may consider that the first-order behavior of the two mean velocity components and the Reynolds shear stress is determined.

We now consider the solution for ${}^{(1)}f(\eta)$, the second term in the expansion of Eq. (8.1.2). From the second of Eqs. (8.1.5) it is convenient to let ξ be the independent variable and to introduce a new dependent variable such that

$$ {}^{(1)}f(\xi) = {}^{(0)}f^2(0)\, F(\xi) $$

since $F(\xi)$ can be determined free of parameters. We obtain the linear equation

$$ F'' + 2\,\xi\, F' + 4\,F = 2\exp(-2\xi^2) \tag{8.1.11} $$

which is to be solved subject to the conditions[†]

$$ F'(0) = F\,(\xi \rightarrow \infty) = 0 $$

Although an analytic solution to Eq. (8.1.11) may be possible, the numerical solution shown in Fig. 8.1.2 is readily found.

[†]Examination of the behavior of Eq. (8.1.11) as $\xi \rightarrow \infty$ indicates that to obtain a unique solution a stronger condition at infinity is required: that $F' \rightarrow \xi^2 \exp(-\xi^2)$. For such an examination it is useful to note that the asymptotic behavior of $y(x)$ as $x \rightarrow \infty$, where $y(x)$ is determined by

$$ y'' + xy' + ny = 0 $$

is

$$ y \approx A_1 \exp\left(-\frac{1}{2}x^2\right) x^{n-1}\,[1 + O(x^{-2})] + A_2 x^{-n}[1 + O(x^{-2})] $$

where A_1 and A_2 are arbitrary constants. When $n > 0$, there are both exponentially decaying and algebraically decaying solutions.

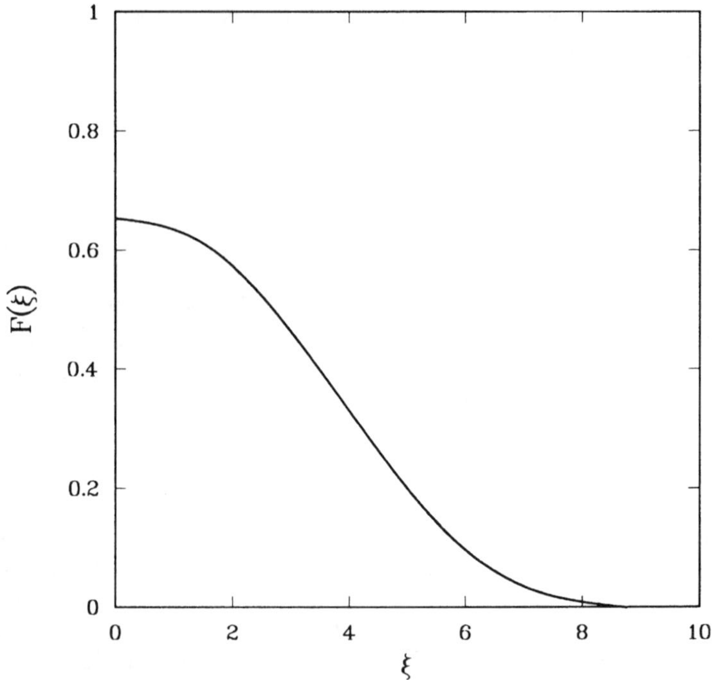

Figure 8.1.2 Second-order solution for the velocity defect in the wake of a cylinder.

The virtual origin can be obtained from experimental data by applying Eq. (8.1.2) at $\eta = 0$. The first approximation to x_0 is obtained from the most downstream data, assumed to correspond to the far wake, with the velocity defect taken to decay as $(x_1 - x_0)^{-1/2}$. A more accurate value of x_0 is obtained if both terms in Eq. (8.1.2) are considered; in this case a least-square determination is indicated.†

Two concluding comments are appropriate. The analysis presented here applies to the far field of a jet discharging into a moving stream provided the sign of ${}^{(0)}f(0)$ is changed. In this case the cylinder is replaced by a two-dimensional slit from which fluid is injected, and the velocity defect becomes a velocity excess. In addition, the nondimensionalizing length d must be identified with the momentum excess of the jet. As noted by Rodi (1975), it is unfortunate that there exist relative to two-dimensional jets in moving streams no data which are sufficiently far downstream to demonstrate the equivalence of these two flows. We know from Section 7.5 that the flow from a two-dimensional slit discharging

†Hariri et al. (1982) described a more accurate means for determining the first approximation to x_0 based on similarity considerations for the second of Eqs. (8.1.1); unfortunately and surprisingly, the additional data needed to validate their perspective are sparse.

into a *quiescent ambient* involves spreading rates and velocity decay rates of x_1 and $x_1^{-1/2}$, respectively. Accordingly, a different but closely related similarity analysis can be readily carried out.

8.2 THE FAR WAKE OF A HEATED CYLINDER

If the cylinder considered in the previous section is heated above ambient temperature, the wake involves both a momentum defect due to the drag and a temperature excess due to heat addition. In the far-wake region the temperature measured relative to the ambient temperature is described by an appropriate specialization of Eq. (4.3.1):

$$\frac{\partial}{\partial x_k}(U_k\,\Theta) = -\frac{\partial}{\partial x_2}\overline{u_2\theta} \qquad (8.2.1)$$

Again we assume that the turbulence Reynolds number is sufficiently large as to permit neglect of transport due to thermal conductivity. Since the velocity and temperature fluctuations are added by the same body, their length scales are essentially equal and the notion of a turbulent Prandtl number applies. Accordingly, we can use our earlier analysis of the momentum wake and consider the turbulent exchange coefficient for heat to be a constant ν_{T0}/σ_T. In this case Eq. (8.2.1) becomes

$$\frac{\partial}{\partial x_k}(U_k\Theta) = \frac{\nu_{T0}}{\sigma_T}\frac{\partial^2\Theta}{\partial x_2^2} \qquad (8.2.2)$$

which is to be solved subject to the boundary conditions

$$\frac{\partial\Theta}{\partial x_2}(x_1, 0) = \Theta(x_1, x_2 \to \infty) = 0$$

Again we restrict attention to the far-wake region and exclude consideration of the flow in the immediate vicinity of the cylinder.

To proceed, we assume a form for Θ analogous to that taken for the velocity distribution [cf. Eq. (8.1.3)]:

$$\Theta(\tilde{x}_1, \xi) \approx \Theta(0)\left[\frac{{}^{(0)}\Theta(\xi)}{\tilde{x}_1^{1/2}} + \frac{{}^{(1)}\Theta(\xi)}{\tilde{x}_1} + \ldots\right] \qquad (8.2.3)$$

where $\Theta(0)$ is a parameter depending on the temperature of the cylinder with the units of degrees, and ${}^{(0)}\Theta(\xi)$, ${}^{(1)}\Theta(\xi)$, etc., give the dimensionless distributions of the mean temperature in terms of $\xi(x_1, x_2)$ [cf. Eqs. (8.1.2) and (8.1.9)]. We implicitly assume here that the virtual origins for the momentum and temperature wakes are the same; to do otherwise complicates the analysis unnecessarily.

Substitution into Eq. (8.2.2) and collection of the terms proportional to $\tilde{x}_1^{-3/2}$ and $\tilde{x}_1^{-2}$ yields the following equations:

$$-(\xi^{(0)}\ \Theta)' = \frac{1}{2\sigma_T}\ {}^{(0)}\Theta''$$
$$-\left({}^{(1)}\Theta + \frac{1}{2}\ \xi^{(1)}\ \Theta' - \frac{1}{2}\ {}^{(0)}f^{(0)}\ \Theta\right) = \frac{1}{4\sigma_T}\ {}^{(1)}\Theta'' \tag{8.2.4}$$

which are to be solved subject to the boundary conditions†

$$ {}^{(0)}\Theta(0) = 1 \qquad {}^{(0)}\Theta(\xi \to \infty) = 0 \qquad {}^{(1)}\Theta'(0) = {}^{(1)}\Theta(\xi \to \infty) = 0$$

Note that the symmetry condition on ${}^{(0)}\Theta(\xi)$ at $\xi = 0$ is automatically satisfied, so ${}^{(0)}\Theta(0)$ can be specified so as to achieve agreement with experiment.

The solution of the first of Eqs. (8.2.4) is

$$ {}^{(0)}\Theta(\xi) = \exp(-\sigma_T\ \xi^2) \tag{8.2.5}$$

In solving the second of Eqs. (8.2.4) we must incorporate the solution for ${}^{(0)}f(\xi)$ given by Eq. (8.1.9). The solutions for the mean temperature distribution in the far wake involve the additional empirical constants $\Theta(0)$ and σ_T. The value of $\Theta(0)$ depends on the extent of the heat addition at the cylinder and therefore varies from one experiment to another.

Experimental data show that the width of the thermal wake is greater than that of the momentum wake and thus that $\sigma_T < 1$ with a value of $\sigma_T = 0.7$ typically assigned. This result is a consequence of rather crude criteria for defining the widths of the momentum and thermal wakes, criteria based on the mean velocity being a specified decrement of U_∞ and Θ being a specified positive increment. Because at the outer edge of the wake the streamwise velocity *within the turbulence* is relatively closer to U_∞ than Θ is to zero, the *apparent thickness* of the thermal wake is greater than that of the momentum wake (cf. Sections 8.5 and 8.6).

The second of Eqs. (8.2.4), written in terms of ${}^{(0)}f(0)\ {}^{(1)}\Theta(\xi)/\Theta(0)$, satisfying the boundary conditions cited earlier and with $\sigma_T = 0.7$, is shown in Fig. 8.2.1.

8.3 THE MIXING LAYER

The two-dimensional mixing layer shown schematically in Fig. 8.3.1 has been subjected to extensive experimental study over many years. Early investigations (e.g., Liepmann and Laufer, 1947) were concerned primarily with the mean velocity distributions and the spreading rate of the layer. As discussed in Section

†Our earlier remark concerning the need for a stronger statement of the condition at infinity applies here as well.

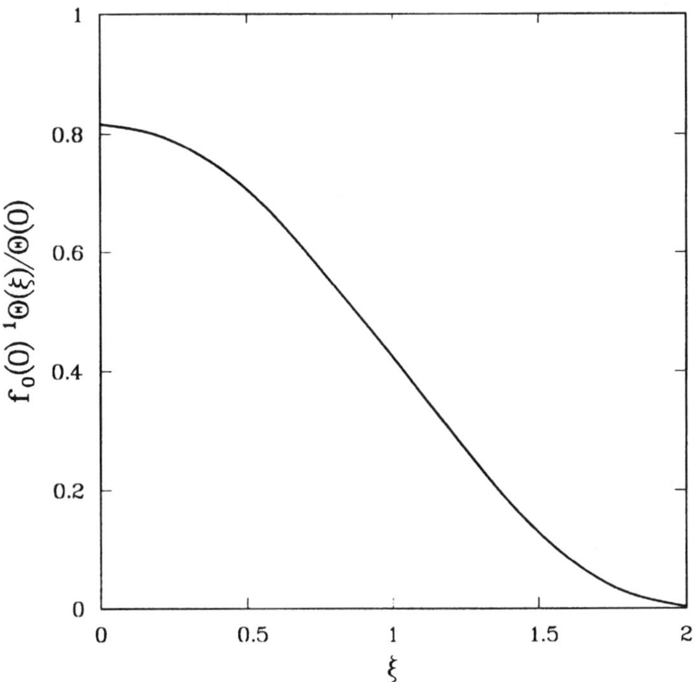

Figure 8.2.1 Second-order temperature distribution in the wake of a heated cylinder.

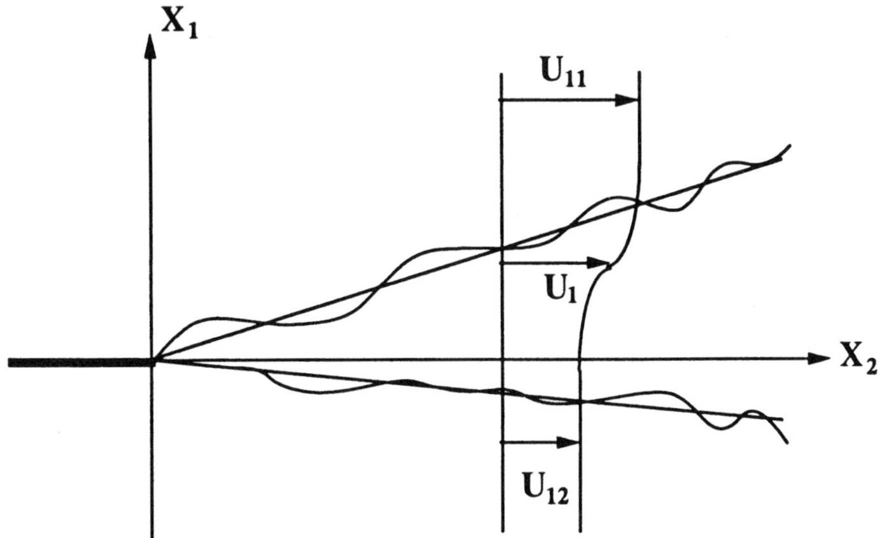

Figure 8.3.1 The turbulent mixing layer.

5.8, in 1974 Brown and Roshko presented dramatic pictures of mixing layers with two-dimensional vortex structures (cf. Fig. 5.6.1); the sequential pairing of these vortices in the downstream direction results in the growth of the layer. The results of Brown and Roshko stimulated intense interest in coherent structures in general as discussed in Section 5.6, and in the two-dimensional mixing layer in particular. The dynamics of two-dimensional vortices and their modification by external excitation attracted special attention (cf. Ho and Huang, 1982). Such excitation can be imposed by mechanically oscillating the edge of the splitter plate, by localized acoustic radiation, or by pulsing the pressure in the settling chambers feeding the two streams. As a result of such disturbances the flow immediately downstream of the splitter plate is altered from its natural state. This is a flow in which conditioning on the phase of the disturbances as discussed in Section 3.6 is effective. Calculations in mixing layers involving either two- or three-dimensional vortices by direct numerical simulation are presented by Metcalfe et al. (1987), and involving two-dimensional vortices by random vortex methods are presented by Ashurst (1979).

Despite this intense interest in two-dimensional vortices and in the two-dimensional mixing layer, it should be kept in mind that in all other turbulent shear flows the turbulent structures are three-dimensional with lengths in the streamwise direction significantly greater than in the two transverse directions (cf. Section 5.1). Thus the two-dimensional mixing layer must be considered unique.

Our discussion of the mixing layer is based on classical turbulence considerations. Within the assumption of constant pressure and a thin mixing layer, Eqs. (8.1.1) and (8.1.2) again apply but with altered boundary conditions. In this case a consistent calculation in agreement with experiment calls for a similarity variable defined by

$$\eta = \frac{\sigma\, x_2}{x_1} \tag{8.3.1}$$

where we follow tradition and take σ as an empirical spreading rate parameter. Experiment shows that $\sigma >> 1$, typically 10 or 12, so that Eq. (8.3.1) is in accord with the thin-layer approximation as may be seen as follows. If the "edges" of the mixing layer are defined by definite values of η of the order of unity, Eq. (8.3.1) shows that they correspond *in physical space* to small values of x_2/x_1 if $\sigma >> 1$. Since the velocity difference across the layer is constant and the thickness grows linearly as indicated by Eq. (8.3.1), we deduce, as discussed in Section 7.5, that the turbulent exchange coefficient increases with the downstream distance x_1. We could introduce a virtual origin in the definition of η by replacing x_1 with $x_1 - x_0$ to reflect the influence of the boundary layers on each side of the splitter plate on the initial growth of the mixing layer. However, our previous discussions of virtual origins provide an adequate exposure to their underlying concept and the means for their determination, and we therefore assume that x_1 in Eq. (8.3.1) is measured from such a suitably chosen origin.

Proceeding as in the analysis of Section 8.1, we assume a form for the mean velocity U_1 suggested by the nature of the flow and take

$$U_1(\eta) = U_{11} f'(\eta) \tag{8.3.2}$$

Substitution into and integration of Eq. (8.1.1) lead to

$$U_2(\eta) = \frac{U_{11}}{\sigma}(\eta f' - f) \tag{8.3.3}$$

where we require that $U_2 = f(0) = 0$.[†]

In the present application involving a similar flow with a thickness growing linearly with the downstream distance x_1, it is convenient to modify Eq. (7.5.2) and take

$$\nu_T = (U_{11} - U_{12})\, \kappa\, x_1 \tag{8.3.4}$$

where κ is an empirical constant. Note that we can always orient our coordinate system so that $U_{11} > U_{12}$ and thus that $\nu_T > 0$. Substitution of Eqs. (8.3.2)–(8.3.4) into Eq. (8.1.1) yields

$$f''' + \lambda f f'' = 0 \tag{8.3.5}$$

where

$$\lambda \equiv \frac{1}{1 - (U_{12}/U_{11})} \geq 1$$

provided $\sigma^2\kappa = 1$, an assumption we may make without loss of generality according to the following argument. Suppose that for a particular velocity ratio U_{12}/U_{11} we solve Eq. (8.3.5) subject to the conditions

$$f(0) = 0 \qquad f'(\eta \to \infty) = 1 \qquad f'(\eta \to -\infty) = \frac{U_{12}}{U_{11}} \tag{8.3.6}$$

We then select σ such that the predicted and measured velocity profiles *at all measuring stations* x_1 agree according to some suitable criterion. To make this selection requires the orientation of the mixing layer to be taken into account.

[†]These conditions beg the practical issue of the orientation of the mixing layer relative to the splitter plate. In reality the x_1 axis identified as the line along which $U_2 = 0$ is not an extension of the splitter plate as shown in Fig. 8.3.1, but rather is along a slightly curved line determined by global features of the flow, e.g., the proximity of one or more walls. Second-order boundary-layer theory resolves this ambiguity (cf. Ting, 1959), but for present purposes we need only call attention to this issue.

The assumption $\sigma^2\kappa = 1$ then implies that the Reynolds number $(U_{11} - U_{12}) x_1/\nu_T = \kappa^{-1}$ is known.

We can calculate the correlation related to the Reynolds shear stress in the mixing layer from Eqs. (8.1.6), (8.3.2), and (8.3.4):

$$-\frac{\overline{u_1 u_2}}{(U_{11} - U_{22})U_{11}} = \kappa\,\sigma\, f''(\eta) \tag{8.3.7}$$

As expected on physical grounds, the Reynolds shear stress vanishes in the two external streams and possesses a maximum in absolute value along a line on which the mean rate of shear strain, $f''(\eta)$, is a maximum.

Figure 8.3.2 shows the distributions of the velocity profiles $f'(\eta) = U_1/U_{11}$ for various values of λ. Note once again that with the simple model for turbulent transport adopted here, Eq. (8.3.5) is identical with that for laminar flow; but because of the differences in ν and ν_T, the spreading rates in physical space are quite different.

8.4 THE HEATED WIRE IN GRID FLOW

A final example of turbulent free shear flows is shown in Fig. 8.4.1. A heated wire with a diameter of the order of the Kolmogorov length is stretched across a turbulent grid flow at x_{1w} measured from a virtual origin determined by the decay of the turbulent kinetic energy. The wire creates a thermal wake in a velocity field with decaying turbulence. Since the source of the temperature fluctuations is different from that of the velocity fluctuations, the length scales of the velocity and temperature fluctuations are different and the concept of a turbulent Prandtl number is inapplicable.

That we might expect different behavior of the thermal wake close to the wire and far downstream is suggested by the following considerations. Close to the wire the heated fluid lies in a thin sheet consisting of fluid which passes in the immediate vicinity of the wire. This sheet is moved transversely by the turbulent velocity field such that a fast-response thermometer close to the wire displays sharp temperature spikes when the sheet crosses the sensor. The resulting *mean temperature* at a particular transverse location thus depends on the number of such spikes encountered per unit time and on their duration and peak value. Their duration is roughly L_B/U_1, where L_B is the Batchelor length (cf. Section 5.7); their number depends on the spectrum of the velocity fluctuations, while their peak value depends on the temperature of the wire. Farther downstream, under the influences of viscous transport and the rates of strain imposed by the velocity field on the sheet, the heated fluid is diffused so the thermometer signal indicates a more continuous temperature, interrupted only when no heated fluid is present at the sensor (cf. Section 8.5). Thus, in the far downstream region we might expect the wake of the wire to resemble that for a heated cylinder.

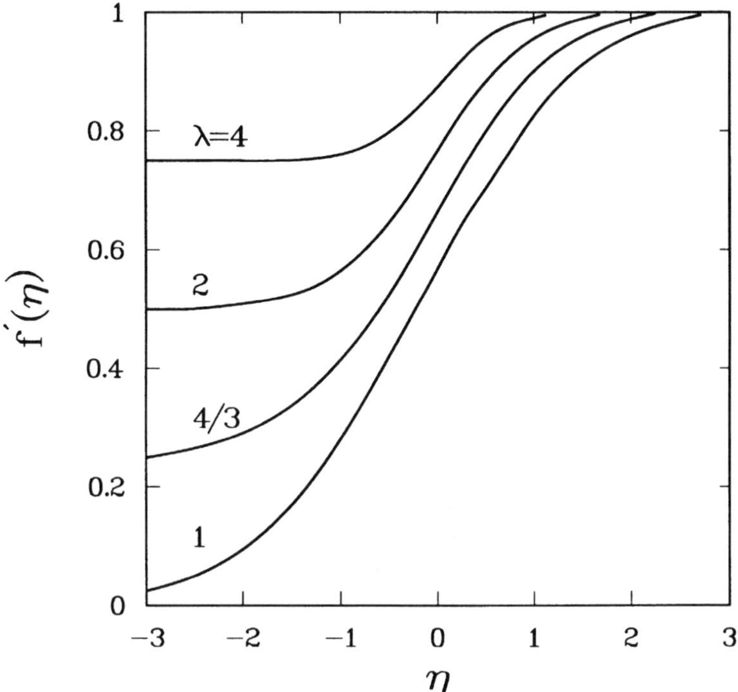

Figure 8.3.2 Mean velocity distribution in turbulent mixing layers.

We now proceed with an analysis based on this picture. Because $U_2 \equiv 0$ and because the wake is suitably thin, the mean temperature is given by

$$\rho\, c_p U_1 \frac{\partial}{\partial x_1} \Theta = -\frac{\partial}{\partial x_2} \overline{u_2 \theta} = \lambda_T \frac{\partial^2 \Theta}{\partial x_2^2} \tag{8.4.1}$$

where U_1 is constant and where λ_T is a turbulent thermal conductivity assumed to be independent of the transverse coordinate x_2. Our task is to determine $\Theta(x_1, x_2)$ and $\lambda_T(x_1)$.

A change of independent variable facilitates the analysis. Let

$$\xi = \frac{1}{\rho\, c_p U_1 L_M} \int_{\bar{x}_{1w}}^{\bar{x}_1} dx\, \lambda_T(x) = \xi(\bar{x}_1) \tag{8.4.2}$$

where L_M is the mesh spacing of the grid, $\bar{x}_1 \equiv x_1/L_M$, $d\xi/dx_1 \propto \lambda_T$. In terms of ξ, Eq. (8.4.1) yields the following simple convection–diffusion balance:

$$\frac{\partial \Theta}{\partial \xi} = \frac{\partial^2 \Theta}{\partial \bar{x}_2^2} \tag{8.4.3}$$

where $\bar{x}_2 \equiv x_2/L_M$. Because of its small diameter, we treat the wire as a point source, with the consequence that the solution to Eq. (8.4.3) is

$$\Theta(\xi, \bar{x}_2) = \Theta(\xi, 0) \exp\left(-\frac{\bar{x}_2^2}{4\xi}\right) \tag{8.4.4}$$

where $\Theta(\xi, 0) = A\xi^{-1/2}$ with A arbitrary.

In this problem it is customary to define the wake width in terms of a transverse length scale according to

$$L_{x_2}^2(\xi) \equiv L_M \frac{\int_0^\infty dx\ \Theta(\xi, x)\, x^2}{\int_0^\infty dx\ \Theta(\xi, x)} \tag{8.4.5}$$

When Eq. (8.4.4) is introduced, this yields

$$L_{x_2}^2 = 2L_M{}^2\, \xi \tag{8.4.6}$$

so that information on the distribution of $L_{x_2}^2$ downstream of the wire permits $\xi = \xi(x_1)$ and hence $\lambda_T(x_1)$ to be determined via Eq. (8.4.2).

If the previous analysis of the thermal wake downstream of a cylinder given in Section 8.2 applies, we have a constant value of λ_T, and from Eqs. (8.4.2) and (8.4.6) $L_{x_2}^2 \propto (x_1 - x_{1w})$. However, experimental data (e.g., Uberoi and

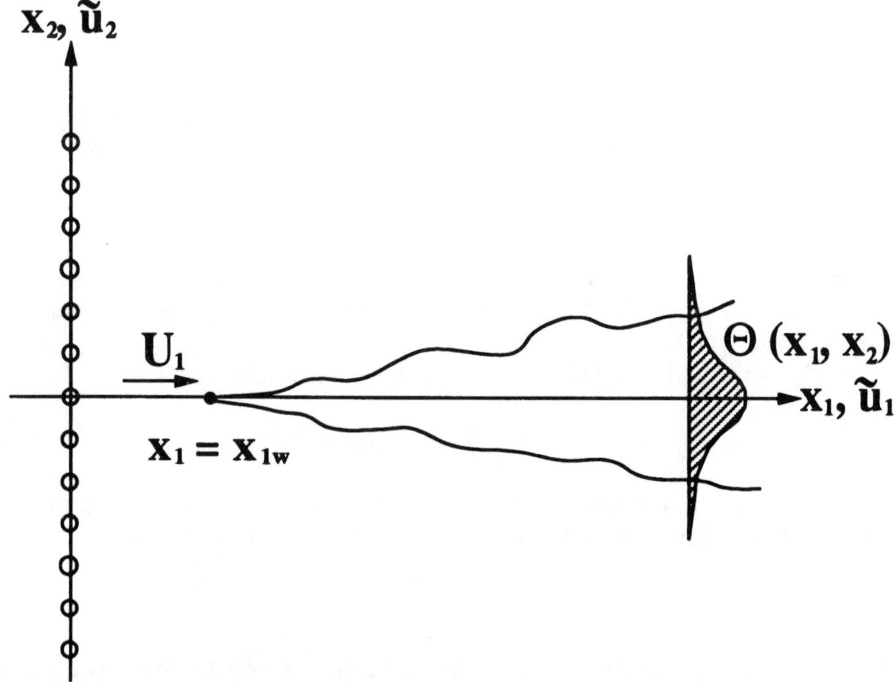

Figure 8.4.1 A heated wire in turbulent grid flow.

Corrsin, 1954; Townsend, 1954) show that $L_{x2}^2 \propto (x_1 - x_{1w})^2$ close to the wire and $L_{x2}^2 \propto (x_1 - x_{1w})$ only far downstream of the wire. This behavior is in accord with our earlier physical reasoning and indicates that λ_T grows initially with $(x_1 - x_{1w})$ before approaching a constant; and further, that L_{x_2} is the correct length scale for the diffusion of heat near the wire while far downstream that length scale is associated with the velocity field and grows with $x_1^{1/2}$.

One simple model which exposes the behavior of the thermal wake downstream of a wire was discussed by Libby and Scragg (1973), who took

$$\lambda_T = \kappa \, \rho \, c_p \, k^{1/2} L_{x2} \tag{8.4.7}$$

where k is the turbulent kinetic energy assumed to be decaying from a virtual origin according to $k^{1/2} = \kappa_0^{-1} x_1^{-1/2}$ with κ_0 an empirical constant. With Eq. (8.4.7), $\xi(x_1)$ and then $\Theta(\xi, \bar{x}_2)$ can be obtained and substituted into Eq. (8.4.7) to yield

$$\frac{L_{x2}^2}{x_{1w} L_M} = \left(\frac{2\,\kappa}{\kappa_0}\right)^2 \left[\left(\frac{x_1 - x_{1w}}{x_{1w}} + 1\right)^{1/2} - 1\right]^2 \tag{8.4.8}$$

Equation (8.4.8) exhibits the near-wire and far-wire characteristics discussed earlier. In Eq. (8.4.8) only the constant κ is available to achieve agreement with experiment. Libby and Scragg (1973) found that $\kappa \approx 1/2$ yields satisfactory agreement with experimental data from several different sources, e.g., from Uberoi and Corrsin (1954) and Townsend (1954). We remark that Anand and Pope (1985) analyzed the dispersion downstream of a line source in grid turbulence in terms of a Langevin equation, i.e., by an entirely different approach to that followed here.

8.5 INTERMITTENCY

All of the statistical quantities discussed in this chapter relate to unconditional averages and disregard the intermittency present in all turbulent free shear flows. In other words, we have ignored the existence in some portions of the flow of two kinds of fluid, turbulent and irrotational. Thus all representations of turbulent flows involving a mean "edge," i.e., a boundary-layer thickness, refer to a mean position of an interface between the two fluids. In reality, a time-resolved photograph of the edge of a shear flow with the turbulent fluid tagged by means of smoke or other tracer shows a highly convoluted surface which separates the two fluids, and which in many turbulent flows oscillates with the large amplitudes characteristic of large-scale fluctuations. Figure 8.5.1 shows a smoke-filled boundary layer illuminated with a sheet of light; the large structures seen there and in most turbulent flows are fully three-dimensional, but under special circumstances they can be two-dimensional (cf. Section 5.8).

Intermittency is most easily understood in terms of the thermal wake of a cylinder, discussed in Section 8.2, with $\tilde{\theta}(\mathbf{x}, t)$ measured relative to the temper-

Figure 8.5.1 Intermittency in a turbulent boundary layer as indicated by smoke. (Courtesy of Robert Falco.)

ature in the external stream. Since momentum is withdrawn from and heat is added to the fluid by the same element, namely, the cylinder, we can associate turbulence with fluid having positive values of $\tilde{\theta}(\mathbf{x}, t)$. Thus, from the output of a temperature sensor with a threshold value $\tilde{\theta}_t$ selected to reflect the accuracy of that sensor and its related instrumentation system, we can construct for a given sensor location within the wake the conditioning function $I(t; \mathbf{x})$ as discussed in Section 3.6, a function with the value unity when $\tilde{\theta} > \tilde{\theta}_t$ and the value zero when $\tilde{\theta} < \tilde{\theta}_t$. It is possible to determine $I(t; \mathbf{x})$ from a careful interpretation of the output from one or more anemometers, i.e., from velocity signals, but the discrimination strategy required to do so is more complex and more subject to ambiguity because of *velocity fluctuations* within the *irrotational fluid.* We discuss these fluctuations later in this section.

Experiment and theory indicate that $\bar{I}(x_1, x_2)$, i.e., the fraction of time that turbulent fluid is present at a point within the far turbulent wake, is well described in terms of the similarity variable of Eq. (8.1.8) by (cf. Libby, 1976)

$$\bar{I}(\xi) = \sin^2\left[\frac{\pi}{2}\exp\left(-\frac{1}{2}\xi^2\right)\right] \tag{8.5.1}$$

where ξ is the wake variable defined by Eq. (8.1.8). The limiting behavior of $\bar{I}(\xi)$ is seen to be

$$\begin{aligned} \bar{I}(\xi \to 0) &= 1 - \frac{\pi^2}{2}\xi^4 \\ \bar{I}(\xi \to \infty) &= \left(\frac{\pi}{2}\right)^2 \exp(-\xi^2) \end{aligned} \tag{8.5.2}$$

According to this representation, the flow is always turbulent on the wake axis and essentially never turbulent at suitably large values of the similarity variable ξ.[†] Similar behavior arises in other turbulent shear flows; for example, in a turbulent boundary layer, the flow at the wall is fully turbulent and irrotational an increasing fraction of the time as locations farther from the wall are considered.

In Section 3.6 we introduced the notion of zone averages. Suppose we consider the zone averages of the temperature within the turbulent fluid in the far turbulent wake. Since the temperature in the irrotational external flow is zero, the zone average of the temperature when $I(t; \mathbf{x}) = 0$ is zero, and in these circumstances there exist simple relations among the unconditional and conditional variables. If we measure the temperature fluctuations from the unconditional mean $\Theta(\mathbf{x})$, then it is easy to show by use of identities such as $\tilde{\theta}(t; \mathbf{x}) \equiv \tilde{\theta}(t; \mathbf{x})I(t; \mathbf{x})$ that

$$
\begin{aligned}
\Theta_1(\mathbf{x}) &= \frac{\Theta(\mathbf{x})}{\bar{I}(\mathbf{x})} \\
\overline{\theta_1^2}(\mathbf{x}) &= \frac{\overline{\theta^2}(\mathbf{x}) - \Theta^2(\mathbf{x})[1 - \bar{I}(\mathbf{x})]}{\bar{I}(\mathbf{x})} \qquad (8.5.3) \\
\overline{\theta_1^3}(\mathbf{x}) &= \frac{\overline{\theta^3}(\mathbf{x} + \Theta^2(\mathbf{x})[1 - \bar{I}(\mathbf{x})]}{\bar{I}(\mathbf{x})}
\end{aligned}
$$

where the subscripts on the quantities on the left side indicate zone averages within turbulent fluid where $I(t; \mathbf{x}) = 1$. We see that in this case knowledge of the intermittency $\bar{I}(t; \mathbf{x})$ and of the unconditional quantities determines the corresponding conditional quantities. Additional relations similar to those of Eqs. (8.5.3) can be readily obtained if needed.

A conceptually important result can be developed from the first of Eqs. (8.5.3). Suppose that we combine this equation with Eqs. (8.5.1) and (8.2.5) and allow $\xi \rightarrow \infty$. We find that

$$
\Theta_1(\xi \rightarrow \infty) = \frac{\Theta(\tilde{x}_1, 0)\exp(-\sigma_T\xi^2)}{(\pi/2)^2\exp(-\xi^2)} \qquad (8.5.4)
$$

Thus the behavior at the outer edge of the far wake of the zone-averaged temperature within turbulent fluid depends on whether σ_T is greater than, equal to, or less than unity. The experimental data of LaRue and Libby (1974) show that these zone averages are constant with $\Theta_1(\xi \rightarrow \infty)/\Theta(\mathbf{x}_1, 0) \approx 0.4$. The implication

[†]In some wake flows with low turbulence intensities, irrotational fluid may occasionally reach the wake axis so that $\bar{I}(0)$ is slightly less than unity; because of the problems of statistical reliability, however, the accuracy of presently available data of $\bar{I}(0)$ is poor. Moreover, in turbulent boundary layers subject to strong favorable pressure gradients, irrotational fluid can reach the wall. Indeed, under the influence of suitably strong favorable pressure gradients, a turbulent boundary layer can be relaminarized so that $\bar{I}(\mathbf{x}, t)$ is identically zero.

of this result is that the approach of the unconditional mean temperature to zero as $\xi \to \infty$ is a consequence of the diminished appearance of turbulent fluid with a constant temperature, namely, the zone-averaged temperature. If we are to achieve agreement between Eq. (8.5.4) and such data, we must set $\sigma_T = 1$ and relinquish the selection of σ_T to achieve agreement with respect to other variables.

It is important to appreciate that the experimental determination of $\Theta_1(\xi \to \infty)$ involves measurement of two quantities, each approaching zero as locations more remote from the wake centerline are considered, and thus each subject to considerable statistical uncertainty.

The situation with respect to zone averages of the velocity components is more complicated because of the existence of velocity fluctuations within the irrotational fluid, fluctuations which themselves are irrotational and due to the pressure fluctuations associated with the turbulence.† For our limited purposes it is sufficient to note that in the far wake the turbulent zone average of the streamwise velocity component defined by

$$U_{11}(\mathbf{x}) = \frac{\lim\limits_{T\to\infty} (1/T) \int_0^T dt\, \tilde{u}_1(t; \mathbf{x})\, I(t; \mathbf{x})}{\bar{I}(\mathbf{x})} \tag{8.5.5}$$

is such that

$$U_{11}(\mathbf{x}) \leq U_1(\mathbf{x}) < U_\infty$$

i.e., the turbulent fluid has an average velocity less than the unconditional velocity and thus less than the external velocity. Because of the experimental difficulties suggested earlier, the velocity decrement $[U_\infty - U_{11}(\xi \to \infty)]/U_\infty$ is uncertain, but values within the range 0.05–0.10 are indicated. It is this relatively small decrement in the streamwise velocity and the relatively large increment of the temperature which accounts for the apparent greater width of the thermal wake relative to the momentum wake, as discussed in Section 8.2. The zone average of the streamwise velocity in the irrotational fluid at a location in the wake involving significant intermittency is less than U_∞. In the case of a jet in a moving stream, the turbulent fluid has a higher streamwise velocity than both the unconditional mean and the external velocity, while the irrotational fluid at locations involving significant intermittency is higher than U_∞ (cf. Anderson et al., 1979).

†It is these fluctuations which make more difficult discrimination between the two fluids on the basis of the output from an anemometer [cf. Bradshaw (1967a) for a discussion of the fluctuations in the irrotational fluid]. In Section 8.6 we discuss the structure of the layer between the rotational and irrotational fluids, the layer within which the temperature increase occurs in the case of a heated wake.

Although we have discussed intermittency within the context of free shear flows, wall-bounded flows such as boundary layers involve intermittency at their edge with the external stream. In this case the distribution of $\bar{I}$ normal to the wall depends on the streamwise pressure gradient.

8.6 THE SUPERLAYER

Although in creating a 0–1 function $I(t; \mathbf{x})$ from a temperature or other signal we have implicitly assumed a discontinuous interface between the two fluids, reality dictates that such a surface must possess a thickness and a structure. The interface is identified as the *superlayer* by Corrsin and Kistler (1954), in contrast with the molecularly dominated layer adjacent to a wall discussed in Section 9.3 and subsequent sections—i.e., the viscous sublayer.

The superlayer can be associated with changes in several fluid mechanical quantities which differ in the two fluids. In the case of the turbulent wake of a heated cylinder there is a temperature change as well as a change in vorticity. In general, we can identify separate structures for these changes, but intuition, experiment, and theory suggest that the thinner of the two superlayers is embedded in the other. We first discuss the vorticity superlayer.

The structure of the layer in which changes in vorticity take place depends on the local rate of strain to which the surface is subjected at a particular time. This can be understood from the following. The velocity relative to a fixed laboratory frame of reference in the neighborhood of a small but arbitrary area of the interface at a particular time involves a translation velocity, a rotation, and a rate of strain. The first two of these motions cannot influence the *structure* of the layer, but the latter does so. Various models of the rate of strain can be assumed in order to analyze the nature of the transition from irrotational fluid on one side of the interface to a vortical flow on the other. The most frequently adopted description involves a compressive rate of strain normal to the surface, i.e., a normal velocity proportional to $a\, n$, where n is the coordinate normal to the area being examined and a is a rate-of-strain parameter with the dimensions of reciprocal time. Accompanying this compressive rate of strain is an extension in one coordinate direction in the plane of the surface and either a compression or an extension in the second direction within the plane. Analysis shows that the thickness of the surface is proportional to $(\nu/\mathrm{a})^{1/2}$, which we can express in more familiar terms according to the following argument. In any flow the vorticity superlayer is subjected to a distribution of rates of strain and therefore possesses a distribution of thicknesses. However, the viscous dissipation ε [cf. Eq. (4.7.1)] in the region of the superlayer with the specified rate-of-strain field can be related by dimensional arguments to the parameter a by $(\varepsilon/\nu)^{1/2}$ (cf. Gibson and Libby, 1972). If we eliminate ε by means of the Kolmogorov length L_K [cf. Eq. (5.3.13)], we find that the mean value of $(\nu/a)^{1/2}$ is L_K; that is, the vorticity superlayer is of the order of a Kolmogorov length, a result which is expected if the superlayer is considered a priori to be dominated by viscous stresses.†

A similar analysis can be carried out for the temperature superlayer. In this case the molecular Prandtl number enters as another parameter and we conclude that the thickness of the temperature superlayer is of the order of the Batchelor length [cf. Eq. (5.7.9)]. Thus, if the Prandtl number is close to unity, the thicknesses of the two superlayers are essentially equal; while if the Prandtl number is large compared with unity, the change in temperature occurs in a layer that is fully embedded in a relatively more extended vorticity layer. On the other hand, if the Prandtl number is small compared to unity, the change in vorticity occurs in a layer that is embedded in a relatively thick temperature superlayer. Measurements of the statistical properties of the superlayer of the heated wake of a cylinder obtained from a fixed thermometer are given by LaRue and Libby (1976).

Although we focus on the nature of the superlayer between the irrotational and turbulent fluids, similar considerations apply to the interface between fluids which are dissimilar in other ways, e.g., between two turbulent fluids of different temperatures. The thermal mixing layer downstream of a turbulence grid with the horizontal rods in the upper half of the tunnel raised to an elevated temperature involves such a temperature interface (cf. LaRue and Libby, 1981).

8.7 SUMMARY

This chapter deals with a class of turbulent flows that is devoid of walls, i.e., jets, wakes, mixing layers, and the like. These flows are easier to treat than those involving walls, since if the turbulence is suitably vigorous, as we assume, viscous and thermal conductivity effects are everywhere negligible. We discuss the statistics of both the velocity and thermal fields. The special case of a small heated wire in a turbulent flow is analyzed; since it involves different length scales for the velocity and temperature fluctuations, the notion of a turbulent Prandtl number is inapplicable to this flow. In connection with the intermittency that is always present in these flows, we discuss the superlayer, the thin interface between turbulent and external, irrotational fluid dominated by viscosity and thermal conductivity.

†We shall see later that the *viscous sublayer* of a turbulent boundary layer also possesses a thickness of the order of L_K, but since for a particular boundary-layer flow the viscous dissipation is considerably higher in the sublayer, its thickness is *considerably less* than that of the corresponding superlayer.

CHAPTER
NINE

WALL-BOUNDED FLOWS

A second class of turbulent shear flows involves one or more solid surfaces and thus wall layers in which the stresses and fluxes due to viscosity and thermal conductivity cannot be neglected no matter how large the turbulent Reynolds numbers of the flow away from the walls. Many flows of practical importance are in this class; examples are flows in pipes, channels, and over wings and bodies in flight and around ships. In this chapter we focus on representative examples which contain the essential features of this class of flows: the symmetric flow in a two-dimensional channel, turbulent Couette flow, pipe flow, and the two-dimensional turbulent boundary layer. The chapter concludes with brief discussions of three-dimensional boundary layers in turbulent flow and the influence of wall curvature on turbulent boundary layers.

9.1 VELOCITY IN TWO-DIMENSIONAL CHANNEL FLOW: GENERAL CONSIDERATIONS

Although we later consider the temperature distribution in a channel, the temperature changes and the size of the channel are taken to be sufficiently small and the velocities sufficiently high so that body forces may be neglected; thus we may consider first the velocity field and subsequently the temperature field. Figure 9.1.1 shows a fully developed turbulent channel flow in which all statis-

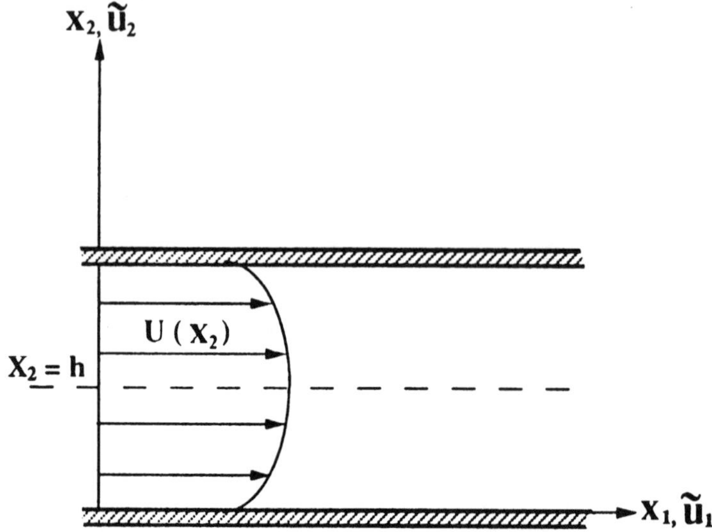

Figure 9.1.1 Turbulent flow in a channel.

tical characteristics of the velocity depend only on the x_2 coordinate. The implication is that we deal with the flow far downstream of the entrance region. Furthermore, we restrict attention to channels with both walls smooth. If the surface finish of the two walls is different—one rough and one smooth, for example—the flow is asymmetric, a situation excluded from our considerations (cf. Hanjalic and Launder, 1972a).

The mean flow is taken to be two-dimensional so that the gradient of all mean quantities in the x_3 direction is zero and $U_3 \equiv 0$. Body forces are neglected, since their inclusion adds nothing essential to our description. Continuity, Eq. (4.1.1), and the fully developed nature of the flow then imply that $U_2 \equiv 0$ and the two momentum equations from Eq. (4.2.1) for $i = 1, 2$ become

$$0 = -\frac{\partial P}{\partial x_1} + \frac{d}{dx_2}\left(\mu \frac{dU_1}{dx_2} - \rho\, \overline{u_1 u_2}\right)$$
$$0 = -\frac{\partial}{\partial x_2}(P + \rho\, \overline{u_2^2}) \qquad (9.1.1)$$

When the second of these equations is integrated with respect to x_2, we find that the mean pressure must be a linear combination of a function of x_1, denoted $P_1(x_1)$, and a function of x_2 equal to $\rho\, \overline{u_2^2}$. Only the former function is of concern here, since the second depends only on x_2 and does not contribute to the $\partial P/\partial x_1$ term in the first of Eqs. (9.1.1); when $P_1(x_1)$ is substituted into that equation, we obtain a result that is analogous to that for fully developed laminar flow, namely, that the mean streamwise pressure gradient in a channel must be constant.

If we integrate the first of Eqs. (9.1.1) with respect to x_2 and impose boundary conditions at $x_2 = 0$, we obtain

$$-\frac{dP_1}{dx_1} x_2 + \mu \frac{dU_1}{dx_2} - \rho\, \overline{u_1 u_2} = \rho\, u_\tau^2 \tag{9.1.2}$$

where $u_\tau \equiv (\tau_w/\rho)^{1/2}$ is the shearing velocity, a convenient replacement for the shear stress τ_w; it arises as a constant of integration and is to be determined as part of the solution. If Eq. (9.1.2) is applied at $x_2 = h$, we obtain

$$-\frac{dP_1}{dx_1} h = \rho\, u_\tau^2 \tag{9.1.3}$$

which suggests a method for determining the shearing velocity from measurements of the distribution of mean pressure along either wall of the channel.

If we nondimensionalize the coordinate x_2 with the channel half-height h and define a new independent variable $\eta \equiv x_2/h$, we obtain

$$\frac{\nu}{h}\frac{dU_1}{d\eta} - \overline{u_1 u_2} = u_\tau^2\,(1 - \eta) \tag{9.1.4}$$

Equation (9.1.4) is central to our analysis of the velocity for fully developed turbulent flow in a channel and forms the basis of several considerations in this chapter and in Sections 10.5 and 10.6.

9.2 THE OUTER FLOW AND THE DEFECT LAW

Our analysis will be facilitated if we recognize that the values of the Reynolds number $u_\tau h/\nu$ of applied interest are large, a number which appears naturally when Eq. (9.1.4) is nondimensionalized further. Typical values are in the range from 50 to 500 so that $\delta \equiv \nu/u_\tau h << 1$. Under these circumstances we are tempted to set the viscous term in Eq. (9.1.4), i.e., the first term on the left side, to zero. The immediate consequence of doing so is the important intermediate result that

$$-\overline{u_1 u_2} = u_\tau^2\,(1 - \eta) \tag{9.2.1}$$

The implication from this equation is that the Reynolds shear stress is distributed linearly across the channel wherever the viscous shear stress is negligible, a result that is free of turbulence modeling. However, we know that this neglect is not valid near the walls of the channel, i.e., that the approximation resulting in Eq. (9.2.1) is not uniformly valid. The degeneration of our basic equation, Eq. (9.1.4), to the algebraic Eq. (9.2.1) when the viscous term is neglected suggests an asymptotic analysis which involves a two-scale description of the flow: an outer flow possessing negligible viscous stresses and a thickness slightly less than $2h$ and thin inner flows near the walls. The thickness of these layers is proportional to the length ν/u_τ, which arises naturally from the only two physical quantities associated with the wall layer.

To make further progress we introduce empiricism, i.e., turbulence modeling, to express the Reynolds shear stress in terms of U_1. As indicated in Chapter 7, various possibilities exist. Here we employ mixing-length theory with an assumed distribution of that length leading to results in accord with experiment. Thus, from Eq. (7.4.5) we have

$$-\overline{u_1 u_2} = \tilde{l}^2 \left(\frac{dU_1}{d\eta}\right)^2 \tag{9.2.2}$$

with

$$\tilde{l} \equiv \frac{l}{h} = \tilde{l}(\eta) \tag{9.2.3}$$

a dimensionless mixing length considered a known function of the indicated variable and independent of δ.

If Eqs. (9.2.2) and (9.2.3) are substituted into Eq. (9.1.4) and we introduce the dimensionless mean velocity $\tilde{U} \equiv U_1/u_\tau$, we obtain†

$$\delta\tilde{U}' + \tilde{l}^2\,\tilde{U}'^2 = (1 - \eta) \tag{9.2.4}$$

where prime denotes differentiation with respect to η. If $\tilde{l}(\eta)$ is known and δ assigned a specific value, Eq. (9.2.4) can be solved by quadrature from $\eta = 0$ with $\tilde{U}(0) = 0$. There results from such a solution a value of $\tilde{U}(1) = U_1(1)/u_\tau$ such that repeated integrations for a range of δ yield a skin friction law of the form $U_1(1)/u_\tau$ versus the Reynolds number $u_\tau h/\nu$; that is, for specified values of $U_1(1)$, h, and ν, we can determine the shearing velocity u_τ and from Eq. (9.1.3) the associated mean pressure gradient. A skin friction coefficient can be determined from

$$c_f \equiv 2\,\frac{\tau_w}{\rho U_1(1)^2} = 2\left(\frac{u_\tau}{U_1(1)}\right)^2 = c_f(\delta) \tag{9.2.5}$$

From this perspective it is useful to consider solutions of Eq. (9.2.4) to be of the form $\tilde{U}(\eta; \delta)$.

An alternative, simpler, and *more illuminating* approach is based on the small magnitude of δ, that is, on $\delta << 1$. In this case we follow standard practice in asymptotic analysis and seek an outer solution by expanding $\tilde{U}(\eta; \delta)$ in powers of δ; that is, we let

†Alternatively, we could use $U_1(1)$ to nondimensionalize the velocities, but the shearing velocity is more convenient.

$$\tilde{U}(\eta; \delta) \approx {}^{(0)}\tilde{U}(\eta) + \delta\ {}^{(1)}\tilde{U}(\eta) + \ldots \tag{9.2.6}$$

where the superscripts identify the various terms in the series expansion. From Eq. (9.2.4) we obtain for the first two terms

$$\begin{aligned} \tilde{l}^2\ {}^{(0)}\tilde{U}'^2 &= (1 - \eta) \\ 1 + 2\ \tilde{l}^2\ {}^{(1)}\tilde{U}' &= 0 \end{aligned} \tag{9.2.7}$$

Several comments are appropriate. Note that the molecular term in Eq. (9.2.4) does not appear in the first of Eqs. (9.2.7) but becomes unity in the second. Note also that if Eq. (9.2.2) is substituted into Eq. (9.2.1), we obtain the first of Eqs. (9.2.7) directly. The advantage of our development is that additional terms in δ, as many as we please, are obtained systematically.

Since we expect the outer solutions to be inapplicable at the wall, where viscous stresses are important, we integrate the first of Eqs. (9.2.7) from $\eta = 1$ to obtain

$${}^{(0)}\tilde{U}(\eta) = {}^{(0)}\tilde{U}(1) - \int_\eta^1 d\eta'\ \frac{(1 - \eta')^{1/2}}{\tilde{l}(\eta')}$$

We thus see that to lowest order in δ the difference between the mean velocity on the channel centerline and at an arbitrary location away from the wall is known, provided the mixing length distribution is given. This equation can be rearranged into the following, more familiar form:

$$\frac{U_1(1) - U_1(\eta)}{u_\tau} = \int_\eta^1 d\eta'\ \frac{(1 - \eta')^{1/2}}{\tilde{l}(\eta')} \tag{9.2.8}$$

where in this context U_1 denotes the mean velocity in the x_1 direction for a suitably large turbulence Reynolds number. This equation is the velocity defect law, widely used for the presentation of experimental data in those portions of channel flow involving negligible molecular stresses. Suppose that the shearing velocity u_τ is determined from a pressure drop measurement and the mean velocity distribution $U_1(\eta)$ is obtained from suitable anemometry. Then the defect law indicates that the velocity difference $[U_1(1) - U_1(\eta)]$ divided by the shearing velocity is independent of the Reynolds number $u_\tau(1)h/\nu$, a result that is in accord with experiments involving suitably high Reynolds numbers and supporting the assumption that $\tilde{l}(\eta)$ is independent of δ.

Although we derive the velocity defect law for a specific turbulence model, its general functional form can be obtained from dimensional arguments. If molecular stresses are negligible, it is reasonable to assume that the difference between the mean velocity on the centerline and at a general location within the channel depends on that location, i.e., on x_2; on the channel half-width; and on the shear at the wall, i.e., on the shearing velocity, u_τ. Since these assumptions imply that a relation $F\,[U_1(x_2 = h) - U_1(x_2), x_2, h, u_\tau] = 0$ exists, dimensional analysis leads to the result that

$$\frac{U_1(1) - U_1(\eta)}{u_\tau} = N(\eta) \tag{9.2.9}$$

where $N(\eta)$ is an unknown function but which, with our a-priori assumption of a turbulence model, is given by Eq. (9.2.8). If we assume that the kinematic viscosity also influences the velocity defect, an extended version of dimensional analysis implies that the right side of Eq. (9.2.9) depends on η and $U_1(1)h/\nu$. In this case the assumption of a turbulence model combined with our asymptotic analysis involving the calculation of ${}^{(1)}\tilde{U}_1(\eta)$, that is, the second term in the expansion of Eq. (9.2.5), again yields an explicit form for the right side as the sum of two terms in Eq. (9.2.6).

We now return to the second of Eqs. (9.2.7), i.e., to the first-order correction to the velocity defect law; its solution is

$${}^{(1)}\tilde{U}(\eta) = \frac{1}{2}\int_\eta^1 \frac{d\eta'}{\tilde{l}^2(\eta')} \tag{9.2.10}$$

Although the analysis of additional terms in the expansion in δ can be carried through without difficulty, our purposes are achieved by limiting attention to the first term, and we therefore do not consider Eq. (9.2.10) further.

Except in the immediate neighborhood of the wall, where a special form is required, we let the distribution of the mixing length be

$$\tilde{l} = \kappa_0[1 - 2(1 - \eta)^2 + (1 - \eta)^4] + \frac{1}{2}\kappa\,(1 - \eta)^2\,[1 - (1 - \eta)^2] \tag{9.2.11}$$

where κ is the von Karman constant, taken here to be 0.40, and κ_0 is a second empirical coefficient whose value is selected to achieve agreement with experimental data relative to the outer flow, i.e., the flow away from the wall. Equation (9.2.11) provides a symmetric mixing-length distribution and leads to the following important behavior as $\eta \to 0$:

$$\tilde{l} \approx \kappa\,\eta + O(\eta^2) \tag{9.2.12}$$

The significance of this result is seen if we use Eq. (9.2.12) in Eq. (9.2.8) to find

$$\begin{aligned} {}^{(0)}\tilde{U}(\eta \to 0) &= {}^{(0)}\tilde{U}(1) + \frac{1}{\kappa}\ln\eta - \left[\int_{\eta^*}^1 d\eta\,\frac{(1-\eta)^{1/2}}{\tilde{l}(\eta)} + \frac{1}{\kappa}\ln\eta^*\right] \\ &= {}^{(0)}\tilde{U}(1) + \left(\frac{1}{\kappa}\ln\eta + C\right) \end{aligned} \tag{9.2.13}$$

where $\eta^* << 1$ is chosen so that the bracketed term on the right side of the first of these equations is essentially independent of η^* and is related in an obvious way to the parameter C.[†] Thus C depends only on the mixing-length distribution $\tilde{l}(\tilde{\eta})$, but κ_0 is at our disposal so that C agrees with experiment. The logarithmic dependence of the mean velocity on η in Eq. (9.2.13) is a manifestation of the singular behavior of the outer solution as the wall is approached. This behavior is a consequence of the neglect of viscous stresses, a neglect which, as has been noted repeatedly, is inapplicable to the wall region.

The mixing-length theory is reasonably tolerant of the assumed distributions of $l(\mathbf{x})$ in wall-bounded flows provided that $l \propto x_2$ near the wall. Thus, for example, in Section 9.4 we employ a considerably simpler distribution than that of Eq. (9.2.11), namely, a linear increase κx_2 away from the wall to a fixed value l_∞, which in the case of the channel would prevail to the centerline. With l_∞ properly chosen, the velocity profile ${}^{(0)}\tilde{U}(\eta)$ would be little altered. If we provisionally take $\kappa_0 = 0.14$ as suggested by later developments related to pipe flow and substitute Eq. (9.2.11) into Eq. (9.2.8), we obtain the results shown in Figs. 9.2.1 and 9.2.2. These have different emphases: In the first the distribution remote from the wall is well resolved, while in the second the distribution near the wall is emphasized. Indeed, in the latter the logarithmic distribution is exposed. As will be noted later, the value of κ_0 used here yields satisfactory agreement between predicted and measured skin friction and a value for the mixing length in the central portion of the channel of roughly one-third the channel width $2h$ as suggested earlier. The associated value of C is -0.75.

9.3 THE WALL LAYER, THE LAW OF THE WALL, AND THE SKIN FRICTION LAW

If the flow close to the wall is to be resolved, we must introduce a new independent variable which expands η by an increasing amount as δ decreases, since the derivative U_1' must increase if the product $\delta U_1'$ which describes the molecular shear stress in Eq. (9.2.4) is to remain significant as $\delta \to 0$. Accordingly, we let

$$\hat{\eta} \equiv \frac{\eta}{\delta} \tag{9.3.1}$$

Several comments are indicated. Note that the variable $\hat{\eta}$ is frequently denoted in the turbulence literature as y^+ because $\hat{\eta} = x_2/(h\,\delta) = u_\tau x_2/\nu \equiv y^+$, a result which confirms our earlier observation that u_τ and ν are the two quantities characterizing the wall layer. We now exploit that observation further. Dimen-

[†]This splitting of an integration range to determine limiting behavior of integrals is used repeatedly in this chapter.

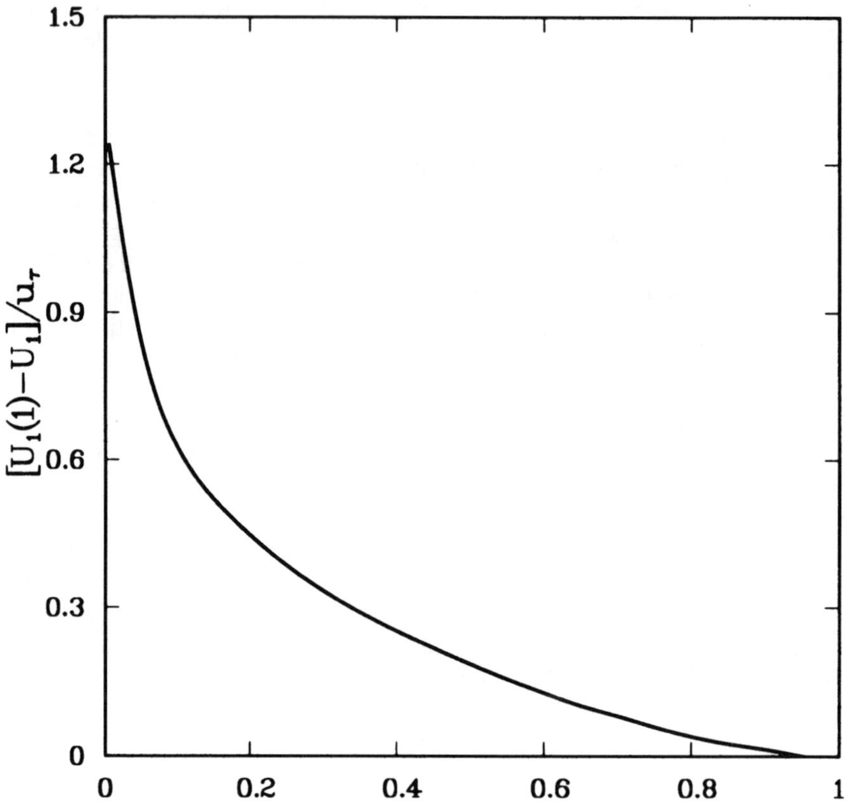

Figure 9.2.1 The velocity defect in a turbulent channel flow, emphasizing the region that is remote from the wall.

sional arguments imply that the the mean viscous dissipation, which has the dimensions of U^3/L, must be proportional to $u_\tau^3/(\nu/u_\tau) = u_\tau^4/\nu$; but from Eq. (5.3.12) we know that the Kolmogorov length $L_K \equiv (\nu^3/\varepsilon)^{1/4}$, so in the inner layer $L_K = \nu/u_\tau$. Thus the coordinate $\hat{\eta}$ that is applicable to the wall layer is a multiple of the Kolmogorov length.

Although not essential, it is convenient to identify as the dependent variable in the wall layer $\hat{U} \equiv U_1/u_\tau = \tilde{U}$. In terms of these new variables, Eq. (9.1.4) becomes

$$\hat{U}' + \hat{l}^2 \hat{U}'^2 = 1 - \delta\hat{\eta} \tag{9.3.2}$$

where

$$\hat{l} \equiv \frac{\hat{l}}{\delta} = \hat{l}(\hat{\eta}) \tag{9.3.3}$$

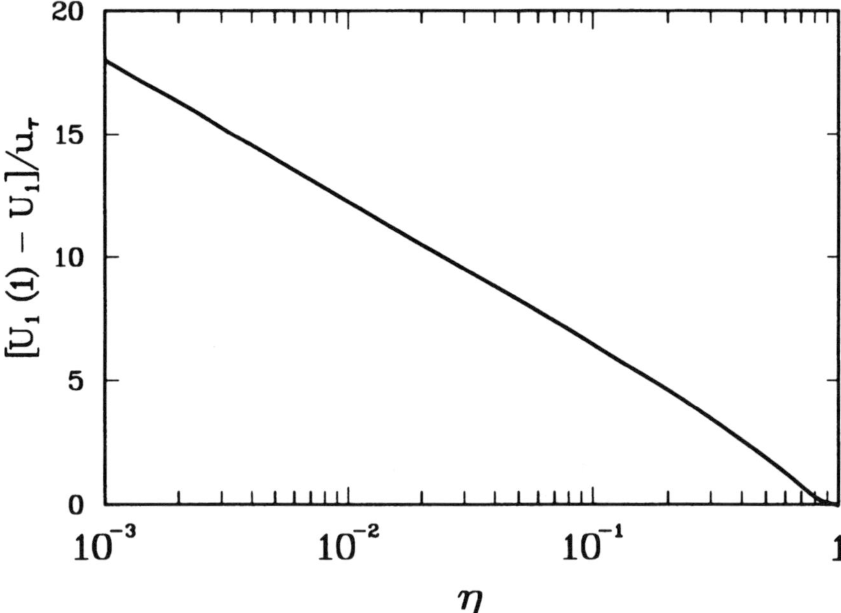

Figure 9.2.2 The velocity defect in a turbulent channel flow, emphasizing the wall region.

is a scaled mixing length. The need to stretch the mixing length in the wall layer is suggested by both physical and mathematical considerations; as the wall is approached we expect the length scales of the turbulence to go to zero, but the second term on the left side of Eq. (9.3.2) must be retained since it describes the turbulent shear stress which is dominant at the *outer edge* of the wall layer. Equation (9.3.1) and this treatment of the mixing length are needed so that *both* viscous and turbulent transport can prevail and can exchange dominance within the wall layer; at the wall only viscous stresses exist, but at the outer edge of the wall layer only turbulent stresses exist. If we refer to Eq. (7.4.6), which relates the ratio of turbulent to viscous shear stresses and the turbulence Reynolds number, we can appreciate that both k and l go to zero as the wall is approached, so the former stress goes to zero. On the other hand, at the outer edge of the wall layer, when the global Reynolds number is large, the turbulence Reynolds number is large and the viscous stresses are small.

Again expanding in a series in δ, we obtain for the first two terms of the inner solution

$$\begin{aligned} {}^{(0)}\hat{U}' + \hat{l}^2\, {}^{(0)}\hat{U}'^2 &= 1 \\ {}^{(1)}\hat{U}' + 2\,\hat{l}^2\, {}^{(0)}\hat{U}'\, {}^{(1)}\hat{U}' &= -\hat{\eta} \end{aligned} \tag{9.3.4}$$

Since Eqs. (9.3.4) apply close to the wall, we impose the boundary conditions

$$^{(0)}\hat{U}(0) = {}^{(1)}\hat{U}(0) = 0 \tag{9.3.5}$$

The solutions are

$$\begin{aligned} ^{(0)}\hat{U}(\hat{\eta}) &= \frac{1}{2}\int_0^{\hat{\eta}} d\hat{\eta}' \, \frac{[1 + 4\,\hat{l}^2(\hat{\eta}')]^{1/2} - 1}{\hat{l}^2(\hat{\eta}')} \\ ^{(1)}\hat{U}(\hat{\eta}) &= -\int_0^{\hat{\eta}} d\hat{\eta}' \, \frac{\hat{\eta}'}{1 + 2\,\hat{U}_0'(\hat{\eta}')\,\hat{l}^2(\hat{\eta}')} \end{aligned} \tag{9.3.6}$$

For our purposes we need not pursue the $^{(1)}\hat{U}$ solutions further. It is important to note that the second term on the right side of Eq. (9.3.2), a term describing the influence of the pressure gradient, does not appear in the lowest-order description of the mean velocity within the wall layer. The implication is that at suitably large Reynolds numbers so that $\delta << 1$, the behavior of the wall layer is independent of the pressure gradient. The first of Eqs. (9.3.4) implies that the sum of the viscous and turbulent shear stresses is constant. Thus the wall layer involves a constant shear stress, zero force with distributions across its thickness of molecular and turbulent shear stresses which add to a constant.

Independent of the form for $\hat{l}(\hat{\eta})$ provided only that $\hat{l}(\hat{\eta} \to 0) \to 0$, the first of Eqs. (9.3.6) yields

$$^{(0)}\hat{U}(\hat{\eta} \to 0) = \hat{\eta} \tag{9.3.7}$$

which implies that, as expected, molecular stresses dominate in the immediate neighborhood of the wall. Again independent of the form of $\hat{l}(\hat{\eta})$ but provided that $\hat{l}(\hat{\eta} \to \infty) \to \kappa\hat{\eta}$, a requirement consistent with the behavior $\tilde{l}(\eta \to 0) \to \kappa\eta$,[†] the first of Eqs. (9.3.6) yields

$$^{(0)}\hat{U}(\hat{\eta} \to \infty) = \frac{1}{\kappa}\ln\hat{\eta} + \left[\frac{1}{2}\int_0^{\hat{\eta}^*} d\hat{\eta}\, \frac{(1 + 4\,\hat{l}^2(\hat{\eta}))^{1/2} - 1}{\hat{l}^2(\hat{\eta})} - \frac{1}{\kappa}\ln\hat{\eta}^*\right] \tag{9.3.8}$$

where $\hat{\eta}^* >> 1$ is selected so that the term in brackets, a quantity we denote as B in accord with tradition, is independent of $\hat{\eta}^*$. Experimental data relative to B or its equivalent in other flows in which it arises are not unambiguous, yielding values between 4 and 6, but specifically for channel flows the value $B = 5.0$ is representative and is adopted here. To evaluate the integral arising in the definition of B we need a form for $\hat{l}(\hat{\eta})$, one containing at least one empirical coefficient to be adjusted so that the desired value of B is obtained.

In view of the restrictions already imposed on $\hat{l}(\hat{\eta})$, consider the form

$$\hat{l}(\hat{\eta}) = \kappa\,\hat{\eta}\left[1 - \exp\left(-\frac{\hat{\eta}}{\kappa_w}\right)\right] \tag{9.3.9}$$

[†]The perspective here follows from the nature of an asymptotic analysis, namely, that with η assigned a small, nonzero value such that $\tilde{l} \approx \kappa\eta$ and with $\delta \to 0$, then $\hat{\eta} \to \infty$.

where κ_w is the requisite empirical coefficient and the exponential term is the van Driest damping factor. For $\hat{\eta} >> \kappa_w$ this term is negligible and we recover the desired asymptotic behavior of $\hat{l}(\hat{\eta} \to \infty)$, but $\hat{l}(\hat{\eta} \to 0) \to (\kappa/\kappa_w)\hat{\eta}^2$, implying that the viscous stresses become rapidly dominant as the wall is approached.

Figure 9.3.1 shows the distribution of ${}^{(0)}\hat{U}(\hat{\eta})$ interpreted as $U_1(\hat{\eta})/u_\tau$ with $\kappa_w = 24.8$, a value that yields $B = 5.0$.‡ This figure indicates the limiting behavior for small and large $\hat{\eta}$, behavior which we discuss in more detail later. Also shown is the ratio of the first and second terms on the left side of Eq. (9.3.4); this ratio is a measure of the relative importance of the viscous and turbulent shear stresses. We see that for $\hat{\eta} > 15$ the viscous stress is negligible, while for $\hat{\eta} < 5$, it is dominant. This figure has important implications. It is a graphical presentation of the law of the wall consisting of a *viscous sublayer* in which $U_1/u_\tau \approx \hat{\eta}$ and a *logarithmic portion,* a division which suggests an idealization of the flow near a wall as follows. The logarithmic portion can be written as

$$\frac{U_1(\hat{\eta})}{u_\tau} = \frac{1}{\kappa} \ln \hat{\eta} + B \tag{9.3.10}$$

The linear variation prevails for $\hat{\eta}$ less than the intersection point of the two approximations, a point $\hat{\eta} = 11$ for the values of the empirical coefficients κ and B used here. The more complete description, of course, involves the smooth transition between these two simple distributions.

‡The value $\kappa_w = 26$ is frequently quoted, but our analysis calls for a precise criterion for its selection and leads to a slightly different value if $B = 5.0$.

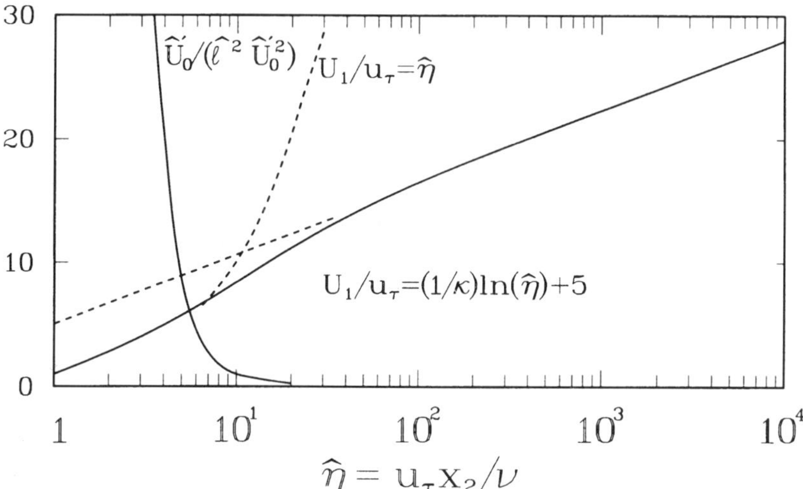

Figure 9.3.1 The mean velocity distribution in the wall layer.

Equation (9.3.10) can be derived in a variety of ways, including by direct dimensional and functional considerations involving neither the conservation and transport equations of fluid mechanics nor turbulence modeling. In this regard it should be noted that the only feature of the mixing-length distribution that is significant for its development is that $l \approx \kappa x_2$ for $\hat{\eta} >> 1$. The logarithmic law represents the outer portion of the wall layer and is widely used in the presentation of experimental data, not only for channel flows, for which the present derivation applies, but for boundary-layer and pipe flows as well. It provides a useful means of obtaining the wall shear from a mean velocity profile measured somewhat away from a wall, since it is difficult to obtain reliable values for the mean velocity within the viscous sublayer. Suppose that experimental data in the form $U_1(x_2)$ are obtained and that u_τ is *estimated.* Then, following Eq. (9.3.10), U_1/u_τ is displayed as a function of $\ln \hat{\eta}$, where $\hat{\eta} = u_\tau x_2/\nu$. If such data fail to agree with Eq. (9.3.10), the value of u_τ is altered until agreement is achieved. Since the pressure gradient term does not enter the determination of ${}^{(0)}\hat{U}(\hat{\eta})$, the law of the wall applies with and without streamwise pressure gradients, provided the Reynolds number is sufficiently high.

To complete the analysis we must demonstrate that the two sets of solutions, those describing the outer and inner flows, are consistent with one another, i.e., that they match. The matching procedure involves approximating the outer solution in the neighborhood of $\eta = 0$ and writing that approximation in terms of the inner variable $\hat{\eta}$. It is then required that for each order in δ the outer and inner solutions be consistent as $\hat{\eta} \to \infty$. Restricting attention to ${}^{(0)}\tilde{U}(\eta)$ and ${}^{(0)}\hat{U}(\hat{\eta})$, i.e., to Eqs. (9.2.13) and (9.3.8), this procedure leads to dominant terms in each solution, namely, to $\kappa^{-1} \ln \hat{\eta}$, which *cancel,* and to remainder terms that yield

$$B = \frac{U_1(1)}{u_\tau} + \frac{1}{\kappa} \ln \delta + C \qquad (9.3.11)$$

Equation (9.3.11) is interpreted as a skin friction law relating either $U_1(1)/u_\tau$ or, as noted in Eq. (9.2.5), a skin friction coefficient to the Reynolds number $u_\tau h/\nu$.

Equation (9.3.11) is conveniently interpreted as follows. With a value of δ arbitrarily assumed, we find $U_1(1)/u_\tau$ from this equation and c_f from Eq. (9.2.5). Then the Reynolds number $U_1(1)h/\nu = (U_1(1)/u_\tau)\delta^{-1}$. The skin friction results obtained in this fashion with the values $B = 5.0$ and $C = -0.75$ found earlier are in good agreement with experimental data, e.g., those of Comte-Bellot (1963), with $\kappa_0 = 0.14$, but these results are relatively insensitive to other values thereof in the range $0.12 \leq \kappa_0 \leq 0.16$.

9.4 TURBULENT COUETTE FLOW: VELOCITY AND TEMPERATURE FIELDS

In Section 6.3 we described the means for establishing turbulent Couette flow and then assumed that an appropriate combination of plate velocity and plate spacing results in homogeneous turbulence subject to a constant rate of strain $S \equiv dU_1/dx_2$ with $U_2 \equiv U_3 \equiv 0$ in the central portion between the plates. We then analyzed that turbulence for a given value of S. In Section 6.4 we carried the analysis further and considered the temperature field between the two plates when they are at different temperatures. In this case we assumed that in the central portion there is a constant gradient of the mean temperature $d\Theta/dx_2$ and then analyzed the combined influence of S and that gradient on the turbulence. Here we apply an asymptotic analysis closely following that in the previous sections to determine the plate velocity U_h and the plate spacing $2h$ required to obtain a specified value of S. We then determine the temperature difference required to obtain a particular mean temperature gradient. With the description of turbulent transport available to us at this point in our discussion, we can deal only with the case of small temperature differences, so the temperature is a passive scalar. Thus the second calculation involves the determination of the mean temperature distribution in a *known* velocity field.

The simplicity of turbulent Couette has resulted in an extensive theoretical literature starting with Burgers and Heisenberg in 1922 and continuing to recent times with Gersten (1985) and Andersson et al. (1992) who provide, among other things, a valuable bibliography of theoretical and experimental studies of this flow. However, the difficulty of establishing a flat surface moving in its plane inhibits experimental study of Couette flow. The obvious alternative, flow in a narrow annulus between two cylinders, one rotating the other fixed, results in turbulence characteristics which differ from those in the planar situation, an indication of the important influence of rotation on turbulence (cf. Section 6.6). Despite these difficulties, Reichhardt (1959), Robertson and Johnson (1970), and El Telbany and Reynolds (1982) provided data on skin friction, rate of shear S, and in some cases the Reynolds stresses from moving belt experiments. An indication of the difficulties connected with such experiments is provided by the questions raised by the direct numerical simulation results of Andersson et al. (1992); the DNS show that within the core region of the flow there is a near balance between turbulence production and dissipation and no need for the diffusion of turbulent kinetic energy suggested by the experimental results of El Telbany and Reynolds. Andersson et al. concluded that a Reynolds stress theory along the lines described in Section 10.10 "should be capable of handling turbulent Couette flow." We are unaware of any experimental results on the temperature field in this flow.

In this chapter we must resort to a gradient transport analysis and must limit our concern to the mean velocity and mean temperature distributions. We deal first with the velocity field. The two viscous sublayers, each adjacent to a wall, and a central portion within which the U_1 velocity varies linearly, or nearly so, with x_2 are shown in Fig. 9.4.1. Viscous transport plays a significant role within the sublayers, with the consequence that they involve large velocity gradients; whereas in the central region, with its relatively large turbulent transport, the gradients of the mean velocity are relatively small. Examination of Fig. 9.4.1

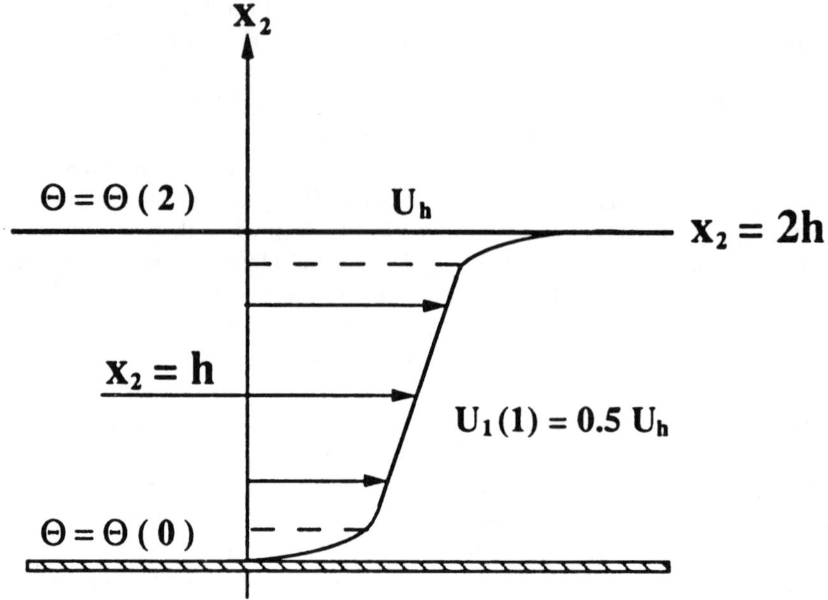

Figure 9.4.1 Turbulent Couette flow.

establishes that $U_1(x_2) - \frac{1}{2}U_h$ is an odd function of $x_2 - h$, that is, $U_1(h) = \frac{1}{2}U_h$. The implication is that we can focus on the lower half of the passage and the solution for that half can be interpreted to apply to the upper half. Indeed, it is convenient to treat $U_1(h)$ as the principal mean velocity characterizing the Couette flow under consideration.

The mean x_1 momentum equation for this flow can be integrated with respect to x_2 to yield

$$\nu \frac{dU_1}{dx_2} - \overline{u_1 u_2} = u_\tau^2 \tag{9.4.1}$$

where $u_\tau \equiv (\tau_w/\rho)^{1/2}$ is again the shearing velocity introduced in Eq. (9.1.2). It arises as a constant of integration and is to be determined as part of the solution.[†] We again adopt mixing-length theory so that as in Eq. (9.2.2) we let

$$-\overline{u_1 u_2} = l^2 \left(\frac{dU_1}{dx_2}\right)^2 \tag{9.4.2}$$

[†]Compare Eqs. (9.4.1) and (9.1.4).

Provided we describe $l(x_2)$, Eqs. (9.4.1) and (9.4.2) can be solved for $U_1(x_2)$ and u_τ subject to the boundary conditions at $x_2 = 0, h$.

If this analysis is to be consistent with Section 6.3, where we require that the correlation $\overline{u_1 u_2}$ *and* the rate of strain dU_1/dx_2 be constant, then we see from Eq. (9.4.2) that in the central portion of the passage the mixing length must be constant with a value denoted l_∞. Accordingly, there is a temptation to assume that the mixing length is constant *throughout* the central portion of the flow and varies only within the wall layer. An asymptotic analysis close to that used in Sections 9.2 and 9.3, however, readily exposes the inappropriateness of that assumption. Briefly, it is not possible to match a solution for the wall layer to that for an outer flow involving a constant mixing length. We thus conclude that at the edges of the outer flow the mixing length must vary as in channel flow, i.e., it must become proportional to η as $\eta \rightarrow 0$. We shall see that such a variation leads to a wall layer identical with that for channel flow, a result which is physically plausible and in accord with the experimental results of Robertson and Johnson (1970) and of El Telbany and Reynolds (1982).

To proceed, we nondimensionalize as in Section 9.2 by letting

$$\eta \equiv \frac{x_2}{h} \qquad \tilde{U} \equiv \frac{U_1}{u_\tau} \qquad \tilde{l} \equiv \frac{l}{h} \tag{9.4.3}$$

so that the combination of Eqs. (9.4.1) and (9.4.2) yields

$$\delta\, \tilde{U}' + \tilde{l}^2\, (\tilde{U}')^2 = 1 \tag{9.4.4}$$

where $\delta \equiv \nu/u_\tau h$ is the inverse of a Reynolds number and where prime denotes differentiation with respect to η. If Eq. (9.4.4) is compared with its counterpart for channel flow, Eq. (9.2.3), we see an essential difference as a consequence of the absence of a streamwise pressure gradient in Couette flow. Equation (9.4.4) indicates that the sum of the viscous and turbulent shear stresses is constant across the *entire* channel.

Since $\delta << 1$ for flows of applied interest, we again carry out an asymptotic analysis involving an outer flow corresponding to the central portion of the passage and inner flows corresponding to the wall layers. The analysis closely follows that of the previous sections, but here we restrict attention to the lowest-order solutions in δ; additional terms, as many as desired, can be systematically obtained, but they are not essential for our purposes. In the outer flow the viscous shear stress is negligible and, as discussed earlier, the mixing length is expected to be nearly constant in the middle portion of the passage but linear in η near the wall. A simple distribution meeting these requirements is shown in Fig. 9.4.2: a linear variation near the wall and a constant beyond a value of $\eta = \eta_m$. Thus we let

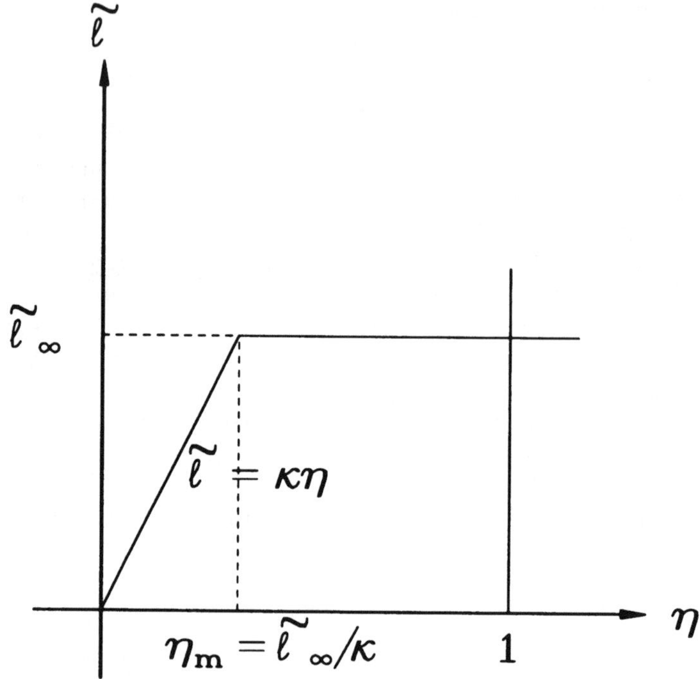

Figure 9.4.2 The distribution of mixing length in terms of $\tilde{l}(\eta)$.

$$\begin{aligned} \tilde{l}(\eta) &= \tilde{l}_\infty \qquad \eta_m \le \eta \le 1 \\ \tilde{l}(\eta) &= \kappa\eta \qquad 0 < \eta \le \eta_m \end{aligned} \tag{9.4.5}$$

where κ is the von Karman constant and $\tilde{l}_\infty$ is as yet arbitrary. Clearly $\eta_m = \tilde{l}_\infty/\kappa$, so either $\tilde{l}_\infty$ or η_m can be considered arbitrary.[†] The parameter η_m determines the edge of the region of uniform velocity gradient and conveniently identifies the various solutions.

The solution to Eq. (9.4.4) with $\delta = 0$ satisfying the boundary conditions at $\eta = 1$ is

$$\tilde{U}(\eta) = \tilde{U}(1) - \int_\eta^1 \frac{d\eta'}{\tilde{l}(\eta')} \tag{9.4.6}$$

This solution can be written in velocity defect form as

[†]As discussed earlier, the distribution of mixing length given by Eq. (9.4.5) suggested by our desire for a region of constant rate of strain in turbulent Couette flow is an alternative to the distribution given by Eq. (9.2.11) for channel and indeed other shear flows involving a wall. In channel flow, for reasonable values of $\tilde{l}_\infty$ the mean velocity profiles in the outer flow differ only slightly from those shown in Figs. 9.2.1 and 9.2.2. The discontinuity in the *slope* of $\tilde{l}(\eta)$ at $\eta = \eta_m$ results in a discontinuity in *curvature* of $\tilde{U}(\eta)$, a discontinuity of no practical significance.

$$\frac{U_1(1) - U_1(\eta)}{u_\tau} = \int_\eta^1 \frac{d\eta'}{\tilde{l}(\eta')} = \frac{1}{\tilde{l}_\infty}(1 - \eta) \qquad \eta_m < \eta \le 1$$

$$= \left[\frac{1}{\tilde{l}_\infty}(1 - \eta_m) + \frac{1}{\kappa}\ln \eta_m\right] - \frac{1}{\kappa}\ln \eta \qquad 0 < \eta < \eta_m \tag{9.4.7}$$

Figure 9.4.3 shows the distributions of the mean velocity as given by Eq. (9.4.7) with η_m as a parameter. If η_m is suitably small, there exist extended regions of uniform rate of strain centered on the plane of symmetry.

It will be of interest for later developments to determine the behavior of $\tilde{U}(\eta \to 0)$. Since $\tilde{l}(\eta \to 0) \approx \kappa\eta$, the second of Eqs. (9.4.7) gives

$$\tilde{U}(\eta) = \tilde{U}(1) + \frac{1}{\kappa}\ln \eta + C \tag{9.4.8}$$

where C is known, being dependent only on η_m.

To resolve the viscous sublayer on the lower, fixed wall, we again must stretch the independent variable by an amount which increases as $\delta \to 0$. Thus we let

$$\hat{\eta} \equiv \frac{\eta}{\delta} \qquad \hat{l} \equiv \frac{l}{\delta} \qquad \hat{U} \equiv \tilde{U} \tag{9.4.9}$$

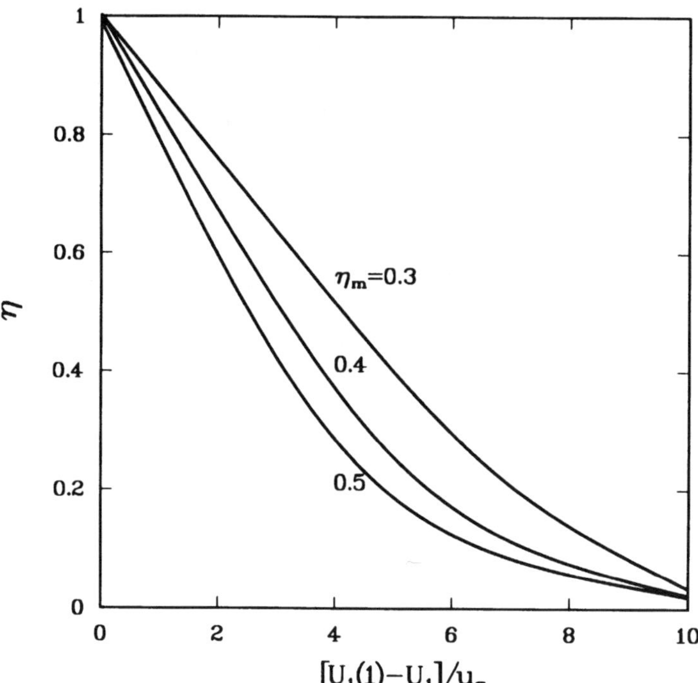

Figure 9.4.3 The outer flow in defect form with η_m as a parameter.

The comments made in connection with Eq. (9.3.2) apply here: As $\delta \to 0$, we stretch the independent variable to retain the viscous term and the mixing length to retain the turbulent term in Eq. (9.4.4). Note that we introduce $\hat{U}(\hat{\eta})$ in order to distinguish the mean velocity within the wall layer from that in the outer flow although no rescaling is involved. When these variables are substituted into Eq. (9.4.4), we obtain

$$\hat{U}' + \hat{l}^2(\hat{U}')^2 = 1 \tag{9.4.10}$$

where prime now denotes differentiation with respect to $\hat{\eta}$. The solution of Eq. (9.4.10) satisfying the condition $\hat{U}(0) = 0$ is

$$\hat{U} = \frac{1}{2}\int_0^{\hat{\eta}} d\eta' \, \frac{[1 + 4\,\hat{l}^2(\eta')]^{1/2} - 1}{\hat{l}^2(\eta')} \tag{9.4.11}$$

Provided only that $\hat{l}(\hat{\eta} \to 0) \to 0$, this solution behaves at the wall as

$$\hat{U} \approx \hat{\eta}$$

which implies that, as expected, viscous stresses prevail at the wall. Equation (9.4.11) is identical with the first of Eqs. (9.3.6). This is in accord with the *absence* of an influence of the streamwise pressure gradient to lowest order in δ within the wall layer in channel flow and, of course, the complete absence of that gradient in Couette flow. The experimental data of Robertson and Johnson (1970) and El Telbany and Reynolds (1982) establish that at suitably high Reynolds numbers, the wall layer in a turbulent Couette flow is essentially identical with that in channel flow and in turbulent boundary layers.

An important result from Eq. (9.3.10) is that at the outer edge of the wall layer, where $\hat{l} \approx \kappa\hat{\eta}$,

$$\lim_{\hat{\eta}\to\infty} \frac{U_1}{u_\tau} = \frac{1}{\kappa}\ln \hat{\eta} + B \tag{9.4.12}$$

where B is a constant that depends on the parameter κ_w in Eq. (9.3.9) and is chosen to achieve agreement with experimental results for the logarithmic portion of the velocity distribution in the wall layer. Robertson and Johnson (1970) give $B = 5.6$,[†] and thus we deduce here that $\kappa_w = 28.0$ in contrast with 24.8 found for channel flow, where experiment suggests $B = 5.0$.

The matching of the inner solution given by Eq. (9.4.11) with that for the outer solutions given by Eq. (9.4.8) results in a skin friction law of the form given by Eq. (9.3.11):

$$B = \frac{U_1(1)}{u_\tau} + \frac{1}{\kappa}\ln \delta + C \tag{9.4.13}$$

[†]They use $\kappa = 0.41$, but we disregard the slight influence on the value of B which results from the difference from our value of 0.40 for the von Karman constant.

The left side of Eq. (9.4.13) is known and if a value for $\tilde{l}_\infty$ is fixed, the parameter C is known so that the right side becomes a relation between $U_1(1)/u_\tau$ and δ. In this case a skin friction coefficient can be defined as

$$c_f \equiv 2\,\frac{\tau_w}{\rho U_1^2(1)} = 2\left[\frac{u_\tau}{U_1(1)}\right]^2 = c_f(\delta;\,\eta_m) \tag{9.4.14}$$

We can now determine the the rate-of-strain parameter of Sections 6.3 and 6.4, namely, $S \equiv dU_1/dx_2$. From Eq. (9.4.7) with $\eta > \eta_m$,

$$\frac{S\,h}{U_1(1)} = \frac{u_\tau}{U_1(1)}\,\frac{1}{\tilde{l}_\infty} \tag{9.4.15}$$

There remains the selection of η_m to bring theory and experiment into agreement. Robertson and Johnson (1970) correlate their data with

$$c_f = \frac{0.072}{[\log(U_1(1)\,h/\nu)]^2} \qquad \frac{S\,h}{U_1(1)} = \frac{0.78}{\log(U_1(1)\,h/\nu)} \tag{9.4.16}$$

Examination of the mean velocity profiles given by Robertson and Johnson (1970) suggests that the mean velocity is nearly linear over the middle half of the passage and thus that $\eta_m \approx 0.5$, a value which gives $\tilde{l}_\infty = 0.2$ and $C = -0.77$. These values lead to the comparison shown in Fig. 9.4.4.† Considering the relative simplicity of the present theory, the agreement must be considered quite satisfactory. As the Reynolds number increases, i.e., δ decreases, the skin friction coefficient and the rate-of-strain parameter both *decrease* and, of course, the viscous sublayers become *thinner*.

Our analysis to this point determines the strain-rate parameter as a function of the Reynolds number $U_1(1)h/\nu$. To complete the connection with the discussion in Section 6.3, we must determine the turbulent kinetic energy in the neighborhood of $\eta = 1$, that is, in the neighborhood of the midplane. There are several ways to proceed, including solution of the equation for $k(\eta)$ [cf. Eq. (10.1.15)]. However, we can use our present solution to obtain a value for the correlation $\overline{u_1 u_2}$ and can then apply Eq. (6.3.8) to obtain an estimate for the turbulent kinetic energy. We have

$$\frac{k}{U_1^2(1)} = 2.8\,\tilde{l}_\infty^2\left(\frac{h\,S}{U_1(1)}\right)^2 = 0.11\left(\frac{h\,S}{U_1(1)}\right)^2$$

Combined with the second of Eqs. (9.4.16) we have a prediction for the variation of the turbulent kinetic energy with Reynolds number.

We now take up the corresponding analysis yielding the connection between the temperature difference of the two plates, $\Theta(2) - \Theta(0)$, and the temperature

†To make this comparison the following procedure is efficient: Pick a value for δ and calculate $U_1(1)/u_\tau$ from Eq. (9.4.13); then $U_1(1)\,h/\nu = [U_1(1)\,h/\nu]\delta^{-1}$ and the values of c_f and $S\,h/U_1(1)$ are found from Eqs. (9.4.14) and (9.4.15).

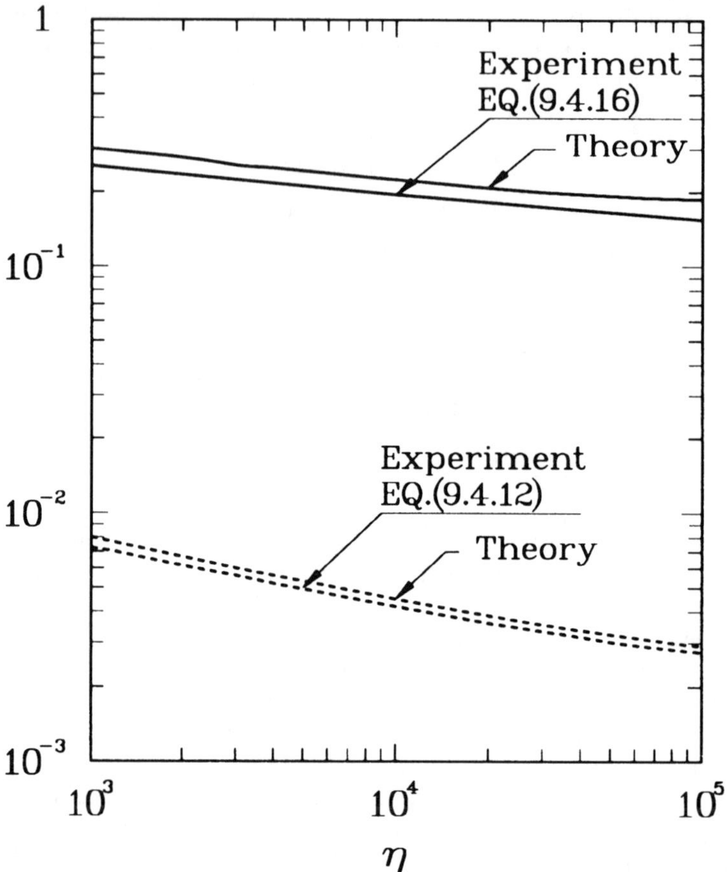

Figure 9.4.4 Variation of the skin friction coefficient and the rate-of-strain parameter with Reynolds number from Eqs. (9.4.13)–(9.4.15) with $\eta_m = 0.5$ and from the correlations of Robertson and Johnson (1970) [cf. Eqs. (9.4.16)].

gradient in the middle region between them, i.e., $d\Theta/dx_2$. Since the simple mixing-length model for turbulent transport that we employ in this section cannot account for the influence of buoyancy, we are restricted to small temperature differences and small plate spacing so that temperature is a passive scalar. Figure 9.4.1 can be reinterpreted to apply to the mean temperature distribution, provided we let $\Theta(2) > \Theta(0)$. Within the two viscous sublayers, where thermal conductivity plays a role, there are large gradients of mean temperature; while in the central region, with its large turbulent transport, those gradients are relatively small. Moreover, from Fig. 9.4.1 it is clear that the mean temperature on the central plane $\eta = 1$ is the arithematic mean of the temperatures of the two plates, that is, $\Theta(1) = \frac{1}{2}[\Theta(2) + \Theta(0)]$. Accordingly, we again can focus on the lower half of the passage and consider $\Theta(1)$ as a characteristic temperature.

The starting point of our considerations is Eq. (4.3.1) specialized to turbulent Couette flow. This equation can be integrated with respect to x_2 to yield

$$\frac{\nu}{N_\sigma}\frac{d\Theta}{dx_2} - \overline{u_2\theta} = -\frac{q_w}{\rho c_p} \tag{9.4.17}$$

where N_σ is the molecular Prandtl number and q_w is the mean heat flux with units of ρU^3, a quantity to be determined. From Eq. (7.4.7) we have for the mixing-length model of turbulent transport of heat

$$-\overline{u_2\theta} = \frac{l^2(dU_1/dx_2)}{\sigma_T}\frac{d\Theta}{dx_2} \tag{9.4.18}$$

where σ_T is the turbulent Prandtl number which applies in this fully developed flow, since the length scales of the velocity and temperature fluctuations are essentially the same. Thus Eq. (9.4.17) becomes, upon introduction of the non-dimensionalization of Eqs. (9.4.3),

$$\left(\frac{\delta}{N_\sigma} + \frac{\tilde{l}^2\tilde{U}'}{\sigma_T}\right)\Theta' = -\frac{q_w}{\rho c_p u_\tau} \tag{9.4.19}$$

With all of the quantities within the factor multiplying Θ' known, either constants or functions of η, Eq. (9.4.19) is to be solved subject to boundary conditions at $\eta = 0, 1$. The quantity q_w is selected so that two conditions can be imposed on a solution to a first-order equation. Note that in those portions of the flow wherein $\tilde{l}$ and $\tilde{U}'$ are constant, we see from Eq. (9.4.19) that Θ' is likewise constant.

We now follow our previous strategy and consider an asymptotic analysis based on $\delta << 1$. The equation for the first-order outer solution is Eq. (9.4.19) with $\delta = 0$, which yields

$$\Theta(\eta) = \Theta(1) + \sigma_T\frac{q_w}{u_\tau}\int_\eta^1 \frac{d\eta'}{\tilde{l}(\eta')} = \Theta(1) + \sigma_T\frac{q_w}{\rho\, c_p\, u_\tau}[\tilde{U}(1) - \tilde{U}(\eta)] \tag{9.4.20}$$

where the second form comes from Eq. (9.4.6). We see that the outer solutions for the mean temperature and the mean velocity are related algebraically, an example of Reynolds analogy connecting the transfer of momentum and heat (cf. Bird et al., 1960). As $\eta \to 0$ we obtain, from Eq. (9.4.8),

$$\Theta(\eta \to 0) = \Theta(1) - \sigma_T\frac{q_w}{\rho\, c_p\, u_\tau}\left(\frac{1}{\kappa}\ln\eta + C\right) \tag{9.4.21}$$

If we stretch the independent variable by introducing $\hat{\eta}$ in order to resolve the mean temperature within the wall layer, then Eq. (9.4.19) becomes

$$\left(\frac{1}{N_\sigma} + \frac{\hat{l}^2\hat{U}'}{\sigma_T}\right)\hat{\Theta}' = -\frac{q_w}{\rho\, c_p\, u_\tau} \tag{9.4.22}$$

where we identify the mean temperature *within the wall layer* as $\hat{\Theta}(\hat{\eta})$. The lowest-order solution of Eq. (9.4.22) satisfying the boundary condition at $\eta = 0$ is

$$\hat{\Theta}(\hat{\eta}) = \hat{\Theta}(0) - \frac{q_w}{\rho\, c_p\, u_\tau} \int_0^{\hat{\eta}} \frac{d\eta'}{(1/N_\sigma) + [(1 + 4\, \hat{l}^2(\eta'))^{1/2} - 1]/(2\sigma_T)} \tag{9.4.23}$$

where we introduce Eq. (9.4.11). The variation of $\hat{l}(\hat{\eta})$ is given by Eq. (9.3.9); in particular, $\hat{l}(\hat{\eta} \to 0) \propto \hat{\eta}^2$, so that from Eq. (9.4.23) we have

$$\hat{\Theta}(\hat{\eta} \to 0) = \hat{\Theta}(0) - \frac{q_w}{\rho\, c_p\, u_\tau} N_\sigma\, \hat{\eta} \tag{9.4.24}$$

We thus see that, as expected, the transport of heat by thermal conductivity dominates at the wall. At the other end of the range of $\hat{\eta}$, as $\hat{\eta} \to \infty$, we have $\hat{l} \approx \kappa \hat{\eta}$, so from Eq. (9.4.23) with our usual splitting of the range of integration,

$$\hat{\Theta}(\hat{\eta} \to \infty) = \hat{\Theta}(0) - \sigma_T \frac{q_w}{u_\tau} \left(\frac{1}{\kappa} \ln \hat{\eta} + B_\theta \right) \tag{9.4.25}$$

where B_θ is a constant independent of $\hat{\eta}^*$ but dependent on N_σ, σ_T, and κ_w in Eq. (9.3.9). Equation (9.4.25) is more general than might be suggested by its genesis here in turbulent Couette flow with heat transfer. Equation (9.4.25) represents the first of several encounters we shall have with a logarithmic distribution of *mean temperature* at the outer edge of the wall layer in nonisothermal turbulent flows; an equivalent equation appears in Section 9.12 in our treatment of a turbulent boundary layer involving heat transfer (cf. Schetz, 1984).

The matching of the distributions of mean temperature given by Eqs. (9.4.21) and (9.4.25) yields a heat transfer law determining the q_w as a function of the Reynolds number, but in the absence of experimental data with which to compare, we do not pursue this possibility further.

Of interest in the context of the analysis of Section 6.4 is the mean temperature distribution in the central portion of the passage. From Eqs. (9.4.15), (9.4.19), and (9.4.26) with $\delta = 0$ we obtain

$$\frac{h}{\Theta(1) - \Theta(0)} \frac{d\Theta}{dx_2} = \frac{1}{\tilde{l}_\infty\, [B_\theta - (1/\kappa) \ln \delta - C]} \tag{9.4.26}$$

which is analogous to Eq. (9.4.15) for the mean rate of strain and yields a variation with Reynolds number similar to that in Fig. 9.4.4.

9.5 PIPE FLOW

Because of its importance in applications, fully developed flow in pipes of circular cross section is the subject of extensive study. However, an analysis closely following that for channel flow in this chapter applies to pipe flow. In fact, some preliminary analysis shows that with suitable reinterpretation and suitable alteration of the empirical coefficients κ_0, κ_w, and B, the previous analysis applies directly to pipe flow.

To start the discussion we introduce axial and radial coordinates denoted x and r, respectively, and their associated mean velocity components U and V. The radius of the pipe is taken to be R. Continuity leads to the result that for fully developed flow the mean radial velocity component is identically zero. By the same arguments used earlier for channel flow, we conclude that the pressure gradient $\partial P/\partial x$ is constant. Thus we are led to the following mean x-wise momentum equation:

$$0 = -\frac{dP}{dx} + \frac{1}{r}\frac{d}{dr}\left[r\left(\mu\frac{dU}{dr} - \rho\,\overline{uv}\right)\right] \tag{9.5.1}$$

Here $\rho\,\overline{uv}$ is the Reynolds shear stress arising from the correlation of the fluctuations of the axial and radial velocity components denoted u and v, respectively.

If Eq. (9.5.1) is integrated with respect to r and the centerline conditions $dU_1/dr = \overline{uv} = 0$ imposed, we obtain

$$0 = -\frac{dP}{dx}\frac{r}{2} + \mu\frac{dU}{dr} - \rho\,\overline{uv} \tag{9.5.2}$$

By applying Eq. (9.5.2) at $r = R$, we can eliminate dP/dx in terms of the shearing velocity u_τ to obtain

$$\begin{aligned} 0 &= u_\tau^2\frac{r}{R} + \nu\frac{dU}{dr} - \overline{uv} \\ &= u_\tau^2\frac{r}{R} + \nu\frac{dU}{dr} - l^2\left(\frac{dU}{dr}\right)^2 \end{aligned} \tag{9.5.3}$$

In developing Eq. (9.5.3) we have again introduced mixing-length theory but have changed the sign in Eq. (9.2.2) to reflect the negative sign of dU/dr.

To facilitate reinterpretation of our previous channel results in terms of pipe flow, we now introduce a new independent variable,

$$\eta \equiv 1 - \frac{r}{R} \tag{9.5.4}$$

so that the centerline corresponds to $\eta = 1$ and the wall of the pipe to $\eta = 0$. Moreover, in anticipation of again finding a two-layer structure to the flow and of employing an asymptotic analysis, we introduce an expansion parameter and a new dependent variable:

$$\delta \equiv \frac{\nu}{u_\tau R} \qquad \tilde{U}(\eta) = \frac{U(\eta)}{u_\tau} \tag{9.5.5}$$

where δ is a small parameter that is completely analogous to the expansion parameter of Sections 9.3 and 9.4. Then Eq. (9.5.3) becomes our final equation, namely,

$$\delta\,\tilde{U}' + \tilde{l}^2\tilde{U}'^2 = (1 - \eta) \qquad (9.5.6)$$

where $\tilde{l} \equiv l/R$ is a dimensionless mixing length.

Equation (9.5.6) is identical with Eq. (9.2.4), although the symbols have slightly but nonessentially altered interpretations. Thus we can solve for the mean velocity distribution with δ as a parameter, but we can also apply asymptotic methods to describe separately the thin viscous sublayer adjacent to the tube wall and a central core devoid of viscous stresses. In particular, our earlier results for channel flows apply to the outer flow of a pipe with revised interpretations of the dependent and independent variables. Thus the discussion of the defect law and the limiting behavior as $\eta \to 0$, Eq. (9.2.13), again apply. The mixing-length distribution $\tilde{l}(\eta)$ of Eq. (9.2.11) is also appropriate for pipe flow, although, as suggested earlier, other reasonable distributions can be chosen without seriously altering the calculated velocity distribution.

The inner solution describing the wall flow is obtained by introducing

$$\hat{\eta} \equiv \frac{\eta}{\delta} \qquad (9.5.7)$$

so that Eq. (9.3.2) is recovered. Thus the solution for the wall layer in pipe flow to lowest order in δ is given by Eq. (9.3.7) and possesses the asymptotic behavior of Eq. (9.3.10). Finally, the skin friction relation arising from matching of the outer and inner solutions, Eq. (9.3.11), also applies, provided the channel half-height h is replaced in the definition of δ with radius R and $U_1(1)$ with $U(1)$. Thus the strategy for exploiting Eq. (9.3.11) to obtain the skin friction as a function of $U(1)R/\nu$ applies to pipe flow.

Given this situation, we seek the value of κ_w so that the value of B for pipe flow is obtained and the value of κ_0 so that the predicted and measured skin friction distributions with Reynolds number agree to the extent possible. Although there is not universal agreement, a value of $B = 5.5$ is recommended for pipe flow by several authors, and we are thus led to $\kappa_w = 27.3$. Moreover, if we take $\kappa_0 = 0.14$, we obtain excellent agreement with various correlations of skin friction with Reynolds number for pipe flow (cf. Bird et al., 1960; Schlicting, 1955).

9.6 HEAT TRANSFER IN A CHANNEL: GENERAL CONSIDERATIONS AND THE SYMMETRIC CASE

We now extend the analysis of Sections 9.1–9.4 to the case of a channel flow with heat transfer either to the fluid from the walls or to the walls from the fluid but subject to the conditions discussed earlier, namely, negligible influence of the temperature distribution on the velocity field, i.e., suitably small rates of heat transfer.

Two different perspectives are possible. If the mean temperature distribution is symmetric with respect to the midplane $x_2 = h$, we must alter our notion of "fully developed" according to the following considerations. The mean energy equation, Eq. (4.3.1), specialized to apply to a two-dimensional flow with $U_1 = U_1(x_2)$, $U_2 \equiv U_3 \equiv 0$ but with $\Theta = \Theta(x_1, x_2)$, is

$$\rho\, U_1 \frac{\partial \Theta}{\partial x_1} = \frac{\partial}{\partial x_2}\left(\frac{\mu}{N_\sigma}\frac{\partial \Theta}{\partial x_2} - \rho\, \overline{u_2\theta}\right) \tag{9.6.1}$$

where it should be recalled that N_σ is the Prandtl number of the fluid. In this equation the streamwise fluxes from both thermal conductivity and turbulent transport are neglected under the assumption that the lateral fluxes are dominant, as is the case in a slender channel. With a symmetric temperature distribution, integration of Eq. (9.6.1) from 0 to h leads after some rearrangement to

$$\frac{d}{dx_1}\int_0^h dx_2\, U_1\, \Theta = -\frac{\nu}{N_\sigma}\frac{\partial \Theta}{\partial x_2}(x_1, 0) \tag{9.6.2}$$

The left side of this equation is the rate of change in the streamwise direction of the total thermal flux through half of the channel; that rate is seen to equal the heat transfer at the wall. If $\partial\Theta/\partial x_2(x_1, 0) > 0$, there is heat transfer from the fluid to the wall and the total thermal flux decreases in the streamwise direction; i.e., the fluid is cooled. If, on the other hand, $\partial\Theta/\partial x_2(x_1, 0) < 0$, the fluid is heated by the wall and the streamwise flux increases. We thus see that in the symmetric case the mean temperature changes in both the normal and streamwise directions; if the streamwise changes are suitably small, our neglect of streamwise transport is justified.

If heat is added to the fluid at the lower wall and withdrawn at the same rate at the upper wall, the temperature is asymmetric and depends only on the normal coordinate. In this case the mean temperature and all statistical quantities involving the temperature can be "fully developed" in the same sense as the velocity field discussed in the previous section. We see from Eq. (9.6.1) that with $\Theta = \Theta(x_2)$ and with transport due to thermal conductivity neglected in the central portion of the channel, the turbulent flux $\overline{u_2\theta}$ is constant. However, we see from Eq. (7.4.7) that the notion of a turbulent Prandtl number cannot apply in this case, since $\partial U_1/\partial x_2 = 0$ at $y = h$ and the turbulent flux cannot be proportional to the velocity gradient. Thus the alternative to Eq. (7.2.1) involving introduction of λ_T, a turbulent thermal conductivity, is indicated; i.e., we might take

$$\rho\, \overline{u_2\theta} \propto k^{1/2}\frac{\partial \Theta}{\partial x_2}$$

However, this calls for calculation of the turbulent kinetic energy $k(x_2)$, so further consideration of this case is deferred to Section 10.7. Accordingly, in this section we consider symmetric temperature distributions which result in a slow streamwise evolution of the mean temperature. Such distributions have the virtue of being readily extended to the practically important case of heat transfer in a pipe.

In the most general case the wall temperature $\Theta(x_1, 0)$ and the heat transfer $(\nu/N_\sigma)\, \partial\Theta/\partial x_2(x_1, 0)$ are related, but special cases, frequently treated because of the simplification they afford, involve either a constant wall temperature or a constant heat transfer. The same fundamental mean conservation equation applies in all cases; accordingly, we nondimensionalize Eq. (9.6.1) as in the previous sections but introduce the streamwise variable $\zeta \equiv x_1/h$ as well. Thus we have from Eq. (9.6.1) the partial differential equation

$$U_1 \frac{\partial\Theta}{\partial\zeta} = \frac{\partial}{\partial\eta}\left(\frac{\delta}{N_\sigma}\frac{u_\tau}{}\frac{\partial\Theta}{\partial\eta} - \overline{u_2\theta}\right) \tag{9.6.3}$$

where, as before, $\delta \equiv \nu/u_\tau h$, and where $U_1(\eta; \delta)$ is known from Eq. (9.1.4). Note that we take ν to be constant, an assumption that is consistent with consideration of *small* temperature changes.

Generally, we must specify both initial and boundary conditions. Consider first

$$\Theta(0, \eta) = \Theta_0(\eta) \qquad \frac{\partial\Theta}{\partial\eta}(\zeta, 1) = 0 \tag{9.6.4}$$

Here $\zeta = 0$ corresponds to some arbitrary station where the velocity distribution is fully developed. In the special cases cited earlier, the second boundary condition is either

$$\Theta(\zeta, 0) = \Theta_w \tag{9.6.5}$$

or

$$\frac{\partial\Theta}{\partial\eta}(\zeta, 0) = \Theta'_w \tag{9.6.6}$$

where Θ_w and Θ'_w are given constants.

The circumstances calling for specification of initial conditions can be understood as follows: Suppose that for a long distance upstream of the station corresponding to $\zeta = 0$ the wall temperature is a constant equal to Θ_i, so that $\Theta_0(\eta) \equiv \Theta_i$ and $\partial\Theta/\partial\eta(0^-, \eta) \equiv 0$. Suppose further that at $\zeta = 0^+$ either a constant wall temperature $\Theta_w \neq \Theta_i$ or a constant heat flux is imposed. Our analysis concerns the evolution of the temperature downstream of this initial station. Other circumstances calling for arbitrary but symmetric distributions of the initial temperature $\Theta_0(\eta)$ can be readily envisaged.

To close Eq. (9.6.3) we must introduce empiricism. Consistent with our earlier analysis, we introduce the mixing-length theory so that from Eq. (7.4.7) we take

$$-\overline{u_2\theta} = \frac{l^2}{\sigma_T}\frac{dU_1}{dx_2}\frac{\partial\Theta}{\partial x_2} \tag{9.6.7}$$

where σ_T is the turbulent Prandtl number, an additional empirical constant, and the mixing length is given by Eq. (9.2.12) in terms of the dimensionless $\tilde{l}(\eta)$. Thus from Eqs. (9.6.3) and (9.6.7) we have an equation suitable for further analysis:

$$\tilde{U}\frac{\partial\Theta}{\partial\zeta} = \frac{\partial}{\partial\eta}\left[\left(\frac{\delta}{N_\sigma} + \frac{\tilde{l}^2}{\sigma_T}\tilde{U}'\right)\frac{\partial\Theta}{\partial\eta}\right] \tag{9.6.8}$$

Here $\tilde{U}(\eta)$ is known from Eq. (9.2.4) and $\tilde{l}(\eta)$ is a dimensionless mixing length given, e.g., by Eqs. (9.2.11) and (9.3.9).

We develop solutions to Eq. (9.6.8) for the case of constant wall temperature by assuming a solution of the form

$$\Theta(\zeta, \eta) = \Theta_w[1 + Z(\zeta)\, N(\eta)] \tag{9.6.9}$$

i.e., a solution involving separation of variables. Then it is easy to determine that

$$Z = Ae^{-\lambda\zeta} \tag{9.6.10}$$

and

$$\left[\left(\frac{\delta}{N_\sigma} + \frac{\tilde{l}^2}{\sigma_T}\tilde{U}'\right)N'\right]' + \lambda\,\tilde{U}\,N = 0 \tag{9.6.11}$$

where λ is the separation parameter. Since the boundary conditions applicable to $N(\eta)$ are

$$N(0) = N'(1) = 0 \tag{9.6.12}$$

we deal with a typical eigenvalue problem. Accordingly, we impose a third, inhomogeneous boundary condition,

$$N(1) = 1 \tag{9.6.13}$$

and select λ so that three conditions can be imposed on a second-order equation.

Equations (9.6.11)–(9.6.13) represent a classical problem with a well-developed supporting theory which establishes that the eigenvalues are positive, real, discrete, and infinite in number. Because $\tilde{l}(\eta)$ and $\tilde{U}(\eta)$ are nonelementary functions, numerical methods must be employed to determine the eigenfunctions $N_n(\eta)$ and their accompanying eigenvalues λ_n. The numerical problem involves treating simultaneously Eqs. (9.6.11) and (9.2.4) with $\tilde{l}(\eta)$ and δ specified and thus with $U_1(1)/u_\tau$ known. Integration is most conveniently initiated at $\eta = 1$, since we know two conditions there; the eigenvalues λ_n are selected so that $N_n(0) = 0$. Standard techniques establish that the eigenfunctions are orthogonal in the sense that

$$\int_0^1 d\eta \, \tilde{U}(\eta) \, N_n(\eta) \, N_m(\eta) = C_n \, \delta_{nm} \tag{9.6.14}$$

The solution to Eq. (9.6.8) is thus of the form

$$\Theta(\zeta, \eta) = \Theta_w \left[1 + \sum_{n=1}^{\infty} A_n e^{-\lambda_n \zeta} N_n(\eta) \right] \tag{9.6.15}$$

where the coefficients A_n are determined in the usual fashion from the initial conditions as

$$A_n = \frac{1}{C_n} \int_0^1 d\eta \, [\Theta_0(\eta) - \Theta_w] \, \tilde{U}(\eta) \, N_n(\eta) \tag{9.6.16}$$

The heat transfer at the wall, a quantity of applied interest, is obtained from Eq. (9.6.15) as

$$q_w = -\frac{\mu \, c_p}{N_\sigma h} \frac{\partial \Theta}{\partial \eta} (\zeta, 0) = -\frac{\mu \, c_p}{N_\sigma h} \Theta_w \sum_{n=1}^{\infty} A_n \, e^{-\lambda_n \zeta} \, N_n'(0) \tag{9.6.17}$$

With $\lambda_n > 0$, $\Theta(\zeta \to \infty, \eta) \equiv \Theta_w$, and $q_w(\zeta \to \infty) \equiv 0$ as called for by physical considerations.

For given distributions of mixing length $\tilde{l}(\eta)$ and the corresponding distributions of mean velocity $\tilde{U}(\eta)$, and for given values of the parameters N_σ and σ_T, the solutions of Eq. (9.6.11) and the evaluation of the integral in Eq. (9.6.16) depend on the Reynolds number via the parameter δ, which appears explicitly in Eq. (9.6.11) and implicitly in $\tilde{U}(\eta)$. Accordingly, for clarity we now write explicitly $N_n(\eta; \delta)$ and $\Theta(\zeta, \eta; \delta)$. The heat transfer, which is proportional to $\partial\Theta/\partial\eta(\zeta, 0; \delta)$, also depends on Reynolds number. Thus a series of solutions of Eq. (9.6.11) covering a range of δ determines, among other things, the dependence of the mean temperature distributions on the Reynolds number and on the initial temperature distribution reflected in the A_n coefficients.

A similar analysis applies for the case of constant heat transfer. In this case we want to solve Eq. (9.6.8) subject to the initial and boundary conditions of Eqs. (9.6.4) and to the boundary condition of Eq. (9.6.6). In anticipation that in due course we shall obtain an eigenfunction solution similar to that defined by Eqs. (9.6.10) and (9.6.11), we replace Eq. (9.6.9) by

$$\Theta(\zeta, \eta) = \Theta_w' \zeta + N_0(\eta) + \sum_{n=1}^{\infty} A_n \, e^{-\lambda_n \zeta} \, N_n(\eta) \tag{9.6.18}$$

Substitution into Eq. (9.6.8) and collection of terms independent of ζ leads to an inhomogeneous equation determining $N_0(\eta)$:

$$\tilde{U}\,\Theta_w' = \left[\left(\frac{\delta}{N_\sigma} + \frac{\tilde{l}^2}{\sigma_T}\tilde{U}'\right) N_0'\right]' \tag{9.6.19}$$

This equation is to be solved subject to the conditions

$$N_0'(0) = \Theta_w' \qquad N_0'(1) = 0 \tag{9.6.20}$$

The remaining terms in Eq. (9.6.18), those involving $N_n(\eta)$, $n = 1, 2, \ldots$, lead to Eqs. (9.6.10) and (9.6.11) with the following homogeneous conditions:

$$N_n'(0) = N_n'(1) = 0 \tag{9.6.21}$$

A third inhomogeneous condition, $N_n(0) = 1$, is imposed and leads to the determination of the eigenvalues λ_n. These eigenvalues differ from those obtained for the case of constant wall temperature, but they have the same characteristics of being positive, real, discrete, and infinite in number. Thus, as $\zeta \to \infty$, the influence of the initial profile decays so that the mean temperature distribution $\Theta(\zeta \to \infty, \eta)$ is given by the first three terms in Eq. (9.6.18).[†] The associated eigenfunctions are orthogonal as in Eq. (9.6.14), so the coefficients A_n are obtained from Eq. (9.6.18) applied at $\zeta = 0$ according to the equation

$$A_n = \frac{1}{C_n}\int_0^1 d\eta\, [\Theta_0(\eta) - N_0]\, \tilde{U}(\eta)\, N_n(\eta) \tag{9.6.22}$$

In this case we see that as $\zeta \to \infty$ the distribution of mean temperature in the fluid continues to change linearly with ζ, although the influence of the initial profile in due course decays.

Our previous discussion of determining the dependence of the solutions on the Reynolds number as reflected in the parameter δ applies in this case as well. Thus a series of solutions with δ as a parameter determines the change in wall temperature with Reynolds number.

Asymptotic methods along the lines employed in Sections 9.2–9.5 and exploiting the small values of δ applied interest simplify both the constant-temperature and constant-heat-transfer cases and remove δ as a parameter. In this method the outer flows involve the expansions

$$N_n(\eta; \delta) = {}^{(0)}N_n(\eta) + \delta\, {}^{(1)}N_n(\eta) + \ldots$$

$$\lambda_n(\delta) = {}^{(0)}\lambda_n + \delta\, {}^{(1)}\lambda_n + \ldots$$

When these expansions are substituted into Eq. (9.6.11) with the known behavior

[†]The requirement that the mean temperature be suitably small restricts Θ_w' and the range of ζ in which our solution applies.

of $\tilde{l}^2\ {}^{(0)}\tilde{U}'$ as $\eta \to 0$ taken into account, the solution for ${}^{(0)}N_n(\eta)$ can satisfy the boundary condition ${}^{(0)}N_n(0) = 0$. The implications of this finding is that the lowest-order eigenvalues are determined by the outer solutions, and the appropriate expansions for the wall layer with our usual $\hat{\eta}$ as the independent variable are

$$\left[\left(\frac{1}{N_\sigma} + \frac{\hat{l}^2}{\sigma_T}\ {}^{(0)}\hat{U}'\right)\ {}^{(1)}\hat{N}_n'\right]' = 0$$

where the prime now denotes differentiation with respect to $\hat{\eta}$. It is unnecessary for our purposes to give further details of this analysis.

9.7 HEAT TRANSFER IN TURBULENT PIPE FLOW

Corresponding to the technologically important skin friction and pressure drop considerations in turbulent pipe flow discussed in Section 9.5, there are equally important thermal considerations. The heat transfer in pipes under Reynolds number sufficiently high so that the flow is turbulent arises in many industrial applications, with the consequence that there is a vast literature involving various correlations and various connections between skin friction and heat transfer. Representative is Bird et al. (1960). Here we are interested in the limited objective of extending the considerations of Section 9.6 to pipe flow.

The equation for the mean temperature $\Theta(x, r)$ in a pipe of radius R when the velocity is fully developed is obtained by averaging the energy equation of Eq. (2.3.1) in cylindrical coordinates and by setting the mean radial velocity $V \equiv 0$. If, as in Section 9.5, statistical quantities are axisymmetric, and if we neglect the mean axial flux of temperature due to both thermal conductivity and turbulent transport, we have

$$\rho\, U \frac{\partial \Theta}{\partial x} = \frac{1}{r}\frac{\partial}{\partial r}\left(\frac{\mu}{N_\sigma}\frac{r\, \partial \Theta}{\partial r} - \rho\, r\, \overline{v\theta}\right) \tag{9.7.1}$$

where $U(r)$ is the axial velocity distributed according to Eq. (9.5.3) and $\rho\, \overline{v\theta}$ is the mean flux of heat in the radial direction due to turbulent transport. If this equation is multiplied by $r\, dr$ and integrated across the radius, we obtain the counterpart of Eq. (9.6.2):

$$\frac{d}{dx}\int_0^R dr\, r\, U(r)\, \Theta(x, r) = \frac{\nu}{N_\sigma} R \frac{\partial \Theta}{\partial r}(x, R) \tag{9.7.2}$$

The left side of this equation is the rate of change of the temperature flux at any streamwise station, a rate that we see depends on heat transfer at the wall of the pipe. If $\partial\Theta/\partial r(x, R) > 0$, heat is added to the fluid and the total flux increases in the streamwise direction; whereas if $\partial\Theta/\partial r(x, R) < 0$, heat is lost by the fluid to the wall of the pipe and the total flux decreases.

In order to expose the connections with the analysis of Sections 9.5–9.7 we now change the independent variable from r to $\eta \equiv 1 - r/R$ [cf. Eq. (9.5.4)]so that $\eta = 0$ and $\eta = 1$ correspond to the wall and axis of the pipe, respectively. Then Eq. (9.7.1) becomes

$$\rho\, U \frac{\partial \Theta}{\partial x} = \frac{1}{R\,(1-\eta)} \frac{\partial}{\partial \eta}\left[\frac{\mu}{N_\sigma R}\frac{\partial \Theta}{\partial \eta} + (1-\eta)\,\rho\,\overline{v\theta}\right] \tag{9.7.3}$$

To close this equation we adapt Eq. (7.4.7) and take

$$\rho\,\overline{v\theta} = \frac{\rho\, l^2}{\sigma_T}\frac{\partial U}{\partial r}\frac{\partial \Theta}{\partial r} \tag{9.7.4}$$

where the sign of the right side reflects the sign of $\partial U/\partial r < 0$. Thus, e.g., if $\partial\Theta/\partial r > 0$, then $\overline{v\theta} < 0$, as expected by physical considerations. If we let $U(1)$ denote the mean velocity on the axis of the pipe and introduce ζ and $\tilde{U}(\eta)$ as in Eq. (9.5.5), Eq. (9.7.3) becomes

$$\tilde{U}\frac{\partial \Theta}{\partial \zeta} = \frac{1}{1-\eta}\frac{\partial}{\partial \eta}\left[(1-\eta)\left(\delta + \frac{\tilde{l}^2}{\sigma_T}\tilde{U}'\right)\frac{\partial \Theta}{\partial \eta}\right] \tag{9.7.5}$$

where $\zeta \equiv x/R$ is a dimensionless streamwise coordinate, prime denotes differentiation with respect to η, and $\tilde{l} \equiv l/R$ is a dimensionless mixing length assumed to be a given function of η.

Equation (9.7.5) is to be solved subject to initial conditions at $\zeta = 0$, a regularity condition $\partial\Theta/\partial\eta = 0$ at the axis, and boundary conditions at $\eta = 0$. Our discussion in Section 9.6 on the various wall conditions which arise and the circumstances calling for specification of initial conditions apply to pipe flow as well. Here we restrict our discussion to the case of a constant wall temperature; the treatment of channel flow involving constant heat transfer in Section 9.6 can be applied with little change to pipe flow. Thus a solution of the form given by Eq. (9.6.9), i.e., involving separation of variables, is indicated, with the consequence that we obtain from Eq. (9.7.5) the following ordinary differential equation for the eigenfunctions $N(\eta)$:

$$\left[(1-\eta)\left(\delta + \frac{\tilde{l}^{2\,\prime}}{\sigma_T}\tilde{U}'\right)' N'\right] + \lambda\,\tilde{U}\,(1-\eta)\,N = 0 \tag{9.7.6}$$

where λ is the usual separation parameter, the eigenvalue in this formulation. The $(1 - \eta)$ factors which appear in this equation as a consequence of the axisymmetry of the flow represent the only difference from Eq. (9.6.11). Accordingly, the analysis of Section 9.6 including the asymptotic treatment for $\delta \to 0$ can be applied with little modification to pipe flow; we do not pursue this analysis further. We note that the mean temperature distribution in the outer portion of the wall layer involves the same logarithmic distribution given by Eq. (9.4.25), another encounter with the logarithmic distribution for the mean temperature. Thus it is found that, to lowest order in δ, the cylindrical curvature of the wall of the pipe is inoperative insofar as the wall layer is concerned.

9.8 THE TURBULENT BOUNDARY LAYER: GENERAL CONSIDERATIONS

We now take up an analysis of a turbulent boundary layer which is isothermal but with a distribution of external velocity. Since such layers are of great applied importance, the literature devoted to them is extensive. Early theories of the turbulent boundary layer including use of the momentum integral method are discussed by Schlichting (1955). Cebeci and Smith (1974) provide a useful survey of theoretical and early computational methods for turbulent boundary layers. For more recent developments, see Reynolds and Cebeci (1971) and Rodi (1980). Finally, we mention the collection of critically assessed experimental data on such boundary layers and the prediction of their characteristics by a variety of methods reported by Kline et al. (1982).

The flow shown schematically in Fig. 9.8.1 is described by the following equations:

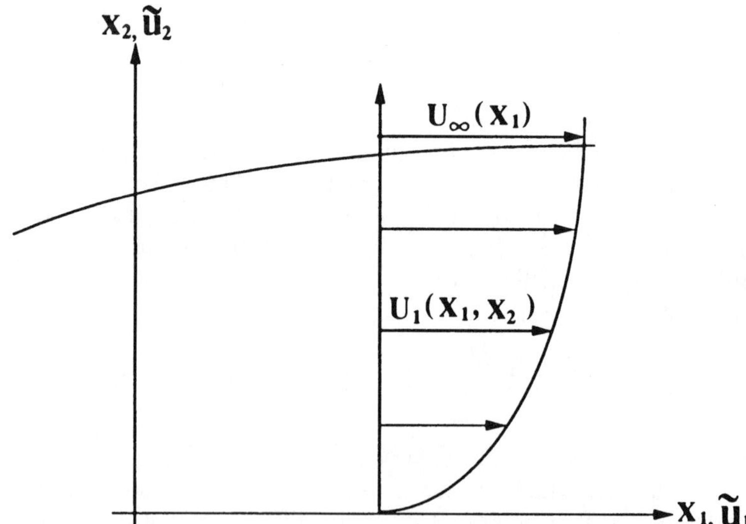

Figure 9.8.1 The turbulent boundary layer.

$$\frac{\partial U_1}{\partial x_1} + \frac{\partial U_2}{\partial x_2} = 0 \qquad (9.8.1)$$

$$U_1 \frac{\partial U_1}{\partial x_1} + U_2 \frac{\partial U_1}{\partial x_2} = U_\infty \frac{dU_\infty}{dx_1} + \frac{\partial}{\partial x_2}\left(\nu \frac{\partial U_1}{\partial x_2} - \overline{u_1 u_2}\right) \qquad (9.8.2)$$

With a model for the Reynolds shear stress $\overline{u_1 u_2}$ and with $U_\infty = U_\infty(x_1)$ given these equations, determine the mean velocity components subject to appropriate initial and boundary conditions. The initial conditions are not relevant to our considerations, but the boundary conditions are:

$$U_1(x_1, 0) = U_2(x_1, 0) = 0 \qquad U_1(x_1, x_2 \to \infty) \to U_\infty(x_1) \qquad (9.8.3)$$

Several comments are appropriate. Since in a "careful" derivation of these equations crucial assumptions must be justified by plausibility arguments, we prefer to draw an analogy with the equations for laminar boundary layers for which rigorous derivation is possible and to add, in an ad-hoc manner, the dominant Reynolds stress. Our earlier discussion of channel flow suggests that the turbulent boundary layer involves a two-layer structure, an outer flow with negligible viscous stresses and a length scale essentially equal to the boundary-layer thickness† identified in some manner, and a wall layer with a length scale measured in terms of $\nu / u_\tau(x_1)$, where u_τ is the local shearing velocity, a quantity which in general and in contrast with channel flow varies in the streamwise direction. The surface on which the boundary layer develops is taken to be essentially flat, so the variation of the external velocity $U_\infty(x_1)$ is assumed to be imposed by other surfaces, such as the other wall of a wind tunnel. We adopt this perspective because if the ratio of turbulent boundary-layer thickness to the longitudinal radius of curvature of the wall is not exceedingly small, the Reynolds stresses are influenced significantly by curvature [cf. Bradshaw (1973) and Section 9.14]. In this regard it is noted that longitudinal curvature is only one of several features of turbulent shear layers considered in the "complex flows" of Bradshaw (1973). Lateral curvature, lateral extension or compression as at a plane of symmetry (cf. Section 9.12), buoyancy as in the atmospheric boundary layer (cf. Section 10.12), and large streamwise acceleration or deceleration are other examples. All these features influence the mean velocity and Reynolds shear stress distributions so as to call for modifications of standard turbulence phenomenology. We emphasize our neglect of the first of these influences by considering the surface to be flat. Finally, we are concerned only with fully turbulent boundary layers, i.e., with stations well downstream of the transition

†Recall from Section 8.5 that the interface between turbulent fluid within the boundary layer and the irrotational outer fluid is highly convoluted and three-dimensional, and that the boundary-layer thickness represents a *mean location* of this interface.

region.[‡] Thus viscous shear stresses are included solely because of their role in the viscous sublayer.

We close Eq. (9.8.2) with a model for the Reynolds shear stress $\rho\ \overline{u_1 u_2}$. If we apply mixing-length theory [cf. Eq. (7.4.5)], we must specify the distribution $l(x_1, x_2)$ so as to achieve agreement with experimentally determined velocity profiles $U_1(x_1, x_2)$. In the wall layer such a distribution is well established; e.g., at the outer edge of that layer $l = \kappa x_2$, where κ is the von Karman constant. However, in the outer flow a simple distribution of the form found appropriate for turbulent channel flow [cf. Eq. (9.2.11)] does not result in satisfactory agreement with experimentally determined velocity profiles, so either a more elaborate representation of the distribution of mixing length or an alternative model for turbulent transport in the outer flow must be sought. Consistent with many previous studies, we follow the latter course and exploit the observation set forth in Section 7.4, that remote from solid walls the turbulent exchange coefficient μ_T is nearly constant in the direction normal to the wall but may vary in the streamwise direction. However, in contrast with previous studies which take μ_T to be constant across the entire outer flow, here matching with the wall layer requires a reduction in the neighborhood of the wall, i.e., at the inner edge of the outer flow. This same requirement was found earlier in Section 9.4 in our discussion of turbulent Couette flow. We take up turbulent transport in more detail later.

The features of the turbulent boundary layer of importance in the present context, the influence of streamwise pressure gradient on the mean velocity profiles and on the relation between the Reynolds number and the skin friction, are exposed by considering special distributions of the external velocity so that the resultant boundary layers exhibit behavior analogous to the similar flows of laminar boundary-layer theory, the Falkner-Skan flows; these are the equilibrium boundary layers. Initiation of the study of these boundary layers is due to Clauser (1956), in a seminal contribution to turbulent boundary-layer theory. Further contributions are due to Townsend (1961), Mellor and Gibson (1966), and Bradshaw (1967b). Under the assumption that equilibrium behavior is established, measurements of the distributions of turbulence quantities at a single streamwise station, when properly interpreted, apply at all other stations, a feature shared with the Falkner-Skan flows. This great advantage in the presentation of experimental data is manifest in the description of the flow in terms of ordinary rather than partial differential equations.

To treat these equilibrium boundary layers we first analyze the outer flow, which is devoid of viscous stresses, and subsequently match the resultant solu-

[‡]The use of turbulence models to *predict* transition from laminar to turbulent flow and the *relaminarization* of a turbulent boundary layer under the influence of a strong favorable pressure gradient are the subject of considerable literature (cf. Pironneau et al., 1990). Analyses of these phenomena require incorporation of viscous and low-Reynolds-number effects throughout the boundary layer and not only in the wall layer as we assume.

tions with those for the wall layer, which, as we saw earlier, is independent of streamwise pressure gradient, provided, as we assume, the Reynolds number is suitably high. In this analysis we restrict attention to the lowest-order terms in a small parameter, the inverse of a Reynolds number.

9.9 THE OUTER FLOW AND APPROXIMATE SIMILARITY

We define similarity for the outer portion of the turbulent boundary layer in terms of the velocity defect law of Eqs. (9.2.9), slightly generalized:

$$\frac{U_\infty(x_1) - U_1(x_1, x_2)}{u_\tau(x_1)} = f'(\eta) \tag{9.9.1}$$

where the similarity variable is

$$\eta \equiv \frac{x_2}{\Delta(x_1)}$$

Here we follow Tennekes and Lumley (1972) and characterize the lateral scale of the boundary layer in terms of an integral thickness $\Delta(x_1)$ defined by

$$\Delta(x_1) \equiv \int_0^\infty dx_2 \frac{U_\infty(x_1) - U_1(x_1, x_2)}{u_\tau(x_1)} = \Delta\,[f(\eta \to \infty) - f(0)] = \Delta \tag{9.9.2}$$

where without loss of generality we take $f(0) = 0$ so that $f(\eta \to \infty) = 1$. Thus two conditions on $f(\eta)$ are fixed, namely, $f(0) = 0$ and $f(\eta \to \infty) = 1$. From Eq. (9.9.1) we recognize that $f(\eta)$ is a streamfunction.

The thickness $\Delta(x_1)$ is related to the usual integral thicknesses of boundary layers as follows. The momentum thickness $\theta(x_1)$ is†

$$\begin{aligned}\theta &\equiv \int_0^\infty dx_2 \frac{U_1}{U_\infty}\left(1 - \frac{U_1}{U_\infty}\right) = \Delta \frac{u_\tau}{U_\infty}\int_0^\infty d\eta\, f'\left(1 - \frac{u_\tau}{U_\infty} f'\right) \\ &= \frac{u_\tau}{U_\infty}\left(1 - \frac{u_\tau}{U_\infty}\int_0^\infty d\eta\, f'^2\right)\Delta\end{aligned} \tag{9.9.3}$$

In addition, the displacement thickness $\delta^*(x_1)$ becomes

$$\delta^* \equiv \int_0^\infty dx_2 \left(1 - \frac{U_1}{U_\infty}\right) = \Delta \frac{u_\tau}{U_\infty}\int_0^\infty d\eta\, f' = \frac{u_\tau}{U_\infty}\Delta \tag{9.9.4}$$

Thus, when we have determined $f(\eta)$, $u_\tau(x_1)$, and $\Delta(x_1)$, we can calculate $\theta(x_1)$ and $\delta^*(x_1)$.

†We use standard notation here with the expectation that there will be no confusion with temperature fluctuations.

If Eq. (9.9.1) is substituted into Eq. (4.1.1), the independent variables changed from x_1, x_2 to x_1, $\eta(x_1, x_2)$, and the result integrated with respect to η, we find the mean velocity normal to the wall as

$$U_2(x_1, \eta) = -[U'_\infty \, \Delta \, \eta - u'_\tau \Delta f + u_\tau \Delta'(\eta \, f' - f)] \tag{9.9.5}$$

where prime denotes differentiation with respect to either x_1 or η as appropriate.

When Eqs. (9.9.1) and (9.9.5) are substituted in Eq. (9.8.2) with the viscous term neglected as called for in the outer flow, there results a cluttered equation with several dimensionless parameters which involve, among others, the derivatives U'_∞, Δ', and u'_τ. If the desired similarity is to prevail so that Eq. (9.9.1) applies, these parameters must be independent of x_1, that is, they must be constants. To achieve tractable results some approximations are called for, and we thus identify the dominant terms in this equation based on the physical notions that throughout the range of x_1 of interest $u_\tau / U_\infty << 1$ and $\Delta' << 1$. To rationalize this latter inequality we imagine that a length L characterizes the range of x_1 of interest and require that

$$\Delta' = O\left(\frac{\Delta}{L}\right) \qquad \frac{\Delta}{L} << 1$$

i.e., the boundary layer is thin. According to this reasoning, terms such as $u'_\tau \Delta / u_\tau$ and Δ' are negligible compared with the following terms:

$$\frac{U'_\infty \, \Delta}{U_\infty} \frac{U_\infty}{u_\tau} \equiv -\Pi \qquad \frac{u'_\tau \Delta}{u_\tau} \frac{U_\infty}{u_\tau} = \Psi \qquad \Delta' \frac{U_\infty}{u_\tau} \equiv \Phi \tag{9.9.6}$$

We shall see later that only one of these three parameters is independent, and thus that we deal with a one-parameter family of flows consistent with the notion of equilibrium boundary layers set forth by Clauser (1956). We select Π as that independent parameter. Note that flows with adverse pressure gradients ($\partial P / \partial x_1 > 0$, $U'_\infty < 0$) correspond to $\Pi > 0$, while those with favorable gradients ($\partial P / \partial x_1 < 0$, $U'_\infty > 0$) to $\Pi < 0$. Note also that our parameter Π is identical with the similarity parameter $\beta = (\delta^* / \tau_w)\,(dP/dx)$ of Clauser (1956).

To facilitate the analysis we extend our definition of similarity and require that $U_\infty(x_1)/u_\tau(x_1)$ be constant. In this case Eqs. (9.9.6) show that $\Psi = -\Pi$, so the number of parameters is reduced, a desirable result. However, we know that the skin friction of a turbulent boundary layer with a uniform external stream, the flat-plate boundary layer which corresponds to $\Pi = 0$, varies slowly in the streamwise direction, roughly as $U_\infty / u_\tau \propto x_1^{1/10}$. This discrepancy is not serious for our purposes, but the implications are that the flat-plate boundary layer is not strictly similar in the sense of Eq. (9.9.1) and that the present analysis should not be used for engineering calculations. Corresponding questions concerning the applicability of our analysis for flows with streamwise pressure gradients do not arise and in any case are again irrelevant for our purposes.

We now take up the modeling of the turbulent transport coefficient ν_T. To do so we anticipate the matching procedure which imposes requirements on ν_T and U_1 at the inner edge of the outer flow and which in due course yields a skin friction law. As discussed earlier, we require the turbulent exchange coefficient to be a constant denoted ν_{TO} away from the immediate vicinity of the wall but to take a form permitting a match with the ν_T prevailing in the outer edge of the wall layer where the logarithmic law exists, i.e., where

$$\frac{U_1}{u_\tau} = \frac{1}{\kappa} \ln \hat{\eta} + B \tag{9.9.7}$$

Here again $\hat{\eta} \equiv \eta/\delta$ is the independent variable for the wall layer, $\delta \equiv \nu/u_\tau\Delta$ is the small parameter appearing in the asymptotic analysis and varying slowly with x_1 because of u_τ and Δ, and κ is the von Karman constant. The use of Eq. (9.9.7) to describe the wall layer with and without streamwise pressure gradients is based on the independence of the wall layer on such gradients if the Reynolds number is suitably high. The constant B is known, e.g., 5.4. If we recall that the sum of the viscous and turbulent shear stresses in the wall layer is constant so that at the outer edge of that layer we have

$$u_\tau^2 = \nu_T \frac{\partial U_1}{\partial x_2}$$

then from Eq. (9.9.7) we find

$$\nu_T = \kappa \, \nu \hat{\eta} \tag{9.9.8}$$

Now we assume that in the outer flow

$$\nu_T = \nu_{TO}[1 - \exp(-\lambda\eta)] \tag{9.9.9}$$

where λ is as yet undefined but assumed positive, so that for $\lambda\eta >> 1$, $\nu_T \approx \nu_{TO}$. Thus the turbulent exchange coefficient far from the wall layer is ν_{TO}. If Eq. (9.9.9) is to match with Eq. (9.9.8), i.e., if $\eta \to 0$ and $\hat{\eta} \to \infty$, then

$$\lambda = \kappa \frac{u_\tau \Delta}{\nu_{TO}} = \kappa \, N_{RO}$$

where N_{RO} is a clearly defined Reynolds number characterizing turbulent exchange as discussed in Section 7.5, i.e., one based on local quantities, the shearing velocity u_τ, the integral thickness Δ, and ν_{TO}. We anticipate that λ is of the order of 10, with the consequence that $\nu_T \approx \nu_{TO}$ for $\eta > 10^{-1}$. Since we match in the outer flow and wall layer the exchange coefficients and later the two mean velocity gradients, the mean Reynolds shear stresses in the two regions will also be matched.

With these considerations the momentum equation, Eq. (9.8.2), becomes

$$2\,\Pi\,f' + (\Phi - \Pi)\,\eta\,f'' = -\left(\frac{\nu_T}{u_\tau \Delta} f''\right)' \tag{9.9.10}$$

The left side of Eq. (9.9.10) arises from convection on the left side of Eq. (9.8.2) and from the streamwise pressure gradient, the first term on the right side of Eq. (9.8.2). If f is to be a function of η alone, Π and Φ in Eq. (9.9.10) must be constants.

Since we deal with an outer solution expected to have a singularity as $\eta \to 0$, Eq. (9.9.10) is integrated from η to ∞ to obtain

$$(3\Pi - \Phi)(1 - f) + (\Pi - \Phi)\,\eta\,f' = \frac{\nu_T}{u_\tau \Delta} f'' \tag{9.9.11}$$

The discussion in Sections 9.2 and 9.3 and the need to match with the velocity U_1 at the outer edge of the wall layer indicate that as $\eta \to 0$ we must require that

$$f'(\eta \to 0) = -\left(\frac{1}{\kappa} \ln \eta + C\right) \tag{9.9.12}$$

In this equation the minus sign arises from the definition of Eq. (9.9.1), and C is a constant to be determined. If this equation is to apply, then $f''(\eta \to 0) = -(\kappa\,\eta)^{-1}$ and with Eq (9.9.9) taken into account the right side of Eq. (9.9.11) is found to approach -1 as $\eta \to 0$. Thus we have the constraint that

$$\Phi = 1 + 3\Pi \tag{9.9.13}$$

Accordingly, the parameter Φ, which determines the growth of the boundary layer in terms of Δ, is related to the external velocity parameter Π. Equation (9.9.13) reduces our similar boundary layers to a one-parameter family. The qualitative behavior of Φ for both positive and negative values of Π is in accord with intuition and experiment: if the pressure gradient is adverse, Π is positive and the rate of growth of the boundary layer is increased; if the pressure gradient is favorable so that Π is negative, that rate is decreased.

When Eq. (9.9.13) is substituted into Eq. (9.9.11) and Eq. (9.9.9) is used to express ν_T, the final equation for $f(\eta)$ is obtained, namely

$$f'' = -N_R\,[(1 + 2\Pi)\,\eta\,f' + (1 - f)] \tag{9.9.14}$$

where $N_R = N_{RO}/[1 - \exp(-\lambda\eta)]$.

We examine next the behavior of $f(\eta \to \infty)$. If we assume that the approach of $f(\eta \to \infty) \to 1$ involves exponential decay of the form suggested by the differential equation,[†]

[†]The behavior of the solutions for turbulent boundary layers as the external stream is approached is the subject of controversy, with some suggesting that there is a singularity at the outer edge of a finite boundary-layer edge (cf. Paullay et al., 1985). Recently Cazalbou et al. (1994) examined the

$$f = 1 + A\eta^n \exp\left(-\frac{\lambda}{2}\eta^2\right)$$

where A is arbitrary. Substitution into Eq. (9.9.11) with $\eta \to \infty$ yields

$$\lambda = N_{R0}(1 + 2\Pi) \qquad n = -2\,\frac{1 + \Pi}{1 + 2\Pi}$$

The implication of these equations is that the analysis applies to favorable pressure gradients for which $\Pi > -\frac{1}{2}$. There appears to be no corresponding restriction for adverse pressure gradients.

Equation (9.9.14) is a linear equation which nevertheless is conveniently solved by numerical means involving integration in two directions: to the right from $\eta_0 << 1$ with C arbitrary, and to the left from a suitably large value of η with A arbitrary. The quantities A and C are selected so that f and f' are continuous at some intermediate value of η. This strategy assures the appropriate approach of the solutions to their external stream values.

For a given value of Π, the single empirical coefficient to be chosen in these calculations is N_{R0}, which in principle might be taken to depend on Π. There are considerable data on the velocity profiles for the flat-plate boundary layer, i.e., for $\Pi = 0$ (cf. Townsend, 1956), data which can guide the choice of this parameter, but conveniently, Cebeci and Smith (1974) show that for the case of ν_T constant *throughout* the outer flow, $\nu_{T0} = \alpha U_\infty \delta^*$ with $\alpha = 0.022$ leads to velocity profiles in the outer flow in good agreement with experiment for such boundary layers. Thus, from Eq. (9.9.4) we find $N_{R0} = \alpha^{-1} = 45.5$.† Although it might be argued that the present analysis with its distribution of ν_T calls for a different value for N_{R0}, our purposes are adequately served by adopting this value not only for the flat-plate boundary layer to which it pertains but for boundary layers with pressure gradient as well. Other values of this coefficient can be selected, but our purposes are adequately dealt with by limiting attention to this representative value.

Figure 9.9.1 shows the velocity defect profiles in terms of $f'(\eta)$ for several values of Π. The corresponding values of $C(\Pi)$ are also given. From this figure

behavior of solutions to the k–ε and related theories (cf. Section 10.1) at such a finite edge. Given the oscillatory nature of the surface bounding the turbulent and irrotational fluids (cf. Section 8.5), the specification of a singular edge is physically unappealing. Here the behavior of the solutions as the external flow is approached depends on the description of turbulent transport employed. For example, the assumption of a constant exchange coefficient as $\eta \to \infty$ leads to an exponential approach to the external stream, behavior which is *required* of laminar boundary layers in order that matching of the boundary layer and external flows can be achieved. In Chapter 10, which involves different models for turbulent exchange, we find a mix of exponential and algebraic decay is called for at infinity.

†Although the value of this Reynolds number compares favorably with those suggested for free shear flows in Section 7.5, the agreement should be considered fortuitous, since all three elements involved are significantly different in the two cases.

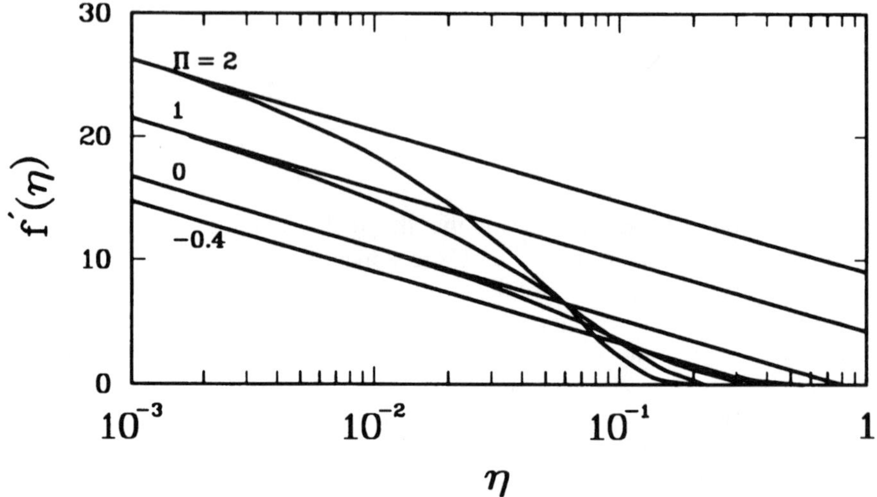

Figure 9.9.1 The velocity defect for similar turbulent boundary layers with emphasis on the wall region; $C(2) = 9.03$, $C(1) = 4.23$, $C(0) = -0.67$, and $C(-0.4) = -2.47$.

we see that the deviation of the velocity profile from the logarithmic distribution applicable to the outer edge of the wall layer is significantly increased as the adverse pressure gradient becomes stronger. Conversely, in favorable gradients the logarithmic law applies over most of the boundary-layer thickness.

9.10 SKIN FRICTION AND STREAMWISE VARIATIONS OF THE EXTERNAL FLOW

The discussion connected with Eqs. (9.9.7), (9.9.8), and (9.9.12) relates to matching of the solutions for the outer and wall flows leading to continuous behavior of the turbulent exchange coefficient, the mean velocity, and the mean turbulent shear stress in their common region. We now complete the matching procedure and thereby establish the skin friction law. From Eq. (9.9.12),

$$\frac{U_\infty(x_1)}{u_\tau(x_1)} - f'(\eta \to 0) = \frac{U_\infty}{u_\tau} + \frac{1}{\kappa} \ln \eta + C$$

$$= \frac{U_\infty}{u_\tau} + \frac{1}{\kappa} \ln \hat{\eta} + \frac{1}{\kappa} \ln \delta + C \tag{9.10.1}$$

Since this distribution must match with Eq. (9.9.7), the $\ln \hat{\eta}$ terms cancel and we are left with

$$B = \frac{U_\infty(x_1)}{u_\tau(x_1)} + \frac{1}{\kappa} \ln \delta + C \tag{9.10.2}$$

which is closely related to Eq. (9.3.11). With both B and C known for a specified value of Π, Eq. (9.10.2) determines a unique value for the quotient $U_\infty(x_1)/u_\tau(x_1)$ in terms of the Reynolds number represented by the reciprocal of δ. The influence of the streamwise pressure gradient on that quotient is contained within $C(\Pi)$. For a fixed value of Π and thus of C, increases in the Reynolds number increase U_∞/u_τ and decrease the skin friction coefficient, as suggested earlier.

With U_∞/u_τ determined for a specified value of δ from Eq. (9.10.2), alternative representations of the velocity profiles are possible. One illuminating alternative involves U_1/u_τ versus the independent variable for the wall layer, i.e., $\hat{\eta}$ as shown in Fig. 9.10.1. Here we take $\delta = 6.25 \times 10^{-4}$, a representative value. For the flat-plate boundary layer this corresponds to a Reynolds number based on momentum thickness $U_\infty\theta/\nu = 5.52 \times 10^3$, again a value representative of those of applied interest. Figure 9.10.1 reinforces our earlier discussion connected with Fig. 9.9.1, namely, that boundary layers in adverse pressure gradients involve relatively small regions with logarithmic distributions of U_1/u_τ, while those in favorable gradients involve relatively extensive regions with such distributions. In the flat-plate boundary layer at the Reynolds number considered in this figure, the wall layer is 3% of the boundary-layer thickness but accounts for 70% of the velocity change from zero at the wall to U_∞/u_τ in the external stream. Both of these percentages are reduced and increased by adverse and favorable pressure gradients, respectively. Indeed, for $\Pi = 2$ with the Reynolds number assumed here, there is little evidence of the logarithmic distribution.

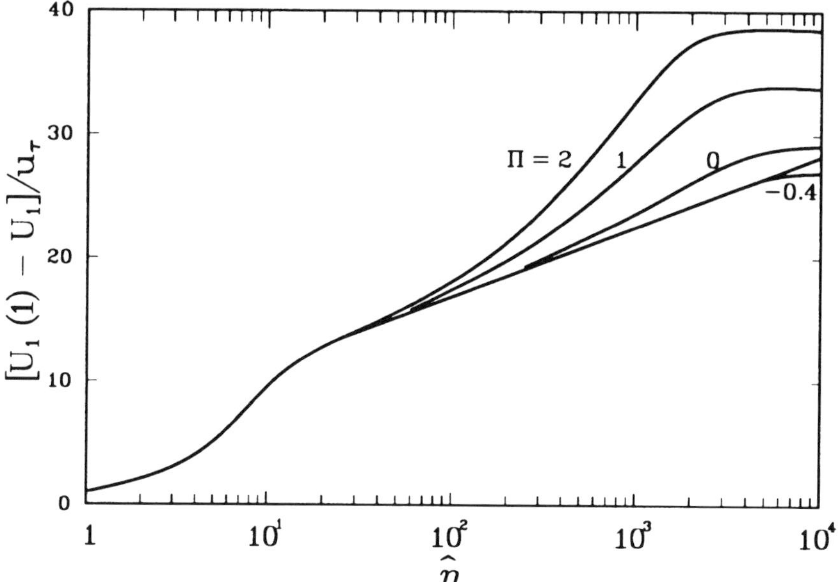

Figure 9.10.1 Velocity profiles in similar boundary layers in terms of the wall variable: $\delta = 6.25 \times 10^{-4}$.

With U_∞/u_τ known, a second representation of the velocity distributions involves the conventional profile U_1/U_∞ as shown in Fig. 9.10.2. For this purpose we define an appropriate boundary-layer thickness, denoted δ_e, defined by the value of η for which $U_1 = 0.99U_\infty$. We then take x_2 divided by δ_e as the independent coordinate in this figure. In this form we again see the significant alteration of the velocity profiles in flows with adverse pressure gradients.

We now turn our attention to other consequences of Eq. (9.10.2). Figure 9.10.3 shows the variation with the Reynolds number $U_\infty\theta/\nu$ of the skin friction coefficient

$$c_f \equiv \frac{2\tau_w(x_1)}{\rho U_\infty^2(x_1)} = \frac{2}{(U_\infty/u_\tau)^2} \tag{9.10.3}$$

where we use Eq. (9.9.3) to determine θ/Δ. We see that favorable and adverse pressure gradients involve increases and decreases, respectively, in skin friction. The results for zero pressure gradient are in good agreement with an expression for c_f due to Falkner (1943), slightly modified:

$$c_f = 0.012\left(\frac{U_\infty\theta}{\nu}\right)^{-1/6} \tag{9.10.4}$$

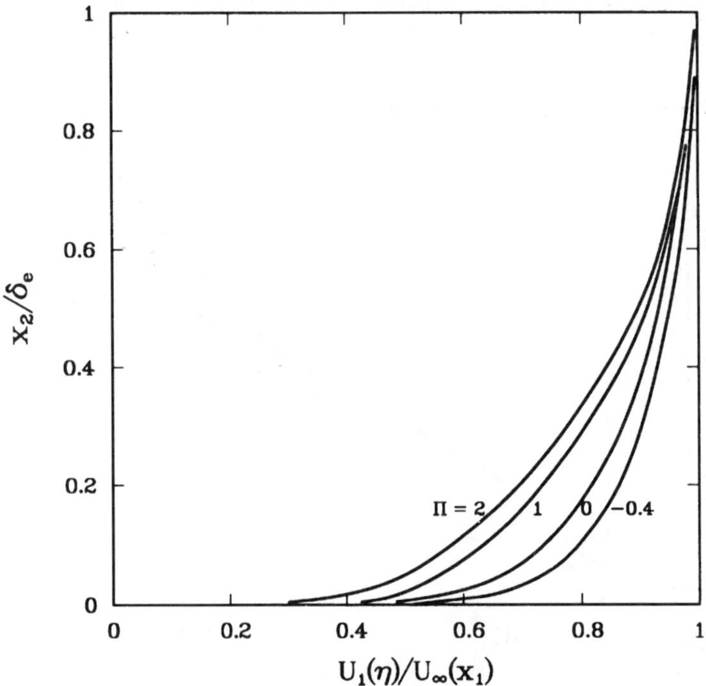

Figure 9.10.2 Velocity profiles in terms of U_1/U_∞ and of the outer flow variable: $\delta = 6.25 \times 10^{-4}$.

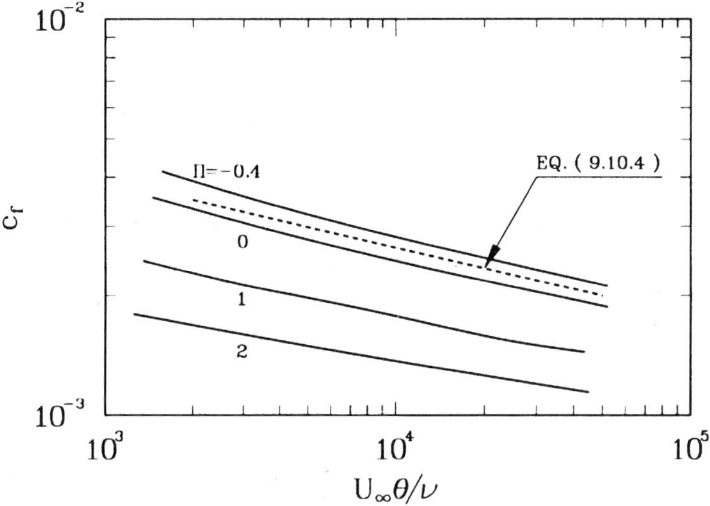

Figure 9.10.3 Skin friction in similar turbulent boundary layers.

We conclude consideration of equilibrium turbulent boundary layers by determining the streamwise distributions of the free-stream velocity and boundary-layer thickness corresponding to specific values of Π. Equations (9.9.6) supplemented with Eq. (9.9.13) yield two differential equations for $U_\infty(x_1)$ and $\Delta(x_1)$:

$$\frac{U'_\infty}{U_\infty} = -\left(\Pi \frac{u_\tau}{U_\infty}\right)\frac{1}{\Delta}$$
$$\Delta' = (1 + 3\,\Pi)\frac{u_\tau}{U_\infty} \tag{9.10.5}$$

The second of these equations can be immediately solved to obtain

$$\Delta(x_1) = \Delta(0)\left[1 + (1 + 3\,\Pi)\frac{u_\tau}{U_\infty}\frac{x_1}{\Delta(0)}\right] \tag{9.10.6}$$

where we place the origin of the x_1 coordinate at the initial station, where $\Delta = \Delta(0)$. We see that the boundary-layer thickness varies linearly with $x_1/\Delta(0)$, increasing if $\Pi > -\frac{1}{3}$ and decreasing if $\Pi < -\frac{1}{3}$. Particularly interesting is the *decrease* in boundary-layer thickness if the pressure gradient is strongly favorable, i.e., if $-\frac{1}{2} < \Pi < -\frac{1}{3}$. The linear increase in thickness for the flat-plate boundary layer, i.e., for $\Pi = 0$, found from this analysis must be compared with the slower development, e.g., $x_1^{4/5}$ (cf. Schlichting, 1955), given by experiment. This discrepancy is consistent with our earlier finding related to the slow streamwise variation of U_∞/u_τ observed experimentally, is associated with our similarity requirements, and is not serious for our purposes. Note that the multiplier of the streamwise variable $x_1/\Delta(0)$ depends on the Reynolds number via the factor U_∞/u_τ.

If Eq. (9.10.6) is substituted into Eq. (9.10.5), we find

$$U_\infty(x_1) = U_\infty(0)\left[1 + (1 + 3\,\Pi)\frac{u_\tau}{U_\infty}\frac{x_1}{\Delta(0)}\right]^{-\Pi/(1+3\Pi)} \tag{9.10.7}$$

For $\Pi > -\frac{1}{3}$ this solution exhibits the expected characteristics associated with adverse, zero, and favorable pressure gradients. For $\Pi = -\frac{1}{3}$, either we can evaluate the indeterminancy 1^∞ or, since for this case Δ is constant as called for by Eq. (9.10.6), we can integrate the first of Eqs. (9.10.5) to find

$$U_\infty(x_1) = U_\infty(0)\exp\left(\frac{1}{3}\frac{u_\tau}{U_\infty}\frac{x_1}{\Delta(0)}\right) \tag{9.10.8}$$

In the far downstream region such that the second terms in the brackets of Eqs. (9.10.7) dominate, we obtain power law distributions of $U_\infty(x_1)$ in accord with Townsend (1961) and Mellor and Gibson (1966).

In a study somewhat related to this but based on the correlation of a large number of experiments connected with turbulent boundary layers, Coles (1956) considered the velocity profile to consist of a logarithmic portion as given by Eq. (9.9.7) plus a universal wake contribution closely approximated by a simple trigonometric function but weighed with a function of x_1. Boundary layers in strong adverse pressure gradients involve large wake contributions, those in strong favorable gradients small and even negative contributions, results in accord with our findings.

Although our analysis applies to similar boundary layers, the results are indicative of the behavior of turbulent boundary layers with arbitrary distributions of free-stream velocity. Thus boundary layers with adverse pressure gradients involve reductions in skin friction and relatively rapid increases in boundary-layer thickness; those with favorable gradients involve increases in skin friction and either diminished rates of growth or actual reductions in the thickness of the layer. The special case of adverse pressure gradient so strong as to result in flow separation warrants special comment. With a fixed Reynolds number, as separation is approached U_∞/u_τ increases indefinitely and our analysis fails. Turbulent boundary-layer separation is of great technological importance relative to the stall of airfoils and diffusers and is the subject of extensive research (cf. Simpson, 1989). It generally involves complex, three-dimensional phenomena including low-frequency, large-amplitude oscillations; i.e., the representation of the reattachment point of a separated boundary layer at a fixed location is highly idealized and ignores large-scale fluctuations. Three-dimensional separation, e.g., on bodies of revolution at angles of attack, is especially complex (cf. Wang, 1974).

9.11 NONISOTHERMAL TURBULENT BOUNDARY LAYERS

Because of their significance in the important applied field of heat transfer, the temperature distributions in turbulent boundary layers are the subject of a considerable literature. Thus all the references cited in Section 9.9 contain discussions of the various methods for calculating the thermal characteristics of turbulent boundary layers and presentations of various heat transfer correlations. These should be consulted for engineering calculations. Here we restrict attention to the fundamental problem of the temperature distributions in the similar turbulent boundary layers dealt with in the previous two sections.

The notion of a two-layer structure applies to such temperature distributions, i.e., there is a relatively thick outer layer dominated by turbulent transport and a thin wall layer in which transport of heat by both thermal conductivity and turbulence are important. To analyze this situation it is expeditious to follow the strategy used earlier, treating first the lowest-order description of the outer portion of the thermal boundary layer and developing subsequently the corresponding description of the temperature distribution in the wall layer. If higher-order Reynolds number effects are desired, a systematic development along the lines described in Sections 9.2–9.4 can be carried out—i.e., additional terms in the series for both the outer and wall layers can be computed. Again we assume that the molecular and turbulent Prandtl numbers are of $O(1)$ and thus that the velocity and temperature distributions within the wall layer have roughly the same thickness. The description of turbulent exchange in terms of a two-layer representation, mixing length within the wall layer and an exchange coefficient in the outer flow given by Eq. (9.9.9), is retained. In implementing this strategy, much of the notation of Sections 9.9 and 9.10 is carried over without change.

Our analysis starts with the equation for the mean temperature $\Theta(x_1, x_2)$ in the outer flow, obtained, e.g., from Eq. (9.6.1), by including an additional convection term on the left side and by incorporating Eq. (7.2.1) to eliminate the mean turbulent flux $\rho\, \overline{u_2\theta}$ with a gradient description of turbulent transport of heat. With transport due to thermal conductivity neglected, as is appropriate for the outer flow,

$$\frac{\partial}{\partial x_1}(U_1\Theta) + \frac{\partial}{\partial x_2}(U_2\Theta) = \frac{\partial}{\partial x_2}\left(\frac{\nu_T}{\sigma_T}\frac{\partial\Theta}{\partial x_2}\right) \tag{9.11.1}$$

If we use Eqs. (9.9.1) and (9.9.5), make the approximations regarding the slow evolution of the boundary layer used to obtain Eq. (9.9.6), and change the independent variables from x_1, x_2 to $\zeta \equiv (u_\tau/U_\infty)(x_1/\Delta)$, $\eta \equiv x_2/\Delta$, we find

$$\frac{\partial\Theta}{\partial\zeta} - (1 + 2\Pi)\,\eta\,\frac{\partial\Theta}{\partial\eta} = \frac{\partial}{\partial\eta}\left(\frac{1}{\sigma_T N_R}\frac{\partial\Theta}{\partial\eta}\right) \tag{9.11.2}$$

where N_R is a turbulent exchange Reynolds number defined in connection with Eq. (9.9.14), namely

$$N_R = \frac{N_{RO}}{1 - \exp(-\lambda\eta)} \qquad N_{RO} \equiv \frac{u_\tau \Delta}{\nu_{TO}} \qquad \lambda = \kappa N_{RO}$$

Note that multiplication by the small constant factor u_τ / U_∞ of the primitive variable x_1/Δ, which can assume large values if streamwise developments over many multiples of the boundary-layer thickness are considered, effectively rescales, i.e., compresses, the streamwise coordinate.

Equation (9.11.2) closely resembles Eq. (9.6.8), arising in channel flow when transport due to thermal conductivity is neglected, i.e., when δ is set to zero. Indeed, the method of solution employed there applies. However, for the present case of a boundary layer with a constant temperature in the free stream and with a constant wall temperature, it is convenient to introduce a new dependent variable such that

$$T(\zeta, \eta) \equiv \frac{\Theta - \Theta_w}{\Theta_\infty - \Theta_w} \tag{9.11.3}$$

which must approach unity as $\eta \to \infty$. Moreover, within the wall layer $T(\zeta, 0) = 0$, but the usual matching of the wall layer and outer flow leads to an appropriate condition for $T(\zeta, \eta \to 0)$.

We now assume a solution in the form

$$T(\zeta, \eta) = 1 + T_0(\eta) + \sum_{n=1}^{\infty} Z_n(\zeta)\, N_n(\eta) \tag{9.11.4}$$

Substitution into Eq. (9.11.2) leads to an equation for $T_0(\eta)$,

$$-(1 + 2\Pi)\, \eta\, T_0' = \left(\frac{1}{\sigma_T N_R} T_0'\right)' \tag{9.11.5}$$

and to an eigenvalue problem with

$$Z_n(\zeta) = A_n\, e^{-\lambda_n \zeta} \tag{9.11.6}$$

and

$$\left(\frac{1}{\sigma_T N_R} N_n'\right)' + (1 + 2\,\Pi)\, \eta\, N_n' + \lambda_n N_n = 0 \tag{9.11.7}$$

where the λ_n are the eigenvalues taken to be real and positive. According to this form of the solution, the first two terms in Eq. (9.11.4) describe the distribution of temperature far downstream from the origin of the x_1, ζ coordinate, where the influence of an arbitrary initial temperature distribution accounted for by the eigenfunctions with their arbitrary constants A_n has effectively disappeared. We now simplify our considerations by focusing on this far-downstream solution and by putting aside the solution of Eq. (9.11.7), which can be treated in the same fashion as the eigenvalue problem outlined in Section 9.6. Thus the mean temperature is now taken to depend only on η, so that variations with respect to x_1 are due to the slow variation of $(\eta/\Delta)\Delta'$.

Since the boundary condition for $\eta \to 0$ must be obtained from matching with the temperature distribution in the wall layer within which the mixing-length model for turbulent transport applies [cf. Eq. (9.6.7)], we analyze next the temperature distribution in the wall layer. The new independent variable needed to resolve that layer is, as before, $\hat{\eta} \equiv \eta/\delta$, where we recall that $\delta \equiv \nu/u_\tau\Delta$ is the small parameter in the asymptotic analysis. Thus Eq. (9.11.1) yields for the lowest-order mean temperature distribution within the wall layer,

$$\left[\left(\frac{1}{N_\sigma} + \frac{\hat{l}^2}{\sigma_T}\hat{U}'\right)\Theta'\right]' = 0 \tag{9.11.8}$$

where $\hat{U} \equiv U_1/u_\tau$ as in Section 9.3, is the dimensionless velocity component in the wall layer, $\hat{l} \equiv (l/\Delta\delta)$ is related to the dimensionless mixing length introduced in connection with Eq. (9.3.2), and the prime now denotes differentiation with respect to $\hat{\eta}$. The implication of this equation is analogous to that deduced from the first of Eqs. (9.3.4): At high Reynolds numbers, the sum of the heat flux due to thermal conductivity and turbulent transport is constant through the wall layer.

Two integrations of Eq. (9.11.9) and some rearrangement lead to the solution for $\hat{T}(\hat{\eta})$ within the wall layer:

$$\hat{T}(\hat{\eta}) = \frac{\Theta'_w}{\Theta_\infty - \Theta_w}\int_0^{\hat{\eta}} \frac{d\hat{\eta}'}{1 + (N_\sigma/\sigma_T)\,\hat{l}^2\hat{U}'} \tag{9.11.9}$$

From this equation we see that the dimensionless temperature distribution in the wall layer is given by quadrature, provided the mixing length in terms of $\hat{l}$ and the mean velocity in terms of $\hat{U}'$ are both known functions of $\hat{\eta}$. If we recall that to lowest order in the reciprocal of a global Reynolds number the wall layer is unaffected by streamwise pressure gradient, the quantity $\hat{U}'$ is given by Eq. (9.3.7). Thus we obtain

$$\hat{T}(\hat{\eta}) = \frac{\Theta'_w}{\Theta_\infty - \Theta_w}\int_0^{\hat{\eta}} \frac{d\hat{\eta}'}{1 + (N_\sigma/2\sigma_T)[(1 + 4\hat{l}^2)^{1/2} - 1]} \tag{9.11.10}$$

so that the mean temperature in the wall layer depends only on the distribution of the mixing length expressed in terms of $\hat{l}(\hat{\eta})$, which can be taken as given by Eq. (9.3.9). As $\hat{\eta}ha \to 0$, the integral in Eq. (9.11.10) becomes $N_\sigma\hat{\eta}$ and we obtain the expected linear variation of the mean temperature as the wall is approached.

To determine the boundary conditions for the outer solutions, the behavior of Eq. (9.11.10) as $\hat{\eta} \to \infty$ is examined; we know that $\hat{l}(\hat{\eta} \to \infty) = \kappa\hat{\eta}$, so

$$\hat{T}(\hat{\eta} \to \infty) = \frac{\Theta'_w}{\Theta_\infty - \Theta_w}\left(\frac{\sigma_T}{\kappa N_\sigma}\ln\hat{\eta} + B_\theta\right) \tag{9.11.11}$$

is the logarithmic law for the mean temperature in the outer portion of the wall layer seen earlier in slightly different form in Eq. (9.4.25). In this equation B_θ is again a known constant which arises from evaluation of the integral in Eq. (9.11.10) at a suitably large value of $\hat{\eta}$.

Figure 9.11.1 shows the solution of Eq. (9.11.10) with $N_\sigma = 0.7$ and $\sigma_T = 1$. It is found that $B_\theta = -14.9$. The close resemblance to the velocity distribution of Fig. 9.3.1 is noted, including the possibility of an idealization in terms of linear and logarithmic distributions at the wall in the viscous sublayer and outer portion, respectively.

Now, if Eq. (9.11.11) is to be matched with the solution to Eq. (9.11.5), we see that as $\eta \to 0$ we must require the outer solution to behave as

$$T_0(\eta \to 0) = \frac{\Theta'_w}{\Theta_\infty - \Theta_w}\left(\frac{\sigma_T}{\kappa N_\sigma}\ln\eta + C_\theta\right) \tag{9.11.12}$$

where C_θ is a constant that depends on Π.

As $\eta \to \infty$, Eq. (9.11.5) becomes

$$T''_0 + (1 + 2\,\Pi)\,\sigma_T\,N_{RO}\,\eta\,T'_0 = 0$$

which has an exponentially decaying solution of the form

$$T_0 \approx A\,\frac{\exp(-\lambda\eta^2/2)}{\eta}$$

where $\lambda = (1 + 2\,\Pi)\,\sigma_T\,N_{RO}$ and A is arbitrary. Thus the strategy discussed earlier is again applicable, namely, integrations to the right from a small value of η with C_θ arbitrary and to the left from a large value of η with A arbitrary are terminated at a common value of $\eta = \eta_m$ and the two parameters adjusted so that the solution is continuous at η_m.

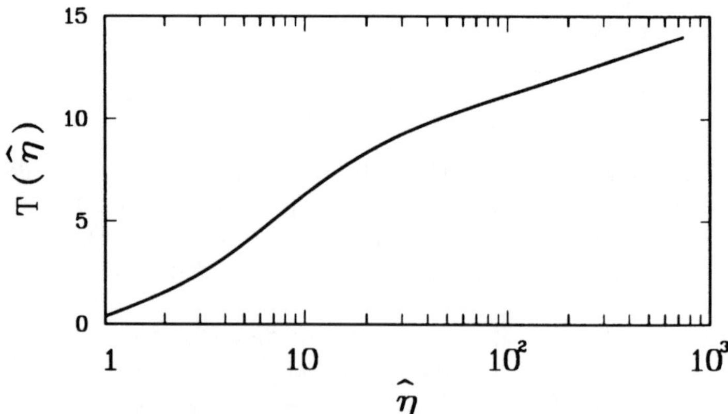

Figure 9.11.1 Dimensionless temperature distribution in the wall layer of a turbulent boundary layer.

Completion of the matching leads, as before, to cancellation of the ln $\hat{\eta}$ terms and to a heat transfer law of the form

$$1 + \frac{\Theta_w'}{\Theta_\infty - \Theta_w}\left(\frac{\sigma_T}{\kappa N_\sigma} \ln \delta + C_\theta - B_\theta\right) = 0 \tag{9.11.13}$$

Figure 9.11.2 shows $T_0(\eta)$ normalized with respect to $T_0(0,1)$ for the values of Π used in Sections 9.9 and 9.10. Also given are the values of $T_0(0,1)$ and C_θ. We see that since $|C_\theta| >> B_\theta$, Eq. (9.11.13) implies that $\Theta_w'/(\Theta_\infty - \Theta_w) > 0$, as called for by physical considerations. We also see from these results that for a given δ, increases in Π result in decreases in $-C_\theta$ and thus increases in $\Theta_w'/(\Theta_\infty - \Theta_w)$.

Our treatment of the characteristics of the thermal boundary layer is confined to the case of constant wall temperature, but a corresponding treatment of the case of constant heat flux can be carried out. The starting point is Eq. (9.11.2) and with the assumption of a solution of the form

$$\Theta(\zeta, \eta) = 1 + \zeta\Theta_w'\overline{\Theta}(\eta) + \sum_{n=1}^{\infty} \overline{Z}_n(\zeta)\overline{N}_n(\eta) \tag{9.11.14}$$

When substituted into Eq. (9.11.2), Eq. (9.11.5) with $T_0(\eta)$ replaced by $\overline{\Theta}(\eta)$ is solved subject to the conditions $\overline{\Theta}'(0) = 1$ and $\overline{\Theta}(\eta \to \infty) = 0$, while the eigenfunctions $\overline{N}_n(\eta)$ are given by Eq. (9.11.7) subject to the conditions $\overline{N}_n'(0) = 0$ and $\overline{N}_n(\eta \to \infty) = 0$. The constant heat flux is related to Θ_w'. The eigenfunctions

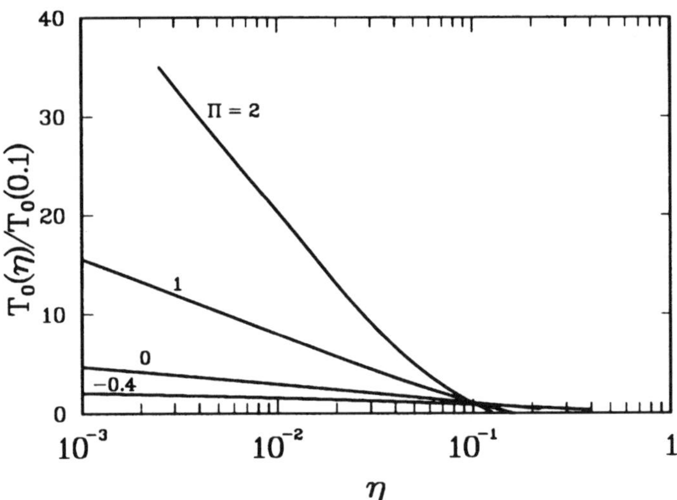

Figure 9.11.2 Mean temperature distribution in the outer flow of an equilibrium turbulent boundary layer. For $\Pi = -0.4$, $T_0(0.1) = -5.08 \times 10^4$, $C_\theta = -1.00 \times 10^6$; for $\Pi = 0$, $T_0(0.1) = -1.33 \times 10^4$, $C_\theta = -5.97 \times 10^4$; for $\Pi = 1$, $T_0(0.1) = -2.79 \times 10^3$, $C_\theta = -4.33 \times 10^4$; and for $\Pi = 2$, $T_0(0.1) = -8.81 \times 10^2$, $C_\theta = -3.73 \times 10^4$.

again account for the influence of the initial temperature distribution on the wall temperature required to maintain the constant heat flux as in Section 9.6. We see again that in the downstream region, where the influence of the initial temperature distribution has vanished, the mean temperature within the boundary layer increases linearly with ζ at a rate that depends on the heat transfer reflected in Θ'_w. We do not pursue this development further.

9.12 THREE-DIMENSIONAL TURBULENT BOUNDARY LAYERS

It is beyond the scope of our considerations to discuss in any detail turbulent boundary layers involving outer flows with *two mean velocity components* parallel to the wall. However, since in many, indeed in most, applications—e.g., in the flow over aircraft, ships, and submarines—the boundary layers are three-dimensional, we call attention to some of their features. Despite their importance, the treatment of these more general boundary layers is not well developed; thus, in the texts cited earlier for engineering calculations, such layers are not discussed. A review of three-dimensional boundary layers in general and turbulent layers in particular is provided by Eichelbrenner (1973). Fannelop and Krogsted (1975) summarize a series of Euromech colloquia dealing with three-dimensional turbulent boundary layers.

Two examples of such layers are shown in Figs. 9.12.1 and 9.12.2. The first is the turbulent boundary layer on an infinite swept body such as a wing. This is a flow which, because of its fundamental nature and applicability to wings of aircraft, is the subject of considerable study. The second is the boundary layer in the vicinity of a plane of symmetry either with inflow, i.e., convergence of the mean streamlines in the outer flow, or with outflow, i.e., divergence of those streamlines. This situation occurs, e.g., on the leeward and windward sides, respectively, of a body of revolution at a small angle of attack. It is also of interest in connection with the alteration by boundary-layer growth on the sidewalls of a nominally two-dimensional boundary layer at the centerplane of a surface between these walls. As mentioned earlier (cf. Section 9.8), turbulent boundary layers at a plane of symmetry represent one of the "complex flows" of Bradshaw (1973). These two flows are sufficiently simple so that we can sketch their analysis and thereby suggest the nature of three-dimensional effects.

Bradshaw and Pontikis (1985) provide data on the turbulent boundary layer on a channel wall with a swept leading edge and with sidewalls arranged to simulate an infinite swept wing, a useful review of the fluid mechanical features of this class of boundary layers, and a discussion of the influence of spanwise flow on the "two-dimensional" boundary layer normal to the leading edge. If the coordinate system is as shown in Fig. 9.12.1, that is, x_1 normal to the leading edge in the surface, x_2 normal to the surface, and x_3 in the spanwise direction,

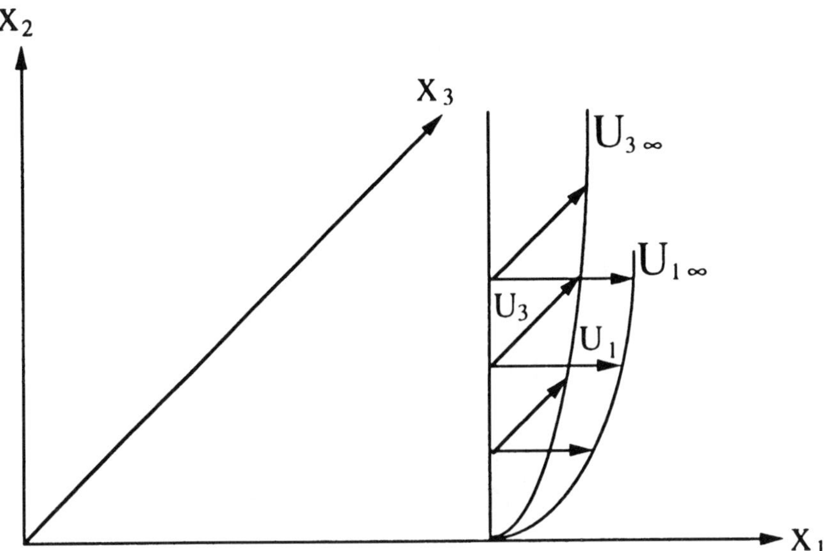

Figure 9.12.1 The turbulent boundary layer on an infinite swept surface. The x_3 axis lies along the leading edge.

then the assumption of an infinite span implies that the x_3 derivative of all statistical quantities is zero. In this case the first-moment equations become

$$\frac{\partial U_1}{\partial x_1} + \frac{\partial U_2}{\partial x_2} = 0$$

$$U_1 \frac{\partial U_1}{\partial x_1} + U_2 \frac{\partial U_1}{\partial x_2} = U_{1\infty} \frac{\partial U_{1\infty}}{\partial x_1} + \frac{\partial}{\partial x_2}\left(\nu \frac{\partial U_1}{\partial x_2} - \overline{u_1 u_2}\right) \tag{9.12.1}$$

$$U_1 \frac{\partial U_3}{\partial x_1} + U_2 \frac{\partial U_3}{\partial x_2} = \frac{\partial}{\partial x_2}\left(\nu \frac{\partial U_3}{\partial x_2} - \overline{u_2 u_3}\right)$$

where all variables except $U_{1\infty}(x_1)$ depend on x_1 and x_2.

The first two of Eqs (9.12.1) are formally identical with those for a two-dimensional turbulent boundary layer in an x_1–x_2 plane; there is therefore a temptation to eliminate the Reynolds shear stress associated with $\overline{u_1 u_2}$ as for the two-dimensional case and to calculate *subsequently* the distribution of the $U_3(x_1, x_2)$ velocity component from the third of Eqs. (9.12.1), i.e., as a second step. Indeed, for laminar flow this is a proper strategy. In turbulent flow, however, such a procedure assumes that the spanwise flow does not alter the turbulence, i.e., the stress $\rho\,\overline{u_1 u_2}$ is unaltered by the U_3 velocity component and its accompanying mean rate of strain $\partial U_3/\partial x_2$. That such an alteration might be expected and thus that the assumed independence may be incorrect is suggested by several

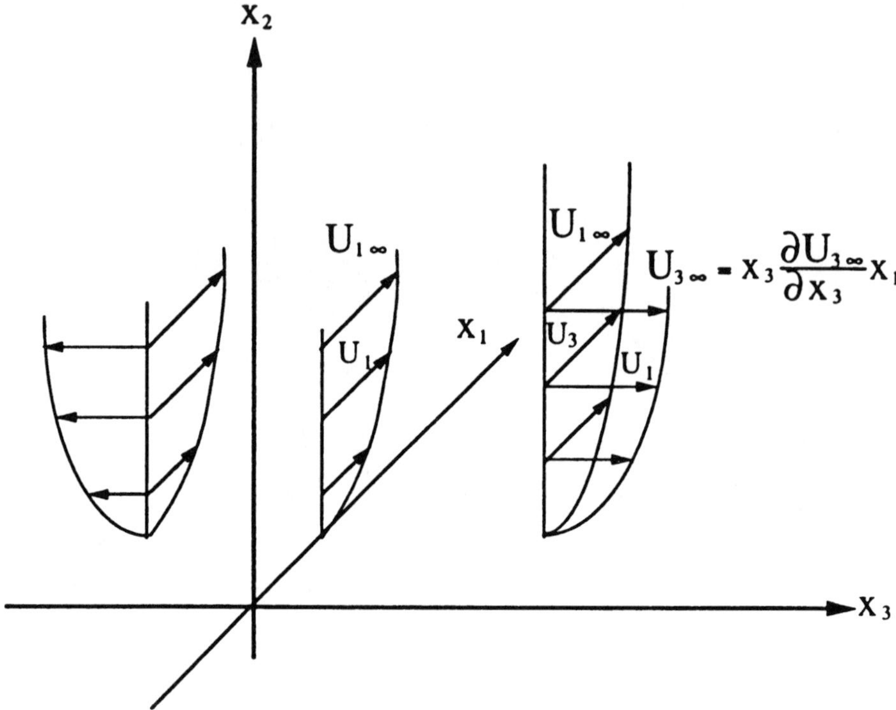

Figure 9.12.2 The turbulent boundary layer near a plane of symmetry.

considerations. If the production terms in the equation for the turbulent kinetic energy, Eq. (4.7.1), are specialized to the three-dimensional boundary layer under consideration, two terms dominate,

$$\overline{u_1 u_2}\,\frac{\partial U_1}{\partial x_2} + \overline{u_2 u_3}\,\frac{\partial U_3}{\partial x_2}$$

and we see that an additional source of turbulence production is associated with the spanwise flow. Furthermore, sweep introduces a second mean vorticity Ω_1 supplementing the usual Ω_3; thus the large turbulent structures in the outer portion of the boundary layer discussed in Section 5.1 and largely accounting for the Reynolds stresses might be expected to be skewed in the spanwise direction and to alter the shear stress $\overline{u_1 u_2}$. Indeed, experimental results confirm the *interdependence* of the chordwise and spanwise boundary layers. It is also found experimentally that the Reynolds shear stress,

$$\mathbf{i}_1\,\rho\,\overline{u_1 u_2} + \mathbf{i}_3\,\rho\,\overline{u_2 u_3}$$

within the boundary layer is not collinear with the vector representing the principal rate of strain, i.e., with

$$\mathbf{i}_1 \frac{\partial U_1}{\partial x_2} + \mathbf{i}_3 \frac{\partial U_3}{\partial x_2}$$

with the implication that classical ideas of turbulent eddy transport, already problematical for simple shear flows (cf. Section 7.3), are even more suspect in these more complex flows. As a consequence, the Reynolds stress modeling outlined in Section 10.10 must be brought to bear in analyzing this three dimensional boundary layer (cf. Rotta, 1979).

We next consider the boundary layer near a plane of symmetry shown in Fig. 9.12.2 but again avoid the complexity of curvature by considering the surface to be flat. Symmetry and antisymmetry considerations at such a plane require that the mean velocities $U_1(\mathbf{x})$ and $U_2(\mathbf{x})$, the Reynolds shear stress $\overline{u_1 u_2}(\mathbf{x})$, and the mean pressure be even functions of x_3 but that the U_3 velocity and the Reynolds shear stress $\overline{u_2 u_3}$ be odd functions of x_3. Accordingly, we let

$$\begin{aligned}
U_1(\mathbf{x}) &\approx {}^{(0)}U_1(x_1, x_2) + O(x_3^2) \\
U_2(\mathbf{x}) &\approx {}^{(0)}U_2(x_1, x_2) + O(x_3^2) \\
U_3(\mathbf{x}) &\approx x_3 \, {}^{(1)}U_3(x_1, x_2) + O(x_3^3) \\
\overline{u_1 u_2}(\mathbf{x}) &\approx {}^{(0)}\overline{u_2 u_2}(x_1, x_2) + O(x_3^2) \\
\overline{u_2 u_3}(\mathbf{x}) &\approx x_3 \, {}^{(1)}\overline{u_2 u_3}(x_1, x_2) + O(x_3^3)
\end{aligned} \tag{9.12.2}$$

The velocity external to the boundary layer is defined in terms of ${}^{(0)}U_{1\infty}(x_1)$ and

$${}^{(1)}U_{3\infty}(x_1) \equiv x_3 \frac{\partial U_3}{\partial x_3} (x_1, x_2 \rightarrow \infty, 0)$$

Outflow corresponds to ${}^{(1)}U_{3\infty} > 0$ and inflow to ${}^{(1)}U_{3\infty} < 0$.

Substitution of these assumed forms into the averaged conservation equations [cf. Eqs. (4.1.1) and (4.2.1)], collection of only the lowest-order terms in x_3, and application of the boundary-layer approximation yields

$$\frac{\partial {}^{(0)}U_1}{\partial x_1} + \frac{\partial {}^{(0)}U_2}{\partial x_2} + {}^{(1)}U_3 = 0$$

$$\begin{aligned}
{}^{(0)}U_1 \frac{\partial {}^{(0)}U_1}{\partial x_1} + {}^{(0)}U_2 \frac{\partial {}^{(0)}U_1}{\partial x_2} &= {}^{(0)}U_{1\infty} \frac{d {}^{(0)}U_{1\infty}}{dx_1} + \frac{\partial}{\partial x_2}\left(\nu \frac{\partial {}^{(0)}U_1}{\partial x_2} - {}^{(0)}\overline{u_1 u_2}\right) \\
{}^{(0)}U_1 \frac{\partial {}^{(1)}U_3}{\partial x_1} + {}^{(0)}U_2 \frac{\partial {}^{(1)}U_3}{\partial x_2} + {}^{(1)}U_3^2 &= {}^{(0)}U_{1\infty} \frac{d {}^{(1)}U_{3\infty}}{dx_1} + \\
&\quad {}^{(1)}U_{3\infty}^2 + \frac{\partial}{\partial x_2}\left(\nu \frac{\partial {}^{(1)}U_3}{\partial x_2} - {}^{(1)}\overline{u_2 u_3}\right)
\end{aligned} \tag{9.12.3}$$

In this case we see that the equations are explicitly coupled, so there is no temptation to assume that the streamwise and crosswise flows are independent.

Given the distribution of the external velocity in terms of ${}^{(0)}U_{1\infty}(x_1)$ and ${}^{(1)}U_{3\infty}(x_1)$, these equations determine ${}^{(0)}U_1(x_1, x_2)$, ${}^{(0)}U_2(x_1, x_2)$, and ${}^{(1)}U_3(x_1, x_2)$. Closure requires description of the two operative Reynolds shear stresses, ${}^{(0)}\overline{u_1u_2}\,(x_1, x_2)$ and ${}^{(1)}\overline{u_2u_3}\,(x_1, x_2)$.

Examination of the production terms in the equation for the turbulent kinetic energy applicable to this boundary layer establishes that again we have two terms:

$$\overline{u_1u_2}\,\frac{\partial U_1}{\partial x_2} + \overline{u_3^2}\;{}^{(1)}U_3$$

In this case we expect neither the skewness of the large turbulent structures in the outer portion of the boundary layer nor the misalignment of the mean shear stress and the mean rate of strain as in the earlier case of the boundary layer on an infinite swept surface, but rather an alteration of the mean velocity and mean shear stress ${}^{(0)}U_1(x_1, x_2)$ and ${}^{(0)}\overline{u_1u_2}(x_1, x_2)$, respectively, as a consequence of the orthogonal rate of strain associated with ${}^{(1)}U_3(x_1, x_2)$. Again, analysis of the flow which involves this complex rate-of-strain field calls for care—perhaps application of the Reynolds stress theory of Section 10.10.

Although we have discussed aspects of three-dimensional turbulent boundary layers within the context of two simple flows, the alteration of the turbulence by rates of strain in orthogonal directions applies more generally and is suggestive of the difficulties associated with the analysis of such layers in more complex geometries.

9.13 TURBULENT CHANNEL FLOW WITH MASS TRANSFER

Turbulent flows sometimes involve porous walls through which fluid is either injected into the stream or withdrawn from the stream. Injected fluid can possess a different composition and/or a different temperature, so in these cases we must deal with both velocity and scalar fields. The cooling of a porous surface exposed to a high-energy stream by injection of a coolant and the application of suction to such a surface in order to suppress boundary-layer separation are practical examples of turbulent flows involving mass transfer, the implication being that a wide variety of such flows can arise. Here we can only illustrate the nature of the changes in the mean velocity distributions resulting from mass transfer while providing entries to the relevant literature.[†]

Early experimental results on turbulent boundary layers with injection by Mickley and Davis (1957) were reanalyzed by Stevenson (1964). Further experimental data on the influence of injection on such layers were given by Collier and Schetz (1984) who provide additional references. Simpson (1970) reviewed

[†]See Antonia et al. (1994) and Mariani et al. (1993) concerning the interesting flow represented by an equilibrium turbulent boundary layer with uniform suction.

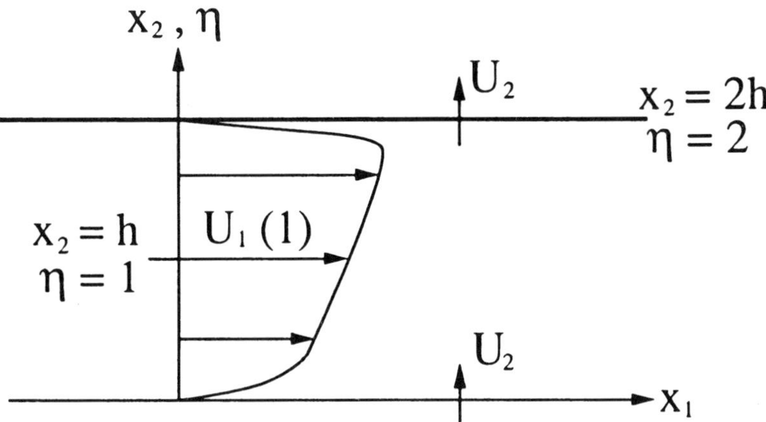

Figure 9.13.1 Schematic representation of the mean streamwise velocity in a channel with mass transfer. Fluid is injected into the flow from the lower wall and withdrawn from the upper wall.

and analyzed the available data on turbulent boundary layers on porous plates subject to injection. More recently, So and Yoo (1987) applied Reynolds stress theory with low-Reynolds-number corrections to shear flows with mass transfer.

Here we consider a simple flow with mass transfer, namely, a modification of the fully developed channel flow treated in Sections 9.1–9.3 involving fluid injection through the lower wall and fluid withdrawn at the same rate through the upper wall.† In this case the mean velocity and turbulence characteristics, although again dependent only on x_2, are not symmetric about the midplane of the channel, i.e., $x_2 = h$. In particular, the mean shear stresses on the two walls are not equal and the two viscous sublayers are different, with the lower one thickened by injection and the upper thinned by suction. Thus, by analyzing the flow across the entire channel we are able to examine the influence of both injection and suction on turbulence behavior in the proximity of walls. Anticipating somewhat the results of our later analysis, Fig. 9.13.1 shows schematically the distribution of the mean streamwise velocity in a channel with mass transfer.

Equations (9.1.1) are altered by the existence of a constant x_2 velocity component to

$$\begin{aligned} \rho U_2 \frac{dU_1}{dx_2} &= -\frac{\partial P}{\partial x_1} + \frac{d}{dx_2}\left(\mu \frac{dU_1}{dx_2} - \rho\, \overline{u_1 u_2}\right) \\ 0 &= -\frac{\partial}{\partial x_2}\left(P + \rho\, \overline{u_2^2}\right) \end{aligned} \tag{9.13.1}$$

By following the steps leading to Eq. (9.1.2) we obtain

†Piomelli et al. (1990) study this flow by means of large eddy simulation.

$$\rho \, U_2 U_1 \;=\; -\frac{\partial P}{\partial x_1}\, x_2 + \mu \,\frac{dU_1}{dx_2} - \rho\, \overline{u_1 u_2} - \rho\, u_{\tau 0}^2 \tag{9.13.2}$$

where $\partial P/\partial x_1$ is constant and $u_{\tau 0} \equiv (\tau_{w0}/\rho)^{1/2}$ is the shearing velocity associated with the *lower wall*. If this equation is applied at the upper wall, i.e., at $x_2 = 2h$, we find

$$-\frac{\partial P}{\partial x_1}\, 2h \;=\; -\tau_{w2} + \rho\, u_{\tau 0}^2 \tag{9.13.3}$$

where τ_{w2} is the mean shear on the upper wall, a negative quantity. If we define $u_{\tau 2} \equiv (-\tau_{w2}/\rho)^{1/2}$, we obtain from Eq. (9.13.3) the analog of Eq. (9.1.3):

$$-\frac{\partial P}{\partial x_1}\, h = \frac{1}{2}\, \rho\, (u_{\tau 0}^2 + u_{\tau 2}^2) \tag{9.13.4}$$

We thus see that the streamwise pressure gradient depends on the mean of the square of the two shearing velocities. Equation (9.13.4) reduces properly if those velocities are equal.

As a further preliminary we introduce $\eta = x_2/h$ and obtain from Eq. (9.13.2) the pivotal equation for channel flow with mass transfer:

$$\frac{\nu}{h}\frac{dU_1}{d\eta} - \overline{u_1 u_2} = u_{\tau 0}^2 \left[1 - \frac{1}{2}\left(1 + \frac{u_{\tau 2}^2}{u_{\tau 0}^2}\right) \eta \right] + U_2 U_1 \tag{9.13.5}$$

which reduces properly to Eq. (9.1.4). If we suppress the viscous stress in Eq. (9.13.5), we see that the Reynolds shear stress is no longer linearly distributed across the channel. Thus fluid elements at various lateral positions in the channel experience shear forces. Since the force on such elements due to the pressure gradient is constant, these shear forces account for the acceleration associated with the $U_2(dU_1/dx_2)$ term in the first of Eqs. (9.13.1).

To close Eq. (9.13.5) we adopt the mixing length model but must take into account that dU_1/dx_2 is negative in some portions of the flow. Accordingly, we rewrite Eq. (9.2.2) as

$$-\overline{u_1 u_2} = \tilde{l}^2 \left| \frac{dU_1}{d\eta} \right| \frac{dU_1}{d\eta} \tag{9.13.6}$$

where $\tilde{l}$ is a dimensionless mixing length which we assume is unaffected by mass transfer so that the representation of Eq. (9.2.12) applies; we argue that the rates of mass transfer of interest are small, typically $U_2/U_1(1)$ of the order of 10^{-2}, and thus that the *mixing length* is not greatly altered by such transfer although $U_1(\eta)$ and $\overline{u_1 u_2}(\eta)$ are significantly influenced.

We now introduce the dependent variable $\tilde{U}_1 \equiv U_1/u_{\tau 0}$ so that our final equation is

$$\delta \frac{d\tilde{U}_1}{d\eta} + \tilde{l}^2 \left|\frac{d\tilde{U}_1}{d\eta}\right| \frac{d\tilde{U}_1}{d\eta} = 1 - R\,\eta + \tilde{U}_2\tilde{U}_1 \tag{9.13.7}$$

where

$$R \equiv \frac{1}{2}\left(1 + \frac{u_{\tau 2}^2}{u_{\tau 0}^2}\right) \qquad \tilde{U}_2 \equiv \frac{U_2}{u_{\tau 0}}$$

are introduced for notational convenience. We consider $\tilde{U}_2$ to be the parameter characterizing the extent of the mass transfer. Without loss of generality we restrict $\tilde{U}_2$ to positive values; if fluid is injected through the upper wall, we simply relocate the origin of the x_2 coordinate at that wall. The parameter $\delta \equiv \nu/u_{\tau 0}h << 1$ is the reciprocal of a Reynolds number, essentially the same small parameter we have encountered repeatedly in this chapter. Equation (9.13.7) is to be solved subject to the conditions $\tilde{U}_1(0) = \tilde{U}_1(2) = 0$ with δ, $\tilde{U}_2$, and a measure of the volumetric flow rate in the channel, e.g., $\tilde{U}_1(1)$, as parameters.

In contrast with our earlier treatment of channel flow without mass transfer, here Eq. (9.13.7) must be integrated numerically. One strategy for doing so involves separate integrations from the centerplane, $\eta = 1$ toward the two walls. With $\tilde{U}_2$ and $\tilde{U}_1(1)$ specified, we select δ and R such that the no-slip condition is satisfied at each wall. By repeating this calculation for a range of $\tilde{U}_1(1)$, we obtain $\delta = \delta(\tilde{U}_1(1); \tilde{U}_2)$ and $R(\tilde{U}_1(1); \tilde{U}_2)$. Again, repetition of this sequence for a range of $\tilde{U}_2$ provides solutions for the entire parameter space associated with this flow.

Rather than pursue this direct treatment of Eq. (9.13.7), we follow our earlier practice of taking advantage of the small magnitude of δ of applied interest and seek an asymptotic solution, one which reflects the relative thinness of the wall layers even with injection if the Reynolds number is sufficiently large, i.e., δ is sufficiently small. In this approach $\tilde{U}_1(\eta)$ is expanded in powers of δ as in Eq. (9.2.6), but here we must do likewise for R so that $R = {}^{(0)}R + \delta\, {}^{(1)}R + \ldots$. It is again assumed that $\tilde{l}(\eta)$ is independent of δ and thus independent of *both* $\tilde{U}_2$ and δ. We restrict our attention to the lowest-order solutions but note that as many additional terms in δ as desired can be systematically considered. In the interest of simplicity we dispense with superscripts on $\tilde{U}_1(\eta)$ and R.

The lowest-order outer solutions are given by Eq. (9.13.7) with $\delta = 0$ and involve $\tilde{U}_1(1)$ and $\tilde{U}_2$ as parameters. Exposition is facilitated if we make the assumption, justified a posteriori, that $d\tilde{U}_1/d\eta > 0$ throughout the outer flow. By this assumption we preclude reduction of the present analysis to that of Sections 9.3 and 9.4, i.e., to the channel without mass transfer.[†] In this case we see from Eq. (9.13.7) with $\delta = 0$ that $1 - R\eta + \tilde{U}_2\tilde{U}_1 > 0$ for all η. Thus we rewrite Eq. (9.13.7) as

[†]See Silva-Freire (1988) for an analysis of the double limit of large Reynolds number and vanishing suction velocity for a turbulent boundary layer.

$$\frac{d\tilde{U}_1}{d\eta} = \frac{(1 - R\eta + \tilde{U}_2\tilde{U}_1)^{1/2}}{\tilde{l}} \tag{9.13.8}$$

It is useful to examine the behavior of the solutions to Eq. (9.13.8) as the walls are approached in order to anticipate the requirements of matching with the inner solutions at each wall. To that end, consider $\eta \to 2$; in this case we see that since $\tilde{l} \to \kappa(2 - \eta)$ [cf. Eq. (9.2.12)], where κ is the von Karman constant, $\tilde{U}_1$ must approach a constant, i.e.,

$$\tilde{U}_1(\eta \to 2) = \frac{2R - 1}{\tilde{U}_2} \tag{9.13.9}$$

The implication of Eq. (9.13.9) is that $2R - 1 > 0$. Thus, in solving Eq. (9.13.8) with $\tilde{U}_2$ and $\tilde{U}_1(1)$ specified, we select R so that Eq. (9.13.9) is satisfied. The complementary integration toward the lower wall is then carried out free of arbitrariness. As $\eta \to 0$, Eq. (9.13.8) becomes

$$\frac{d\tilde{U}_1}{d\eta} = \frac{(1 + \tilde{U}_2\tilde{U}_1)^{1/2}}{\kappa\eta} \tag{9.13.10}$$

Suppose that Eq. (9.13.10) applies for $\eta < \eta^* << 1$; then we have

$$\frac{2}{\tilde{U}_2}(1 + \tilde{U}_2\tilde{U}_1)^{1/2} = \frac{1}{\kappa}\ln\eta + C \tag{9.13.11}$$

where $C = C[\tilde{U}_1(1), \tilde{U}_2]$ is known from the integration from the midplane and is independent of η^*. If this behavior is compared with that of the outer solutions at an impermeable wall [cf. Eq. (9.2.13)], the significant influence of injection on the turbulence is evident.

In developing the equations for the wall layers we follow the analysis of Section 9.3. The dependent variable in each layer remains unchanged, that is, $U_1/u_{\tau 0}$, but to facilitate identification of the inner solutions we introduce $\hat{U}_1$. Without ambiguity, $\tilde{U}_2$ is retained as the parameter characterizing the extent of mass transfer.

For the upper wall we introduce the new independent variable $\bar{\eta} \equiv (2 - \eta)/\delta$. Again exposition is facilitated if we assume that $d\hat{U}_1/d\bar{\eta} > 0$ throughout the layer adjacent to the upper wall. In this case we thus obtain from Eq. (9.13.7) to lowest order in δ the equation

$$\frac{d\hat{U}_1}{d\bar{\eta}} + \hat{l}^2\left(\frac{d\hat{U}_1}{d\bar{\eta}}\right)^2 = (2\,R - 1 - \tilde{U}_2\,\hat{U}_1) \tag{9.13.12}$$

where $\hat{l} \equiv \tilde{l}/\delta$ is a stretched dimensionless mixing length applicable to the wall layer [cf. Eq. (9.3.2)] and where R is known from the outer solutions for specified values of $\tilde{U}_2$ and $\tilde{U}_1(1)$. Consistent with our earlier assumption that the mixing length is not altered by mass transfer, we use Eq. (9.3.9) to describe

$\hat{l}(\overline{\eta})$. However, κ_w can no longer be selected so that the constant in the log law has a specified value; rather, we consider κ_w to be an empirical coefficient. Since as $\overline{\eta} \to 0$ the turbulent transport term vanishes, we find from Eq. (9.13.12) that

$$\hat{U}_1 \approx \frac{2R - 1}{\tilde{U}_2} (1 - e^{-\tilde{U}_2 \overline{\eta}}) \approx (2R - 1)\, \overline{\eta} \tag{9.13.13}$$

If the shearing velocity $u_{\tau 2}$ is computed from Eq. (9.13.13), this equation is consistent with the definition of R given by Eq. (9.13.7). As $\overline{\eta} \to \infty$, turbulent transport dominates with $\hat{l} \propto \overline{\eta}$ at the outer edge of the wall layer, with the consequence that

$$\hat{U}_1(\overline{\eta} \to \infty) = \frac{2R - 1}{\tilde{U}_2} \tag{9.13.14}$$

Comparison of Eqs. (9.13.9) and (9.13.14) establishes that the matching requirements in the neighborhood of the upper wall are automatically met.

We turn next to the inner solution adjacent to the lower wall. In this case we introduce as the independent variable $\hat{\eta} \equiv \eta/\delta$ and assume that $d\hat{U}_1/d\hat{\eta} > 0$ throughout the wall layer so that Eq. (9.13.7) becomes

$$\frac{d\hat{U}_1}{d\hat{\eta}} + \hat{l}^2 \left(\frac{d\hat{U}_1}{d\hat{\eta}}\right)^2 = 1 + \tilde{U}_2 \hat{U}_1 \tag{9.13.15}$$

which can be rewritten in the more convenient form

$$\frac{d\hat{U}_1}{d\hat{\eta}} = \frac{-1 + [1 + 4\, \hat{l}^2(1 + \tilde{U}_2 \hat{U}_1)]^{1/2}}{2\, \hat{l}^2} \tag{9.13.16}$$

Consider next the behavior of the solutions to Eq. (9.13.16) in the two limits, $\hat{\eta} \to 0$ and $\hat{\eta} \to \infty$. In the former limit with the turbulent transport term absent the solution is

$$\hat{U}_1 \approx \frac{1}{\tilde{U}_2} (e^{\tilde{U}_2 \hat{\eta}} - 1) \approx \hat{\eta} \tag{9.13.17}$$

As $\hat{\eta} \to \infty$, it is readily found that with $\hat{l} \propto \hat{\eta}$ we have

$$\frac{d\hat{U}_1}{d\hat{\eta}} \approx \frac{1}{\kappa \hat{\eta}} (1 + \tilde{U}_2 \hat{U}_1)^{1/2}$$

which has a solution

$$\frac{2}{\tilde{U}_2} (1 + \tilde{U}_2 \hat{U}_1)^{1/2} \approx \frac{1}{\kappa} \ln \hat{\eta} + B \tag{9.13.18}$$

where B is a constant determined as part of the solution. Note that Eq. (9.13.18) indicates an $\ln^2 \hat{\eta}$ dependence at the outer edge of the wall layer.

As suggested earlier, the matching of the inner and outer solutions at the upper wall provides no new information, but matching at the lower wall utilizing the procedure leading to Eq. (9.4.1) yields

$$B = \frac{1}{\kappa} \ln \delta + C \tag{9.13.19}$$

Although it is not apparent, Eq. (9.13.18) is indeed an implicit skin friction law because while B depends only on $\tilde{U}_2$, the parameter C depends on both $\tilde{U}_2$ and $\tilde{U}_1(1)$. Thus, for a fixed rate of mass transfer determined by $\tilde{U}_2$, the variation of Reynolds number reflected in δ with $\tilde{U}_1(1)$ is obtained from Eq. (9.13.18).

Figure 9.13.2 shows the solutions for the outer flow for a representative value of $\tilde{U}_1(1)$, namely, 25, and for two values of $\tilde{U}_2$, namely, 0.25 and 0.50, values which correspond to injection velocities of 1% and 2% of $U_1(1)$. The associated values of R are found to be 4.89 and 9.87, while those of C are 20.3 and 14.2 for the values 0.25 and 0.50, respectively. We thus deduce from the definition of R [cf. Eq. (9.13.7)] that these small rates of injection at the large Reynolds numbers under consideration result in significant differences in the mean shear stress at the two walls.

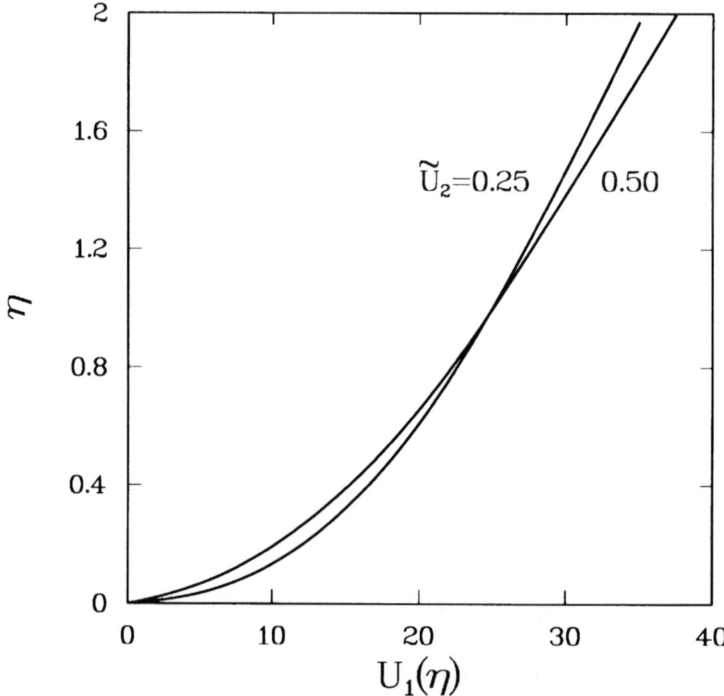

Figure 9.13.2 Distribution of the mean streamwise velocity in the outer flow with two rates of dimensionless mass transfer.

The numerical solutions for the two wall layers require a value for κ_w in the representation of the mixing length $\hat{l}$. As noted in Sections 9.3 and 9.4, we choose this coefficient to obtain the experimentally observed constant B in the log portion of the law of the wall, but here we have no such criterion: If experimental data on the variation of skin friction with Reynolds number and injection rate were available, we could use Eq. (9.13.19) to guide the choice of κ_w. Absent such data we set $\kappa_w = 10$, a value which results in a consistent calculation but which could be altered if data on channel flow with mass transfer were available. With this value assumed for the layers adjacent to each wall, we obtain from Eqs. (9.13.12) and (9.13.16) the results shown in Figs. 9.13.3 and 9.13.4. For these calculations we use the two values of $\tilde{U}_2$ cited earlier. Also shown in each figure are the mean velocity distributions absent mass transfer taken from Fig. 9.3.1. Figure 9.13.3 pertains to the upper wall and shows that suction alters the law of the wall profoundly; indeed, there is no similarity with the usual law of the wall. In Fig. 9.13.4 we see the $\ln^2 \hat{\eta}$ distribution of the mean velocity under the influence of injection, again a profound alteration of the wall layer. The values of B from Eq. (9.13.18) for $\tilde{U}_2$ equal to 0.25 and 0.50

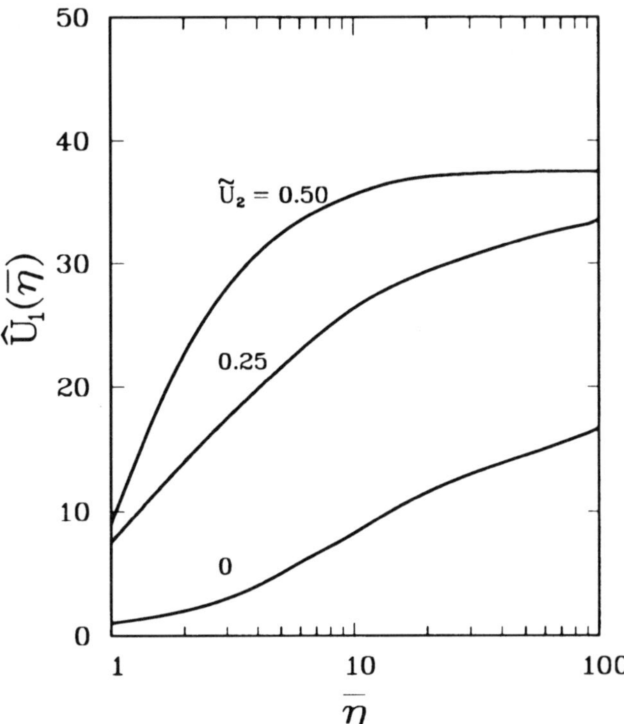

Figure 9.13.3 Distribution of the mean streamwise velocity in the wall layer with suction at two dimensionless rates.

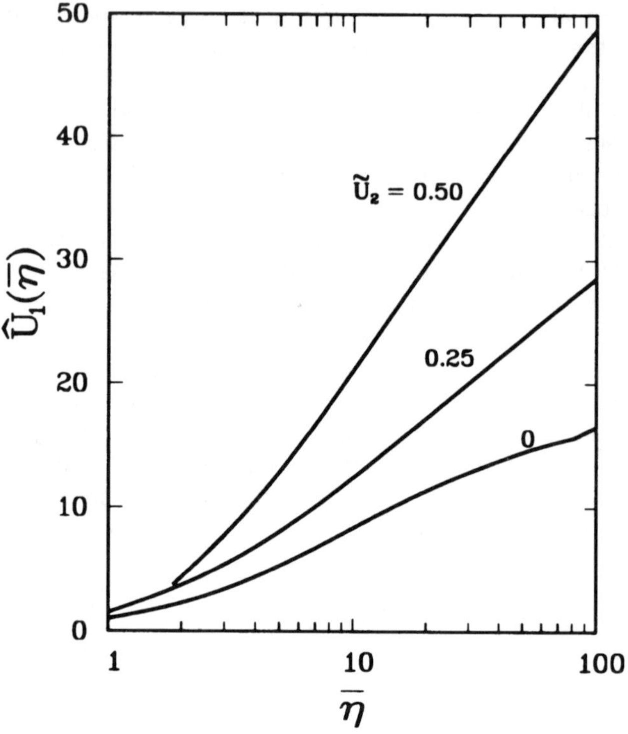

Figure 9.13.4 Distribution of the mean streamwise velocity in the wall layer with injection at two dimensionless rates.

are 11.2 and 8.61, respectively. The results in Figs. 9.13.3 and 9.13.4 are in good qualitative agreement with the calculations of Piomelli et al. (1990).

We can calculate from Eq. (9.13.19) the Reynolds number corresponding to $\tilde{U}_1(1)$ for the two rates of mass transfer. We find for $\tilde{U}_2$ equal to 0.25 and 0.50 that δ is 2.63×10^{-2} and 1.07×10^{-1}, respectively. Thus the increase in mass transfer for the same mean velocity on the midplane corresponds to a decrease in Reynolds number based on $u_{\tau 0}$.

9.14 THE INFLUENCE OF LONGITUDINAL CURVATURE

We emphasized earlier that the boundary layers dealt with in Sections 9.8–9.12 are assumed to occur on flat surfaces, so curvature effects are absent. Here we discuss briefly and qualitatively the important and interesting problem of the influence of longitudinal curvature on such layers. The boundary layer on the wings of aircraft or on turbine blades can be altered significantly by curvature of the mean streamlines if the ratio of thickness to the radius of curvature of the surface is not suitably small. A study of the effect of longitudinal curvature in homogeneous shear flows was made recently by Gatski and Savill (1989) and Tselepidakis et al. (1992), the former publication provides an extensive listing of the literature on curvature effects, including the early seminal contribution of Bradshaw (1969, 1973). Despite a long history of studies of the effect of longitudinal curvature on boundary layers and shear flows, an entirely satisfactory predictive method does not appear to be available (cf. Tselepidakis et al., 1992).

A sense of the influence of longitudinal curvature on turbulence is given by considering the conservation and transport equations in terms of s–n coordinates as shown in Fig. 9.14.1. The two velocity components $\tilde{u}$ and $\tilde{v}$ are in the s and n coordinate directions, respectively. There is a third coordinate z normal to the s–n plane with associated velocity $\tilde{w}$, but we need not consider the flow in this third direction. The surface defined by $n \equiv 0$ has a distribution of curvature described by $\kappa(s)$ with a convex surface as shown in Fig. 9.13.1 corresponding to $\kappa > 0$. Inspection of this figure suggests that if $\kappa(s) < 0$, the range of permissible n is restricted, but in this case we limit n to suitably small values. Although we think of the curve $n \equiv 0$ as being a wall, it can also be treated as a mean streamline in a free shear flow.

The continuity equation and the momentum equations in the s and n directions are (van Dyke, 1969)

$$\frac{\partial \tilde{u}}{\partial s} + \frac{\partial}{\partial n}(h\, \tilde{v}) = 0$$

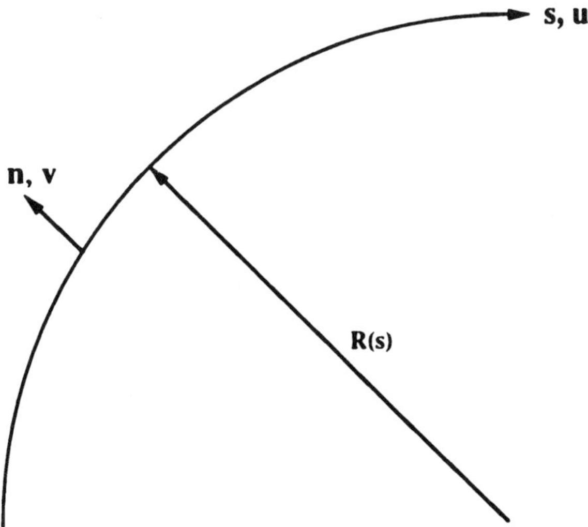

Figure 9.14.1 A curvilinear coordinate system to describe the boundary layer on a curved wall.

$$\frac{\partial \tilde{u}}{\partial t} + \frac{\partial \tilde{u}^2}{\partial s} + \frac{\partial}{\partial n}(h\,\tilde{v}\,\tilde{u}) + \kappa\,\tilde{u}\,\tilde{v} = -\frac{1}{\rho}\frac{\partial \tilde{p}}{\partial s} + \nu\,h\,\frac{\partial}{\partial n}\left\{\frac{1}{h}\left[\frac{\partial}{\partial n}(h\,\tilde{u}) - \frac{\partial \tilde{v}}{\partial s}\right]\right\}$$

$$\frac{\partial \tilde{v}}{\partial t} + \frac{\partial}{\partial s}(\tilde{u}\tilde{v}) + \frac{\partial}{\partial n}(h\,\tilde{v}^2) - \kappa\,\tilde{u}^2 = -\frac{h}{\rho}\frac{\partial \tilde{p}}{\partial n} + \nu\,h\,\frac{\partial}{\partial s}\left\{\frac{1}{h}\left[\frac{\partial \tilde{v}}{\partial s} - \frac{\partial}{\partial n}(h\,\tilde{u})\right]\right\} \tag{9.14.1}$$

where $h \equiv 1 + n\,\kappa(s)$ is a dimensionless curvature function. If we decompose the flow variables into mean and fluctuating components, we obtain the following first-moment equations:

$$\frac{\partial U}{\partial s} + \frac{\partial}{\partial n}(h\,V) = 0$$

$$\frac{\partial U^2}{\partial s} + \frac{\partial}{\partial n}(h\,U\,V) + \kappa\,U\,V = -\frac{\partial}{\partial s}\left(\frac{P}{\rho} + \overline{u^2}\right) - \frac{\partial}{\partial n}(h\,\overline{uv}) - \kappa\,\overline{uv} \tag{9.14.2}$$

$$\frac{\partial}{\partial s}(U\,V) + \frac{\partial}{\partial n}(h\,V^2) - \kappa\,U^2 = -\frac{\partial}{\partial n}\left(\frac{P}{\rho} + \overline{v^2}\right) + \kappa\,(\overline{u^2} + \overline{v^2})$$

where we drop the viscous terms as being unimportant for our considerations; i.e., when we consider the curve $n = 0$ to be a wall, we deal with the turbulence *external* to the wall layer. If $\kappa(s) \equiv 0$ so that $h = 1$, we recover our earlier first-moment equations [cf. Eq. (4.2.1)], but we see various additions to the conservation and transport equations when the wall is curved.

The corresponding second-moment equations are obtained in the usual fashion from Eqs. (9.14.1), i.e., by multiplication and/or cross-multiplication, addition, and averaging. Of interest in the present discussion are the following equations:

$$\frac{1}{2}\frac{\partial}{\partial s}(U\,\overline{u^2} + \overline{u^3}) + \frac{1}{2}\frac{\partial}{\partial n}\,[h\,(V\,\overline{u^2} + \overline{uv^2})] + \overline{u^2}\,\frac{\partial U}{\partial s} + \overline{uv}\,\frac{\partial U}{\partial n} + \kappa\,(U\,V + \overline{uv})$$

$$= -\frac{1}{\rho}\,\overline{u\,\frac{\partial p}{\partial s}} - \frac{1}{2}\,\varepsilon_{uu}$$

$$\frac{1}{2}\frac{\partial}{\partial s}\,(U\,\overline{v^2} + \overline{uv^2}) + \frac{1}{2}\frac{\partial}{\partial n}[h\,(V\,\overline{v^2} + \overline{v^3})] + \overline{v^2}\,\frac{\partial V}{\partial s} + \overline{uv}\,\frac{\partial V}{\partial n}$$

$$- \kappa\,(2\,U + \overline{uv} + \overline{u^2v}) = -\frac{h}{\rho}\,\overline{v\,\frac{\partial p}{\partial n}} - \frac{1}{2}\,\varepsilon_{vv}$$

$$\frac{\partial}{\partial s}(U\,\overline{uv} + \overline{u^2v}) + \frac{\partial}{\partial n}[h\,(V\,\overline{uv} + \overline{uv^2})] + \overline{u^2}\,\frac{\partial V}{\partial s} + \overline{uv}\left(\frac{\partial U}{\partial s} + h\,\frac{\partial V}{\partial n}\right)$$

$$+ \overline{v^2}\,\frac{\partial}{\partial n}(h\,U) - \kappa(2\,U\,\overline{u^2} + \overline{u^3} - \overline{uv}\,V - \overline{uv^2}) = -\frac{1}{\rho}\left(\overline{v\,\frac{\partial p}{\partial s}} + \overline{h\,u\,\frac{\partial p}{\partial n}}\right) - \frac{1}{2}\,\varepsilon_{uv}$$

(9.14.3)

If $\kappa = 0$ so that $h = 1$, these equations reduce to those derivable from Eq. (4.4.1). Here viscous effects enter implicitly via the dissipation terms ε_{uu}, ε_{vv}, and ε_{uv}. If these equations are to form the basis for quantitative predictions of the influence of longitudinal curvature, they must be closed by appropriate modeling, e.g., as discussed in Chapter 10. This modeling must reflect the influence of curvature on the pressure and thus on the pressure–rate-of-strain terms in the Reynolds stress equations (cf. Sections 10.10 and 10.13). This influence and the explicit role of curvature contained in Eqs. (9.14.3) alters the turbulent structure of shear layers. For example, it is found that streamwise longitudinal vortices tend to be strengthened in boundary layers on concave surfaces and to be suppressed on convex surfaces.

Our purposes are served by deducing from Eqs. (9.14.3) the qualitative influence of longitudinal curvature. If we consider a turbulent boundary layer on a slightly curved surface such that $h \approx 1$ and $U >> V$, we see that the dominant influence of curvature is reflected in the $-2\ \kappa\ U\ \overline{uv}$ and $-2\ \kappa\ U\ \overline{u^2}$ terms in the $\overline{v^2}$ and $\overline{uv}$ equations, respectively. The implication is that on a convex surface ($\kappa > 0$) the intensity $\overline{v^2}$ and shear stress $\rho\ \overline{uv}$ are *decreased* by curvature, while on a concave surface ($\kappa < 0$) they are both *increased.*

9.15 SUMMARY

This chapter concerns the second class of turbulent shear flows, those involving one or more walls. In these flows viscosity and thermal conductivity are important in a layer adjacent to the wall, the viscous sublayer, no matter how large the global Reynolds number of the flow. When that Reynolds number is large, the sublayer is thin and the flow is amenable to analysis by asymptotic methods; in this case description of the flow separates into an inner solution within which viscous and turbulent transport exchange dominance and an outer solution devoid of viscous effects. A variety of wall-bounded flows have been treated in this chapter: channel flow, Couette flow, pipe flow, and boundary layers. The problem of the three-dimensional boundary layer is sketched. The dramatic alteration of the sublayer by small rates of mass transfer has been discussed within the context of a channel with porous walls permitting the injection of fluid through the lower wall and withdrawal at the upper wall. The chapter concluded with a brief discussion of the influence of longitudinal curvature.

CHAPTER

TEN

HIGHER-ORDER MOMENT AND RELATED METHODS

Chapters 8 and 9 focused on the mean velocity and the Reynolds shear stress distributions and on their temperature counterparts in free shear and wall-bounded flows. If information concerning the other Reynolds stresses in these flows, particularly the intensities $\overline{u_i^2}$, $i = 1, 2, 3$, is desired, the second-moment equations, Eqs. (4.4.1), must be considered. To do so, however, requires significant additional modeling to achieve a closed system of equations. This chapter discusses such considerations and reviews some contemporary moment methods for the treatment of turbulent flows. There are currently available a variety of second-moment methods involving various levels of complexity and generality and undergoing continuous improvement. Useful recent reviews are by Launder (1990) and Hanjalic (1994).

Our purposes are served by emphasizing the widely used k–ε method and its application to one or more simple turbulent flows and by then moving on to a discussion of the Reynolds stress closure, which is generally considered to be the most complete moment formulation of applied significance, one providing the framework for future developments in moment methods. In the course of this discussion, reference is made to other theories so that a general perspective on the more advanced moment methods is set forth. The concluding section calls attention to several recent developments in second-moment methods and to an alternative to such methods.

Initially we consider only isothermal flows and neglect body forces, but subsequently we shall incorporate such forces to deal with buoyancy effects. Similarly, we initially consider free shear flows with high turbulence Reynolds numbers and thus neglect molecular transport. Subsequent applications to wall-bounded flows call for modifications to incorporate the effects of low turbulence Reynolds numbers, which always arise in the viscous sublayer.

10.1 THE k–ε THEORY

To set the stage for our considerations, we rewrite the averaged transport equations, Eqs. (4.1.1), (4.2.1), and (4.7.1). One motivation for considering Eq. (4.7.1), for the turbulent kinetic energy, resides in our earlier observation in connection with Eq. (7.1.3), that if gradient transport is assumed, then information concerning the turbulent exchange coefficient μ_T, the turbulent kinetic energy $k(\mathbf{x})$, and the mean velocity components $U_i(\mathbf{x})$ determines *all* of the Reynolds stresses. Accordingly, we take up the following equations:

$$\frac{\partial U_k}{\partial x_k} = 0$$

$$\frac{\partial}{\partial x_k}(U_k U_i) = -\frac{\partial}{\partial x_i}\left(\frac{P}{\rho}\right) - \frac{\partial}{\partial x_k}(\overline{u_i u_k}) \tag{10.1.1}$$

$$\frac{\partial}{\partial x_k}(U_k k) + \overline{u_k u_l}\,\frac{\partial U_l}{\partial x_k} = -\frac{\partial}{\partial x_k}\left(\frac{\overline{pu_k}}{\rho} + \frac{1}{2}\,\overline{u_k u_l u_l}\right) - \varepsilon$$

where we have neglected for the time being molecular stresses.† These five equations are closed by appropriate modeling so as to permit determination of the five principal dependent variables: U_i, $i = 1, 2, 3$, P, and k. To achieve closure, however, many additional quantities must be expressed in terms of these principal variables, namely, the six Reynolds stresses, $\rho\,\overline{u_i u_j}$; the three turbulent diffusion terms,

$$\frac{\overline{pu_i}}{\rho} + \frac{1}{2}\,\overline{u_i u_l u_l} \tag{10.1.2}$$

and the mean viscous dissipation ε.

The gradient transport model for the Reynolds stresses given by Eq. (7.1.3) yields a significant advance toward closure:

$$-\overline{u_i u_j} = \nu_T\left(\frac{\partial U_i}{\partial x_j} + \frac{\partial U_j}{\partial x_i}\right) - \frac{2}{3}\,k\,\delta_{ij} \tag{10.1.3}$$

†Later we discuss a low-Reynolds-number version of the k–ε equations.

Since the terms of Eq. (10.1.2) appear differentiated in the last of Eqs. (10.1.1) and can therefore be interpreted as representing diffusive effects, by analogy with the model for the turbulent flux of a scalar, Eq. (7.2.1), they are described by a gradient model,

$$\frac{\overline{pu_i}}{\rho} + \frac{1}{2}\overline{u_i u_k u_k} = -\frac{\nu_T}{\sigma_k}\frac{\partial k}{\partial x_i} \tag{10.1.4}$$

where σ_k is the first of several empirical coefficients.[†] Notice that the influence of pressure fluctuations is modeled in terms of local quantities, whereas we know from Eq. (2.4.1) that the pressure at a given point in space and time depends on the instantaneous velocity throughout all the space occupied by fluid at that time. We discuss modeling of the pressure in more detail in Section 10.10. With Eqs. (10.1.3) and (10.1.4), attention focuses on the single quantity ν_T, the turbulent exchange coefficient.

The dimensions of ν_T are UL and, as noted earlier, the various models for this coefficient differ in the "velocity" and "length" they involve. With k treated as a principal dependent variable, an obvious model is

$$\nu_T = \kappa\, k^{1/2} L \tag{10.1.5}$$

where κ is an empirical coefficient and L is expressed by an algebraic equation either in terms of a global length of the flow, e.g., the thickness of the turbulent wake or jet, or in some other relatively simple fashion. This is the Prandtl-Kolmogorov model for turbulent exchange. At the present time it is occasionally adopted either when a form for L is considered more reliable than an alternative model for ν_T or when the resulting simplification more than compensates for the possibly diminished accuracy (cf. Donaldson, 1972; Hunt et al., 1988). When this model is used, the mean dissipation ε in the third of Eqs. (10.1.1) is replaced by the product of an empirical coefficient times $k^{3/2}/L$ so that Eqs. (10.1.1)–(10.1.4) determine U_i, $i = 1, 2, 3$, P, and k as functions of $\mathbf{x}$. However, a widely accepted alternative model for ν_T is based on k and ε. Implicit in this approach is the recognition that $k^{3/2}/\varepsilon$ *defines* a length L, which we know from the discussion in Sections 5.2–5.4 characterizes the large, energy-containing fluctuations. According to the spectral cascade notions discussed in those sections, the fluctuations *determine* viscous dissipation. Implementation of such a model calls for an equation for ε the $(\mathbf{x})$.

A formal derivation of a partial differential equation for the mean viscous dissipation can be carried out by appropriate operations applied to the momentum equations, Eqs. (2.2.1) (cf. e.g., Hanjalic and Launder, 1972b), but the complexity resulting from this formal attack and the extensive and uncertain modeling needed to reduce the resulting equation to usable form call for a pragmatic

[†]We adopt the generally accepted notation for the empirical constants arising in the k–ε method, but in doing so we violate in this instance our rule reserving the subscript k for summation.

alternative. Thus there is introduced an equation based largely on modeling. Guidance is provided by other equations, e.g., Eqs. (4.4.1) and (4.7.1). From this perspective, consider the following equation:

$$\frac{\partial}{\partial x_k}(U_k\,\varepsilon) - c_{\varepsilon 1}\frac{\varepsilon}{k}\overline{u_k u_l}\frac{\partial U_l}{\partial x_k} = \frac{\partial}{\partial x_k}\left(\frac{\nu_T}{\sigma_\varepsilon}\frac{\partial \varepsilon}{\partial x_k}\right) - c_{\varepsilon 2}\frac{\varepsilon^2}{k} \tag{10.1.6}$$

where we arrange the terms in the same order as they appear in the cited transport equations so that from left to right we have convection, production, turbulent diffusion, and mean dissipation. The extent of the modeling in Eq. (10.1.6) is clearly suggested by the presence of three additional empirical constants, $c_{\varepsilon 1}$, σ_ε, and $c_{\varepsilon 2}$. A partial rationale for the production and dissipation terms is as follows. Suppose that $c_{\varepsilon 1} = c_{\varepsilon 2}$ and that in the third of Eqs. (10.1.1) and (10.1.6) there is a balance between production and dissipation (cf. Section 6.3 for a discussion of this balance in simple shear flows); then both of these equations reduce to

$$\overline{u_k u_l}\frac{\partial U_l}{\partial x_k} - \varepsilon = 0 \tag{10.1.7}$$

Thus, under these circumstances the transport equations for the turbulent kinetic energy and the mean dissipation are consistent. In applications the empirical constants $c_{\varepsilon 1}$ and $c_{\varepsilon 2}$ are assigned somewhat different values to provide for deviations from strict equilibrium between production and dissipation in a variety of flows. In this regard it should be recalled that in Sections 6.3–6.5 we found it necessary to restrict these coefficients and indeed to assign nonstandard values in order to achieve certain desired results.

Discrepancies which arise between prediction and experiment in some applications of the k–ε method are usually attributed to shortcomings of the dissipation equation, Eq. (10.1.6), with its heavy modeling. Accordingly, alternative equations are periodically suggested (cf., e.g., Speziale et al., 1990a), and indeed, as indicated earlier, Eq. (10.1.5) with an algebraic form for L is sometimes a preferred approach to the representation of ν_T. It should also be noted that in this discussion a single length scale $k^{3/2}/\varepsilon$ is assumed to characterize all turbulent processes. This is the same assumption permitting introduction of a turbulent Prandtl number to relate the turbulent flux of temperature to the Reynolds stresses [cf. Eq. (7.2.1)], but our earlier warnings of its inapplicability in some cases, e.g., the heated wire in turbulent grid flow discussed in Section 8.4, apply in this case as well. Thus, in some applications a second dissipation equation characterizing a second turbulent length scale is incorporated into the set of describing equations (cf. Hanjalic et al., 1980; Jones and Musonge, 1983).

With k and ε treated as principal dependent variables, a model for ν_T is clearly suggested:

$$\nu_T = c_\mu \, k^{1/2}\left(\frac{k^{3/2}}{\varepsilon}\right) = c_\mu \, \frac{k^2}{\varepsilon} \tag{10.1.8}$$

where c_μ is a final empirical coefficient. Equation (10.1.8) provides an alternative to the classical models of turbulent exchange discussed in Chapters 7–9, one which does not require the assumption of a mixing length or equivalent but which is still based on a gradient transport assumption. It possesses the further virtue of permitting *all* the Reynolds stresses to be determined once the mean velocities and the turbulent kinetic energy are found.

With the substitution of all of the models into the second and third of Eqs. (10.1.1) and (10.1.6) and with a slight rearrangement to achieve standard forms, we have the general equations for the k–ε method:

$$\begin{aligned}
&\frac{\partial}{\partial x_k}(U_k U_i) = -\frac{\partial}{\partial x_i}\left(\frac{P}{\rho} + \overline{u_i^2}\right) + c_\mu \frac{\partial}{\partial x_k}\left[\frac{k^2}{\varepsilon}\left(\frac{\partial U_i}{\partial x_k} + \frac{\partial U_k}{\partial x_i}\right)\right] \\
&\frac{\partial}{\partial x_k}(U_k k) - \left[c_\mu \frac{k^2}{\varepsilon}\left(\frac{\partial U_k}{\partial x_l} + \frac{\partial U_l}{\partial x_k}\right) - \frac{2}{3} k \, \delta_{kl}\right]\frac{\partial U_l}{\partial x_k} \\
&\quad = \frac{c_\mu}{\sigma_k}\frac{\partial}{\partial x_k}\left(\frac{k^2}{\varepsilon}\frac{\partial k}{\partial x_k}\right) - \varepsilon \\
&\frac{\partial}{\partial x_k}(U_k \varepsilon) - c_{\varepsilon 1}\frac{\varepsilon}{k}\left[c_\mu \frac{k^2}{\varepsilon}\left(\frac{\partial U_k}{\partial x_l} + \frac{\partial U_l}{\partial x_k}\right) - \frac{2}{3} k \, \delta_{kl}\right]\frac{\partial U_l}{\partial x_k} \\
&\quad = \frac{c_\mu}{\sigma_\varepsilon}\frac{\partial}{\partial x_k}\left(\frac{k^2}{\varepsilon}\frac{\partial \varepsilon}{\partial x_k}\right) - c_{\varepsilon 2}\frac{\varepsilon^2}{k}
\end{aligned} \tag{10.1.9}$$

When supplemented with the mean continuity equation, i.e., with the first of Eqs. (10.1.1), these rather formidable equations define the mean velocity components U_i, $i = 1, 2, 3$, the mean pressure P, the turbulent kinetic energy k, and the mean dissipation ε. In the first of Eqs. (10.1.9) we include the contributions to the normal stresses from the intensities of the velocity fluctuations, but these are frequently neglected as being unimportant. The appropriate initial and boundary conditions to be imposed on these equations are discussed as specific applications are taken up.

As they stand, these partial differential equations are elliptic and thus require specification of conditions on a closed boundary. Accordingly, when extended to include the low-Reynolds-number effects arising at solid surfaces, they can be applied to cavity and separated flows, e.g., the flow over a backward-facing step. When specialized to the description of thin shear layers such as boundary layers, mixing layers, wakes, and jets, they reduce to the usual parabolic form; thus in these flows, which are furthermore two-dimensional in mean quantities, we obtain the following reduced set of equations:

$$U_1 \frac{\partial U_1}{\partial x_1} + U_2 \frac{\partial U_1}{\partial x_2} = - \frac{\partial}{\partial x_1}\left(\frac{P}{\rho}\right) + c_\mu \frac{\partial}{\partial x_2}\left(\frac{k^2}{\varepsilon}\frac{\partial U_1}{\partial x_2}\right)$$

$$U_1 \frac{\partial k}{\partial x_1} + U_2 \frac{\partial k}{\partial x_2} - c_\mu \frac{k^2}{\varepsilon}\left(\frac{\partial U_1}{\partial x_2}\right)^2 = \frac{c_\mu}{\sigma_k}\frac{\partial}{\partial x_2}\left(\frac{k^2}{\varepsilon}\frac{\partial k}{\partial x_2}\right) - \varepsilon \qquad (10.1.10)$$

$$U_1 \frac{\partial \varepsilon}{\partial x_1} + U_2 \frac{\partial \varepsilon}{\partial x_2} - c_{\varepsilon 1}\, c_\mu\, k \left(\frac{\partial U_1}{\partial x_2}\right)^2 = \frac{c_\mu}{\sigma_\varepsilon}\frac{\partial}{\partial x_2}\left(\frac{k^2}{\varepsilon}\frac{\partial \varepsilon}{\partial x_2}\right) - c_{\varepsilon 2}\frac{\varepsilon^2}{k}$$

In these equations $P(x_1)$ is given so that when they are combined with the mean continuity equation we have four equations for the two mean velocity components U_1 and U_2, the turbulent kinetic energy k, and the viscous dissipation ε. Appropriate initial and boundary conditions again complete the formulation. Since we make no special provision for curvature effects, these equations should be considered applicable to flows with planar or nearly planar mean streamlines. It is worth noting that the production term in the second and third of Eqs. (10.1.10) arise from mean shear, i.e., from $\partial U_1/\partial x_2$; although by far the greatest number of applications of the k–ε theory relate to shear flows with this production mechanism, alternative flows such as that in a contraction section discussed in Section 6.5 are of interest, flows in which the production mechanism involves extension and compression and in which Eqs. (10.1.9) must be applied. Turbulence stagnating against a wall or bluff body is another example of production due to these rates of strain (cf. Champion and Libby, 1991, 1994).

There are alternative but closely related formulations to the k–ε theory discussed here. The k–τ theory replaces the mean dissipation by a turbulence time scale $\tau \equiv k/\varepsilon$, with the benefit of improved behavior in the low-Reynolds-number region at the wall (cf. Speziale et al., 1990a). Wilcox and co-workers extend and apply to various flows the Saffman k–ω formulation, where $\omega = \tau^{-1} = \varepsilon/k$ (cf. Wilcox, 1993). Although these gradient transport methods and indeed the Reynolds stress theory discussed later are under continuous development and assessment, at present the k–ε theory remains the most commonly applied procedure beyond those based on the classical closures discussed in Chapters 7–9.

In the relatively simple flows we are able to discuss, the distributions of the mean velocity, the Reynolds shear stress, and the Reynolds flux obtained on the basis of the k–ε theory differ inconsequentially from those given by the classical theories discussed in Chapters 8 and 9. However, we are able to calculate the normal Reynolds stresses and the length scales of the turbulence. In addition, we illustrate application of the k–ε theory and thereby suggest the means for treating more complex flows for which the classical theory, requiring as it does the assumption of either an exchange coefficient or a distribution of mixing length, may be problematical.

10.2 THE EMPIRICAL COEFFICIENTS

We now take up the means for determining the five empirical coefficients and their universality, i.e., their independence of the specific flow being treated. Two means are adopted, both based on the implicit assumption that once they are determined, these coefficients are applicable to all flows, i.e., they are universal—an assumption found not to be completely valid, as suggested by our discussion in Sections 6.3–6.5. In one method the general equations are specialized to simple flows so that the resultant equations involve only a few of the coefficients, preferably only one. As a consequence, comparison with experimental data determines those coefficients which are thereafter applied to more complex flows.

An example of this approach is provided by grid flows discussed in Section 6.1; in this case Eqs. (10.1.9) and (10.1.10) reduce to

$$U_1\frac{dk}{dx_1} = \frac{c_\mu}{\sigma_k}\frac{d}{dx_1}\left(\frac{k^2}{\varepsilon}\frac{dk}{dx_1}\right) - \varepsilon \approx -\varepsilon$$
$$U_1\frac{d\varepsilon}{dx_1} = \frac{c_\mu}{\sigma_\varepsilon}\frac{d}{dx_1}\left(\frac{k^2}{\varepsilon}\frac{d\varepsilon}{dx_1}\right) - c_{\varepsilon 2}\frac{\varepsilon^2}{k} \approx -c_{\varepsilon 2}\frac{\varepsilon^2}{k} \tag{10.2.1}$$

As on several previous occasions, we neglect turbulent diffusion in the streamwise direction. By differentiating the first of Eqs. (10.2.1) and eliminating $d\varepsilon/dx_1$ and ε between the resulting three equations, we obtain a single equation for $k(x_1)$:

$$\frac{k\,(d^2k/dx_1^2)}{(dk/dx_1^2)} = c_{\varepsilon 2} \tag{10.2.2}$$

which is solved to obtain[†]

$$k = C\,x_1^{1/(c_{\varepsilon 2}-1)} \tag{10.2.3}$$

where C is an arbitrary constant determined by comparison of the rate of decay dk/dx_1 with experimental data. If we accept the not fully realistic arguments of Section 6.1 in favor of $k \propto x_1^{-1}$, we are led to $c_{\varepsilon 2} = 2$. The recommended value is $c_{\varepsilon 2} = 1.92$, which gives a slightly nonlinear rate of decay that is more in accord with experimental data. However, if we retain $c_{\varepsilon 2} = 2$, then $\varepsilon \propto x_1^{-2}$ from Eq. (10.2.2) and Eq. (10.1.9) implies that ν_T is constant, an analytically attractive result but one not fully in accord with experiment.

The constant c_μ can be determined as follows. In thin, free shear flows the dominant terms are frequently found to involve a balance between production

[†]Notice that the independent variable x_1 is absent, so that with k considered a new independent variable, one integration is immediately possible. Moreover, the resultant first-order equation can then be readily integrated.

and mean viscous dissipation. This notion introduced into Eqs. (10.1.10) implies that

$$c_\mu \frac{k^2}{\varepsilon}\left(\frac{\partial U_1}{\partial x_2}\right)^2 = \varepsilon$$

but from Eqs. (10.1.3) and (10.1.8) for these flows we know that

$$c_\mu \frac{k^2}{\varepsilon}\frac{\partial U_1}{\partial x_2} = -\overline{u_1 u_2}$$

If we eliminate ε between these two equations, we have

$$c_\mu = \left(\frac{\overline{u_1 u_2}}{k}\right)^2 \tag{10.2.4}$$

In the flows under consideration—and indeed in a variety of turbulent shear flows—experiment shows that to a close approximation $\overline{u_1 u_2} \approx -0.30\, k$. Thus, according to these considerations, $c_\mu = 0.09$, a recommended value.[†]

Finally, in the logarithmic portion of the wall layer in wall-bounded flows, a portion in which viscous stresses are negligible, production of turbulent kinetic energy again closely balances mean viscous dissipation, the mean Reynolds shear stress $\overline{u_1 u_2} = u_\tau^2$ is constant where u_τ is the shearing velocity, and

$$U_1 = u_\tau\left(\frac{1}{\kappa}\ln\frac{u_\tau x_2}{\nu} + B\right) \tag{10.2.5}$$

Thus

$$\frac{\partial U_1}{\partial x_2} = \frac{u_\tau}{\kappa x_2}$$

where κ is the von Karman constant. The second and third of Eqs. (10.1.10) are consistent if, as $x_2 \to 0$ within the logarithmic portion of the wall layer, $k \to u_\tau^2/c_\mu^{1/2}$ and $\varepsilon \to 1/(\kappa x_2)$ (cf. Section 10.6 for a further discussion of this behavior). In this case Eq. (10.1.16) leads to a restraint among several coefficients:

$$c_{\varepsilon 1} = c_{\varepsilon 2} - \frac{\kappa^2}{\sigma_\varepsilon c_\mu^{1/2}} \tag{10.2.6}$$

Despite these considerations, the recommended values of the coefficients appearing in this equation respect Eq. (10.2.6) only approximately.

[†]As noted on several earlier occasions, Bradshaw et al. (1967) provide a theory for turbulent boundary layers involving, among other things, the elimination of the Reynolds shear stress $\overline{u_1 u_2}$ by means of Eq. (10.2.4) and the inclusion of the second of Eqs. (10.1.10) somewhat modified. However, a shortcoming of Eq. (10.2.5) *as used in that theory* arises at planes of symmetry and at similar points where $\overline{u_1 u_2}$ is zero but $k \neq 0$. Here we use Eq. (10.2.5) only to determine the value of c_μ so that this shortcoming does not arise in the k–ε theory.

The remaining empirical coefficients in the k–ε method are chosen by a second method, numerical experimentation, involving comparison with a variety of turbulent flows. Thus a recommended set of coefficients is established:

$$c_\mu = 0.09 \qquad c_{\varepsilon 1} = 1.44 \qquad c_{\varepsilon 2} = 1.92 \qquad \sigma_k = 1 \qquad \sigma_\varepsilon = 1.3 \tag{10.2.7}$$

Note that if the values of the first four of these coefficients are adopted, Eq. (10.2.6) is satisfied with $\sigma_\varepsilon = 1.11$, a value close to 1.3. Somewhat altered values of these coefficients are suggested, indeed *required,* in some applications, as we know from the discussion in Sections 6.3–6.5. For example, in some cases it is essential that $c_{\varepsilon 1} = c_{\varepsilon 2}$ in order that production and dissipation be precisely balanced in the second and third of Eqs. (10.1.10). Later we shall encounter an example in which the asymptotic approach of the solutions to the external stream for the far wake of a cylinder restricts these coefficients. Finally, we note that the empirical coefficients are selected largely on the basis of turbulent shear flows, so alternative values may be required in applications to flows that are dominated by extensive or compressive rates of strain. However, the values of Eqs. (10.2.8) should be adopted at least provisionally in applications of the k–ε method, unless there are known reasons to do otherwise.

10.3 THE TURBULENT FAR WAKE OF A CYLINDER

In Section 8.1 we discussed as a prototypical free shear flow the turbulent far wake of a cylinder and employed a classical gradient transport model to achieve closure. We now repeat this calculation using the k–ε theory. The early considerations in Section 8.1, those leading to Eqs. (8.1.1)–(8.1.5), apply here. Thus we have a single dependent variable of similarity form,

$$\eta = \frac{x_2}{d\,\tilde{x}_1^{1/2}} \tag{10.3.1}$$

while the x_1 variable is replaced by the dimensionless $\tilde{x}_1 \equiv x_1/d$, where the quantity d is a length characterizing the size of the body, e.g., the diameter of a circular cylinder. If we restrict our attention to the lowest-order terms for the far wake, we must consider only the U_1 velocity component as given by

$$U_1 = U_\infty\left(1 - \frac{{}^{(0)}f(\eta)}{\tilde{x}_1^{1/2}} + \ldots\right) \tag{10.3.2}$$

where ${}^{(0)}f(\eta)$ is one of three dependent variables to be calculated.

We next adopt suitable similarity forms for the turbulent kinetic energy and the viscous dissipation, forms which are suggested by dimensional and fluid mechanical considerations. Our earlier analysis indicates that the turbulent exchange coefficient is constant throughout this wake flow and that the wake thickness grows as $\tilde{x}_1^{1/2}$. Accordingly, we must restrict the $\tilde{x}_1$ dependence of k and ε so that k^2/ε and $k^{3/2}/\varepsilon\, \tilde{x}_1^{1/2}$ are independent of $\tilde{x}_1$. We are thus lead to

$$k = \frac{U_\infty^2}{\tilde{x}_1}\,[{}^{(0)}K(\eta) + \ldots] \qquad \varepsilon = \frac{U_\infty^3}{\tilde{x}_1^2 d}\,[{}^{(0)}E(\eta) + \ldots] \tag{10.3.3}$$

If Eqs. (10.3.2) and (10.3.3) are substituted into Eqs. (10.1.14)–(10.1.16) and the independent variables changed from x_1, x_2 to $\tilde{x}_1$, $\eta(x_1, x_2)$, we find after some algebra the following three equations:

$$-\frac{1}{2}(\eta\,{}^{(0)}f)' = c_\mu\left(\frac{{}^{(0)}K^2}{{}^{(0)}E}{}^{(0)}f'\right)'$$

$$-{}^{(0)}K - \frac{1}{2}\,\eta\,{}^{(0)}K' - c_\mu\,\frac{K_0^2}{E_0}\,({}^{(0)}f')^2 = \frac{c_\mu}{\sigma_k}\left(\frac{K_0^2}{{}^{(0)}E}{}^{(0)}K'\right)' - {}^{(0)}E \tag{10.3.4}$$

$$-2\,{}^{(0)}E - \frac{1}{2}\,\eta\,{}^{(0)}E' - c_{\varepsilon 1}\,c_\mu\,{}^{(0)}K\,{}^{(0)}f'^2 = \frac{c_\mu}{\sigma_\varepsilon}\left(\frac{{}^{(0)}K^2}{{}^{(0)}E}{}^{(0)}E'\right)' + c_{\varepsilon 2}\,\frac{{}^{(0)}E^2}{{}^{(0)}K}$$

which determine ${}^{(0)}f(\eta)$, ${}^{(0)}K(\eta)$, and ${}^{(0)}E(\eta)$ provided appropriate boundary conditions are specified. Those conditions are:

$$\begin{gathered} {}^{(0)}f'(0) = {}^{(0)}K'(0) = {}^{(0)}E'(0) = 0 \\ {}^{(0)}f(\eta \to \infty) = {}^{(0)}K(\eta \to \infty) = {}^{(0)}E(\eta \to \infty) = 0 \end{gathered} \tag{10.3.5}$$

With these conditions Eqs. (10.3.4) are satisfied by the uninteresting solution $f \equiv K \equiv E \equiv 0$. Therefore, we are led to an inverse approach as follows. We impose an additional condition assuring a nontrivial solution and consider one of the empirical coefficients to be a parameter at our disposal. To this end we specify ${}^{(0)}f(0)$ and consider $c_{\varepsilon 1}$ as a quantity to be determined, but the former value can be systematically altered so that the *calculated* value of $c_{\varepsilon 1}$ approximates its recommended value of 1.44. As noted earlier, experimental data indicate that $1.12 < {}^{(0)}f(0) < 1.17$.

The behavior of ${}^{(0)}K(\eta)$ and ${}^{(0)}E(\eta)$ as $\eta \to \infty$ requires discussion. Since both of these quantities approach zero as $\eta \to \infty$, and since the quotient ${}^{(0)}K^2/{}^{(0)}E$ appears in Eqs. (10.3.4), it is clear that some attention to their rate of approach to zero is indicated. If we assume that this quotient approaches a constant asymptotically as $\eta \to \infty$, that is, that the turbulent exchange coefficient approaches a constant as in the analysis of Section 8.1 based on classical closure, a consistent calculation results provided the empirical coefficients σ_k and σ_e are constrained. As discussed in Section 9.10, the treatment of the boundary conditions at the outer edge of a shear layer is subject to some controversy. However, the various means for imposing these conditions probably do not result in significantly different solutions except as the free stream is approached. Our treatment avoids specification of an "edge" and results in solutions which approach the free stream exponentially, features we consider to represent significant advantages.

To implement this strategy we assume that beyond a value of η denoted η^*, $c_\mu{}^{(0)}K^2/{}^{(0)}E \approx T_\infty$, a constant. We seek solutions to Eqs. (10.3.4) for $\eta > \eta^*$. From the first of these equations we readily find

$$^{(0)}f = A_f \exp\left(-\frac{1}{4T_\infty}\eta^2\right) \tag{10.3.6}$$

where A_f is an arbitrary constant. In the second and third of these equations the production and dissipation terms, which are quadratic in the dependent variables, decay faster than the linear convection and diffusion terms. Accordingly, we are led to a convective-diffusive balance as the external stream is approached and to the following linear equations:

$$\begin{aligned} -{}^{(0)}K - \frac{1}{2}\eta{}^{(0)}K' &= \frac{T_\infty}{\sigma_k}{}^{(0)}K'' \\ -{}^{(0)}E - \frac{1}{2}\eta{}^{(0)}E' &= \frac{T_\infty}{\sigma_e}{}^{(0)}E'' \end{aligned} \tag{10.3.7}$$

When $\eta \to \infty$, there are both algebraically and exponentially decaying solutions to Eq. (10.3.7). For the purpose of imposing boundary conditions on Eqs. (10.3.4), we require exponential decay and assume that these asymptotic solutions apply with sufficient accuracy for $\eta > \eta^*$, that is, wherever the turbulent exchange coefficient is approximately constant. Thus we take

$$^{(0)}K = A_K \exp\left(-\frac{\alpha_K}{2}\eta^2\right) \qquad ^{(0)}E = A_E \exp\left(-\frac{\alpha_E}{2}\eta^2\right) \tag{10.3.8}$$

where A_K and A_E are arbitrary coefficients. Substitution into the second and third of Eqs. (10.3.4) and collection of the η^2 terms yield

$$\alpha_K = \frac{\sigma_K}{2T_\infty} \qquad \alpha_E = \frac{\sigma_e}{2T_\infty} \tag{10.3.9}$$

and since T_∞ is constant, $2\,\alpha_K = \alpha_E$ and thus

$$2\,\sigma_k = \sigma_e \tag{10.3.10}$$

for $\eta > \eta^*$. We thus see that consistent behavior of the variables describing the turbulent flow as the outer edge of the wake is approached places a restriction on σ_k and σ_e. If for simplicity we retain the standard value $\sigma_k = 1$, we can assume a distribution of σ_e of the form

$$\sigma_e = 2 - 0.7 \exp\left(-4\frac{\eta}{\eta^*}\right) \tag{10.3.11}$$

where the multiplier in the argument of the exponential function is arbitrary but not critical in determining the nature of the solutions.

The possibility of computing each of the Reynolds stresses from Eq. (7.1.3) provided ν_T and k are known was noted earlier. When this possibility is exploited it is found that to lowest order in $\tilde{x}_1^{-1/2}$ the turbulence is isotropic, that is, $\overline{u_i^2} = \frac{2}{3}k$, $i = 1, 2, 3$, and

$$\overline{u_1u_2} = c_\mu \, U_\infty^2 \frac{{}^{(0)}K^2}{{}^{(0)}E} \frac{1}{\tilde{x}_1} {}^{(0)}f' \tag{10.3.12}$$

Since wake flow is dominated by simple shear, the prediction is consistent with the discussion in Section 7.3, namely, that in all turbulent flows with simple shear, gradient descriptions of turbulent transport yield equal contributions to the turbulent kinetic energy from the intensities of the fluctuations in each coordinate direction. In the present case, as in many other shear flows, this equipartioning of the energy is not in accord with experiment (e.g., Townsend, 1956). If accurate distributions of the contributions to the turbulent kinetic energy are desired, the Reynolds stress theory must be employed (cf. Section 10.11). From Eq. (10.3.12) we see that the requirement of ${}^{(0)}K^2/{}^{(0)}E$ approaching a constant as $\eta \to \infty$ results in $\overline{u_1u_2}$ decaying exponentially fast at the outer edge of the wake, which is an attractive feature.

The numerical treatment of the equations arising in the k–ε theory, e.g., Eqs. (10.3.4), is not straightforward, although most applications of the theory in the literature acknowledge no numerical difficulties. The troubles appear to be related to the near balance in some portions of the flow of production and dissipation, so that the derivative terms are given by the small differences of relatively large algebraic quantities, a characteristic of stiff equations. Finite-difference methods combined with linearization and iteration to deal with the two-point boundary-value problem and utilizing a quite accurate initial estimate of all dependent variables at all grid points is generally required to achieve a successful calculation. The point to be made is that the numerical solution of the equations encountered in the k–ε theory should not be treated casually, despite the absence of comments concerning numerical difficulties in the literature.

Figures 10.3.1 and 10.3.2 show the distributions of ${}^{(0)}f(\eta)$ and ${}^{(0)}K(\eta)$ and ${}^{(0)}E(\eta)$, respectively, as obtained from Eqs. (10.3.4). There is nothing noteworthy about the monotonic decline of the velocity defect reflected in ${}^{(0)}f(\eta)$ seen in Fig. 10.3.1. However, the distributions of turbulent kinetic energy and mean viscous dissipation in Fig. 10.3.2 exhibit a maximum off axis that calls for comment. This behavior is a consequence of the nonmonotonicity of the production of the turbulent kinetic energy in the wake. Such production is associated with the $({}^{(0)}f')^2$ term in the second and third of Eqs. (10.3.4), a term which is

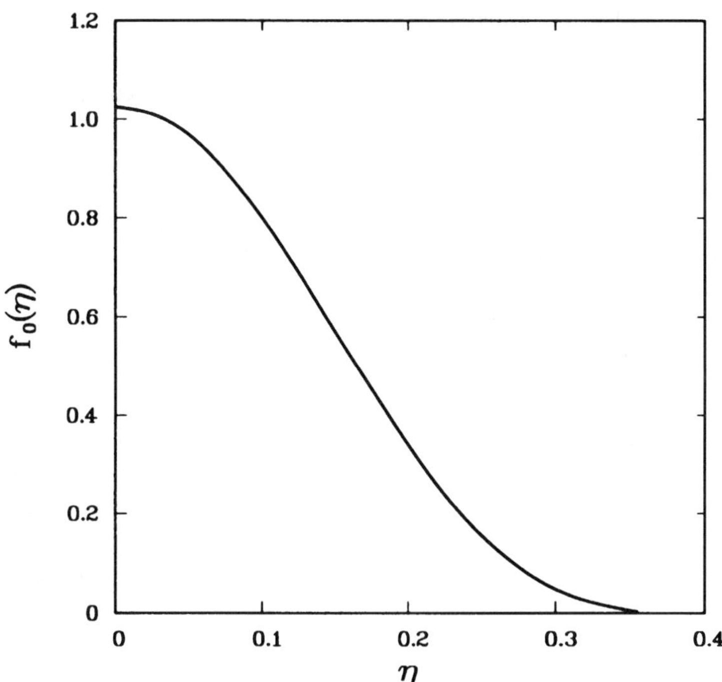

Figure 10.3.1 The velocity defect for the far wake of a cylinder according to the k–ε theory.

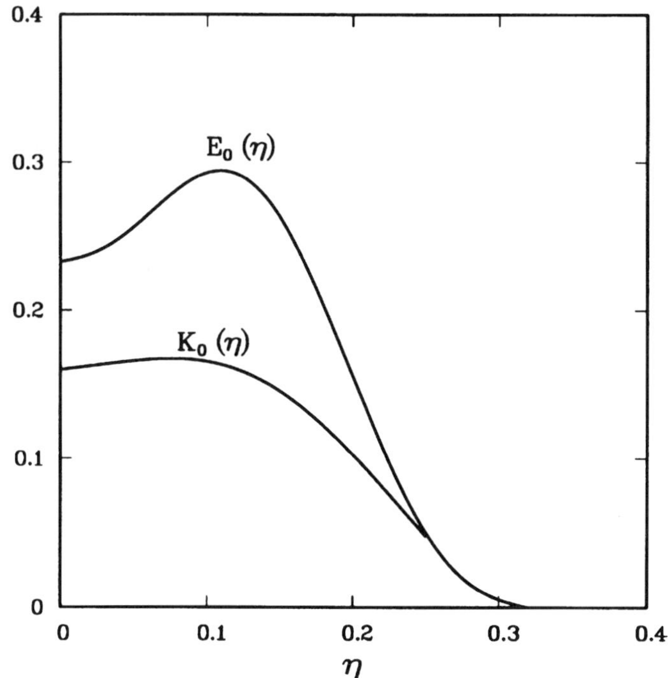

Figure 10.3.2 Turbulent kinetic energy and mean viscous dissipation for the far wake of a cylinder according to the k–ε theory.

zero on axis and in the ambient stream and which thus possesses a maximum off axis.

At several points in our discussion we have interpreted the various mean equations of turbulence in terms of a balance among competing influences. In the context of the present analysis we can make such a balance more quantitative as follows. Consider the second of Eqs. (10.3.4) and let all of the terms be brought to the right side. We can then identify the contribution of diffusion via (c_μ/σ_k) [ ]'; of production via $c_\mu({}^{(0)}K^2/{}^{(0)}E)({}^{(0)}f')^2$; of dissipation via ${}^{(0)}E$; and of convection via $K_0 + (\eta/2)K_0'$. Figure 10.3.3 shows the distribution of these contributions across the wake; positive values imply contributions *increasing* turbulent kinetic energy and vice versa. This is a convenient procedure for examining the various contributors to the behavior of turbulence, which we shall use on several occasions later. We can see immediately that while the production and dissipation terms assume the largest absolute values anywhere in the wake, there are regions where other contributions are dominant. For example, in the neighborhood of the wake axis, the balance of turbulent kinetic energy is largely between convection and dissipation, while, as noted earlier, at the outer edge there is a convective-diffusive balance. We thus see in this relatively simple flow

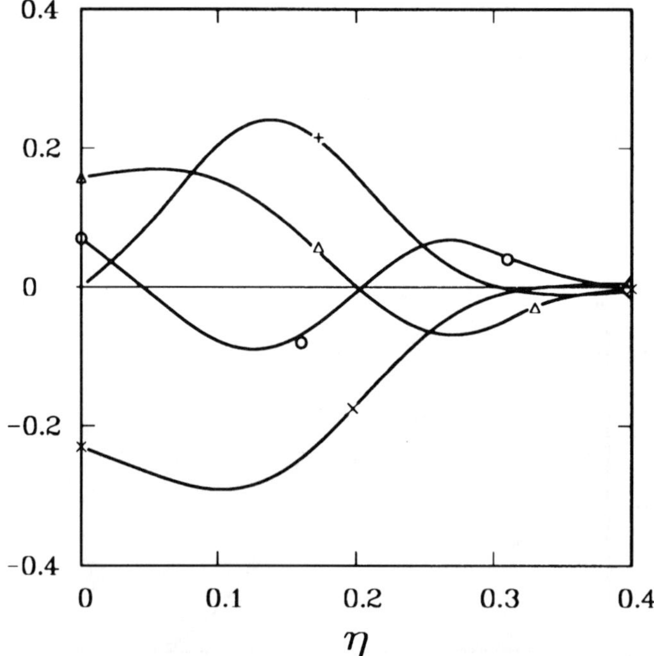

Figure 10.3.3 Distributions of the contributions to the balance of the turbulent kinetic energy in the far wake of a cylinder according to the k–ε theory. Δ = diffusion; + = production; × = dissipation; ○ = convection.

a complex interplay among competing influences determining the distribution of the turbulent kinetic energy. Similar considerations apply to the other equations in this system, and indeed to other turbulent flows.

Several further points can be made with respect to these results. The specified value of ${}^{(0)}f(0) = 1.15$ is in good agreement with the available data. However, comparison with classical theory [cf. Eq. (8.1.8) with $U_\infty d/\nu_{T0}$ selected to yield agreement with experimental data] shows that the present calculation yields a wake width roughly one-half the correct value even if the largest of the experimental values of $U_\infty d/\nu_{T0}$ is adopted. This result was also found by Patel and Scheuerer (1982) and is a shortcoming of the k–ε theory for wake flows [see also Launder (1989a) for a discussion of the shortcomings of the basic model in the Reynolds stress theory with respect to the plane wake flow]. Moreover, the value of ${}^{(0)}K(0)$ is found to be 0.17, while experiment gives values between 0.25 and 0.31. The implication of this discrepancy is that the empirical coefficients must be adjusted—i.e., nonstandard values must be adopted, to achieve a higher spreading rate—if quantitative agreement with experiment is to be achieved. For example, if $c_{\varepsilon 1} = 1.33$ and $c_{\varepsilon 2} = 2.0$, the agreement between our k–ε calculation and the classical solution is greatly improved; the wake width is essentially brought into agreement and ${}^{(0)}K(0) = 0.28$, a value that is within the range of experimental data. We thus see another example of the adjustment of empirical coefficients to achieve a desired result.

10.4 THE TEMPERATURE IN THE FAR WAKE OF A HEATED CYLINDER

If in Eq. (8.2.1), the equation for the mean temperature in the wake of a heated cylinder, we employ the gradient model of Eq. (7.2.1) to eliminate the mean turbulent flux $\rho\overline{u_2\theta}$, we have

$$\frac{\partial}{\partial x_1}(U_1\Theta) + \frac{\partial}{\partial x_2}(U_2\Theta) = \frac{\partial}{\partial x_2}\left(\frac{\nu_T}{\sigma_T}\frac{\partial\Theta}{\partial x_2}\right) \tag{10.4.1}$$

In extending the analysis of the previous section to the determination of the temperature distribution in the wake, we follow closely the treatment in Section 8.2 but with the k–ε model for the turbulent exchange coefficient and with σ_T a specified constant having values in the range 0.5–1.5 depending on the application. Thus restriction to the far wake, as in the previous section, calls for

$$\Theta(\tilde{x}_1, \eta) \approx \frac{\Theta(0)}{\tilde{x}^{1/2}}\,[{}^{(0)}\Theta(\eta) + \ldots] \tag{10.4.2}$$

where $\Theta(0)$ is a dimensional constant that depends on the temperature of the cylinder. From Eqs. (10.4.1) and (10.4.2) we obtain the equation for the mean temperature,

$$-\frac{1}{2}(\eta^{(0)}\Theta)' = \frac{c_\mu}{\sigma_T}\left(\frac{^{(0)}K^2}{^{(0)}E}\,{}^{(0)}\Theta'\right)' \tag{10.4.3}$$

which is to be solved subject to the conditions $^{(0)}\Theta(0) = 1$, $^{(0)}\Theta(\eta \to \infty) = 0$. The solution can be obtained by quadrature as

$$^{(0)}\Theta(\eta) = \exp\left(-\frac{\sigma_T}{2\,c_\mu}\int_0^\eta d\eta'\eta'\,\frac{^{(0)}E}{^{(0)}K^2}\right) \tag{10.4.4}$$

Figure 10.4.1 shows this solution based on the results of Fig. (10.3.2) with $\sigma_T = 1$ and $c_\mu = 0.09$; smaller values of σ_T result in wider thermal wakes in terms of η. The distribution of the turbulent flux, $\overline{u_2\theta}$, which is proportional to $-(^{(0)}K^2/^{(0)}E)^{(0)}\Theta'$, can be deduced from Figs. 10.3.2 and 10.4.1; it is zero on the wake axis and in the ambient stream and thus possesses a maximum off axis.

10.5 THE WALL LAYER AND LOW-REYNOLDS-NUMBER CORRECTIONS WITH APPLICATION TO CHANNEL FLOW

When they are applied to wall flows, all second-moment methods must be modified to account for the alteration of turbulence within the viscous sublayer, no matter how large the global Reynolds number. In particular, the viscous stresses and transport of heat by thermal conductivity can no longer be neglected, and the difference noted earlier in the character of the mean viscous dissipation with high and low turbulence Reynolds numbers must be taken into account. In addition, we show later that as the wall is approached, the turbulence becomes highly anisotropic. A variety of such modifications is available. Patel et al. (1985) provide a valuable review of the low-Reynolds-number versions of the k–ε and equivalent methods, including comparison of predictions with experimental data from a large number of test cases. Mansour et al. (1989) and Speziale et al. (1990a) also give valuable critiques of near-wall modeling for the k–ε theory. For our purposes it is sufficient to introduce one such version and discuss its application to turbulent channel flow.

As a preliminary to a discussion of turbulent wall flows within the context of second-moment methods, it is useful to consider the behavior as the wall is approached of second-moment quantities, the turbulent intensities, the turbulent shear stress, the turbulent flux of heat, and the mean dissipation. To that end, consider the expansion of the fluctuations of the three velocity components about their wall values with the usual orientation of the coordinate system. The no-slip condition implies that at $x_2 = 0$ for any time,

$$\frac{\partial u_1}{\partial x_1} \equiv \frac{\partial u_3}{\partial x_3} \equiv 0 \tag{10.5.1}$$

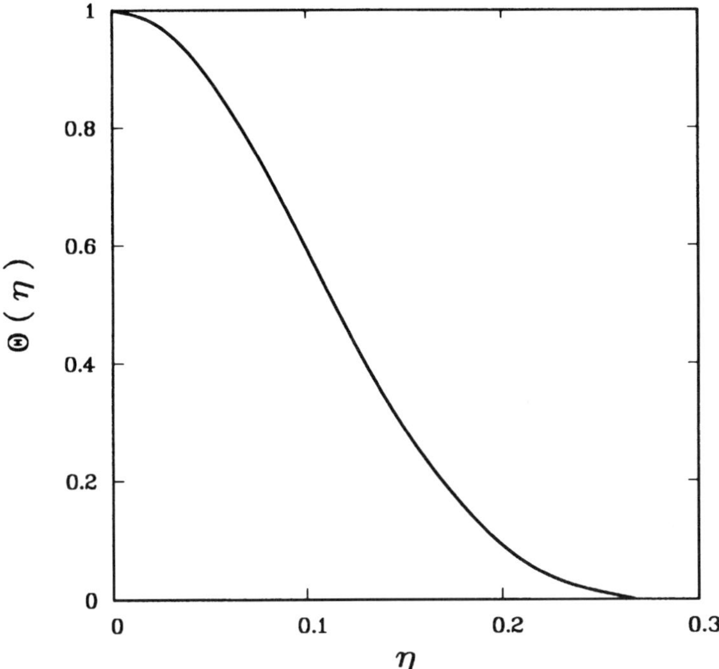

Figure 10.4.1 Temperature distribution in the far wake of a heated cylinder according to the k–ε theory.

and thus we find from the continuity equation that $\partial u_2/\partial x_2 \equiv 0$ at the wall. Accordingly, the first terms in the desired expansions are

$$u_1(\mathbf{x}, t) \approx \frac{\partial u_1}{\partial x_2}(x_1, 0, x_3, t)\, x_2 + \frac{\partial^2 u_1}{\partial x_2^2}(x_1, 0, x_3, t)\, \frac{x_2^2}{2} + \ldots$$

$$u_2(\mathbf{x}, t) \approx \frac{\partial^2 u_2}{\partial x_2^2}(x_1, 0, x_3, t)\, \frac{x_2^2}{2} + \ldots \tag{10.5.2}$$

$$u_3(\mathbf{x}, t) \approx \frac{\partial u_3}{\partial x_2}(x_1, 0, x_3, t)\, x_2 + \frac{\partial^2 u_3}{\partial x_2^2}(x_1, 0, x_3, t)\, \frac{x_2^2}{2} + \ldots$$

One implication of these equations is that the turbulence becomes highly anisotropic as the wall is approached, resembling two-dimensional turbulence in planes parallel to the wall (cf. Section 4.7 and Fig. 4.7.1). Furthermore, as $x_2 \to 0$, the turbulent kinetic energy behaves such that

$$k(\mathbf{x}) \approx \frac{1}{2}(\overline{u_1^2} + \overline{u_3^2}) \approx a_2(x_1, x_3)x_2^2 + \ldots \tag{10.5.3}$$

where, from Eqs. (10.5.2), we see that the $a_2(x_1, x_3)$ coefficient depends on the sum of the mean square values of the gradients of the u_1 and u_3 velocity components at the wall.

These considerations can be extended. The variation of the Reynolds shear stress in the neighborhood of the wall is readily found from Eqs. (10.5.2) to be

$$-\rho\, \overline{u_1 u_2} \approx \rho\, b_3(x_1, x_3)\, x_2^3 + \ldots \tag{10.5.4}$$

where $b_3(x_1, x_3)$ depends on the mean value of the product of the gradient and the curvature of the u_1 and u_2 velocity components, respectively. Furthermore, for a fixed wall temperature, the fluctuations of the temperature in the fluid adjacent to the wall can be approximated by

$$\theta(\mathbf{x}, t) = \frac{\partial \theta}{\partial x_2}(x_1, 0, x_3, t)\, x_2 + \ldots \tag{10.5.5}$$

so that the mean flux of temperature behaves as

$$-\rho\, \overline{u_2 \theta} \approx \rho\, c_3(x_1, x_2)\, x_2^3 + \ldots \tag{10.5.6}$$

where $c_3(x_1, x_3)$ depends on the mean of the product of the gradient of the temperature and the curvature of the u_2 velocity at the wall. We know from our earlier discussion (Section 9.2) that in shear flows, as the wall is approached, the reduction in the turbulent shear stress and the turbulent heat flux according to Eqs. (10.5.4) and (10.5.6), respectively, are compensated by corresponding increases in molecular transport so that the total stress and total flux are constant in the viscous sublayer.

Equations (10.5.3)–(10.5.6) clearly expose the important role played by the intensities and correlations of the time-dependent viscous shear stresses τ_{12} and τ_{23} and the heat flux q_2 associated with thermal conductivity in determining the behavior of *turbulence quantities* as the wall is approached. Accurate measurements of the time-resolved shear stresses at a wall in turbulent flow are not easily made, but there is an extensive literature concerned with such measurements (cf. Alfredsson et al., 1988).

The vanishing of the various second-moment correlations as the wall is approached contrasts with the behavior of the mean dissipation, as may be seen as follows. In Eq. (4.7.2) the derivatives with respect to x_1 and x_3 of *all* velocity components are identically zero at $x_2 = 0$, but only $\partial u_2/\partial x_2(x_1, 0, x_3, t) = 0$, with the consequence that

$$\varepsilon(x_1, 0, x_3) = \nu \overline{\left[\left(\frac{\partial u_1}{\partial x_2}\right)^2 + \left(\frac{\partial u_3}{\partial x_2}\right)^2\right]}(x_1, 0, x_3) \neq 0 \tag{10.5.7}$$

Several comments are appropriate. Again we see the importance of fluctuations of the viscous shear stresses at the wall. Note that the [ ] term in this equation equals the coefficient $a_2(x_1, x_3)$ of Eq. (10.5.3), a result connecting the behavior of the turbulent kinetic energy and the mean dissipation at the wall according to

$$\frac{\partial^2 k}{\partial x_2^2}(x_1, 0, x_3) = \frac{\varepsilon(x_1, 0, x_3)}{\nu} \tag{10.5.8}$$

An additional term in the expansion of k about $x_2 = 0$, that is, the x_2^3 term, involves quantities whose signs are uncertain, so the rate of change of $\varepsilon(x_1, 0, x_3)$ with x_2 cannot be simply deduced. Equation (10.5.7) supports our earlier observation (cf. Section 9.3) that the Kolmogorov length characterizes the wall layer; we see that $\varepsilon \propto \nu\, (\tau_w/\mu)^2 = u_\tau^4/\nu$, and thus $L_K = \nu/u_\tau$, and finally that the independent variable for the wall layer $y^+ = x_2/L_K$.

Equations (10.5.1)–(10.5.8) describe the exact behavior of turbulence quantities at the wall. Although it might be expected that turbulence models accounting for low-Reynolds-number effects within the wall layer would respect all of these results, many of those currently in the literature do not do so. The implication appears to be that detailed behavior of the solutions in the immediate vicinity of the wall does not critically influence the solutions in the outer portion of the wall layer, the portion subject to comparison with experimental data.

An example of such disregard is provided by the the mixing-length theory. We know from Eq. (7.4.5) that

$$-\rho\, \overline{u_1 u_2} = \rho\, l^2 \left(\frac{\partial U_1}{\partial x_2}\right)^2$$

As $x_2 \rightarrow 0$ [cf. Eqs. (9.3.11)], this becomes

$$-\rho\, \overline{u_1 u_2} \rightarrow \rho \left(\frac{\kappa}{\kappa_w}\right)^2 \hat{\eta}^4 \propto x_2^4$$

rather than x_2^3 as called for by Eq. (10.5.4).

Patel et al. (1985) discuss other examples of flaws in the modeling of low-Reynolds-number effects and state "most modifications to the basic high Reynolds number turbulence models lack much physical basis." In recent years, improved models of these effects have become available.

We now return to the treatment of turbulent channel flow within the context of a second-moment method. We do so in terms of a slight modification of the low-Reynolds-number version of the k–ε theory due to Myong and Kassagi (1990). In this theory the changes to the describing equations are formally rather minimal. Viscous effects are incorporated by adding the kinematic viscosity to the $c_\mu k^2/\varepsilon$ coefficient of turbulent exchange in each of the diffusion terms in Eqs. (10.1.10). In addition, this exchange coefficient is modified by a multiplicative function f_μ, which depends on distance from the wall, the dependence being such that far from the wall $f_\mu \rightarrow 1$, that is, the usual $c_\mu\, k^2/\varepsilon$ is recovered. Furthermore, there is incorporated an additional function f_2 which multiplies the dissipation term in the dissipation equation and which approaches unity asymptotically when far from the wall, an addition accounting for the change in mean viscous dissipation with turbulence Reynolds number as discussed in Section 5.3. The advantage of the Myong and Kassagi (1990) model is that in contrast with some other low-Reynolds-number corrections, f_μ and f_2 yield the proper turbulence characteristics as $x_2 \to 0$. Details of these two functions are given later.

With these preliminaries completed, our analysis of turbulent channel flow starts with Eq. (9.1.4) as modified by a gradient description of the Reynolds shear stress, i.e., by Eq. (7.1.3). We have

$$\frac{1}{h}(\nu + \nu_T)\, U_1' = u_\tau^2 (1 - \eta) \tag{10.5.9}$$

where we recall that $\eta \equiv x_2/h$ is the dimensionless independent variable and prime denotes differentiation with respect to η. From Eq. (10.1.8), modified as indicated earlier to account for low-Reynolds-number effects, we have for the turbulent exchange coefficient

$$\nu_T = c_\mu f_\mu \frac{k^2}{\varepsilon} \tag{10.5.10}$$

For channel flow the equations for the turbulent kinetic energy and viscous dissipation, the second and third of either Eqs. (10.1.9) or (10.1.10), become

$$\begin{aligned} 0 &= \left[\left(\nu + \frac{\nu_T}{\sigma_k}\right) k'\right]' + \nu_T (U_1')^2 - h^2 \varepsilon \\ 0 &= \left[\left(\nu + \frac{\nu_T}{\sigma_\varepsilon}\right) \varepsilon'\right]' + c_{\varepsilon 1}\, \nu_T \frac{\varepsilon}{k} (U_1')^2 - c_{\varepsilon 2}\, f_2\, h^2 \frac{\varepsilon^2}{k} \end{aligned} \tag{10.5.11}$$

Note that convective terms are absent from Eqs. (10.5.9) and (10.5.11), as expected in fully developed flow.

In the k–ε method the turbulence Reynolds number is defined in terms of k, ε, and ν, as

$$N_{RT} \equiv \frac{k^2}{\varepsilon \nu} = \frac{k^{1/2}(k^{3/2}/\varepsilon)}{\nu} \tag{10.5.12}$$

where the second form emphasizes the turbulent length scale $k^{3/2}/\varepsilon$. Note that from Eqs. (10.5.3) and (10.5.7), as the wall is approached, N_{RT} goes to zero rapidly, in particular as x_2^4. Following Myong and Kassagi (1990) as slightly modified by Speziale et al. (1990a), we take

$$f_\mu = \left(1 + \frac{3.45}{N_{RT}^{1/2}}\right) \tanh\left(\frac{\hat{\eta}}{70}\right) \qquad f_2 = \left[1 - \exp\left(\frac{\hat{\eta}}{4.9}\right)\right]^2 \tag{10.5.13}$$

where $\hat{\eta} \equiv \eta/\delta$ with $\delta \equiv \nu/u_\tau h$ is the independent variable for the wall layer introduced in Sections 9.2–9.4. Recall that δ is the reciprocal of a Reynolds number based on the channel half-height, the shearing velocity, and the kinematic viscosity.

We now examine the behavior of these functions as $x_2 \to 0$. From Eq. (10.5.4) and the gradient description of Reynolds stresses [cf. Eq. (7.3.4)], we see that $\nu_T(x_1, x_2 \to 0, x_3)$ must be proportional to x_2^3. Thus, if k and ε have the physically correct behavior as $x_2 \to 0$, then from Eq. (10.5.10) we see that $f_\mu(x_2 \to 0)$ must be proportional to x_2^{-1}. The first of Eqs. (10.5.13) has this behavior, since $\tanh(x \to 0) = x$ and $N_{RT} \propto x_2^4$ as discussed earlier. With this behavior for ν_T, k, and ε, the first of Eqs. (10.5.11) reduces to a diffusive-dissipative balance. Furthermore, if the second of Eqs. (10.5.11) is likewise to describe such a balance as $x_2 \to 0$, then in this limit f_2 must be proportional to x_2^2 to nullify the k^{-1} factor in the dissipation term. From the second of Eqs. (10.5.13) we see that f_2 has the proper behavior. We thus see that the present low-Reynolds-number functions assure physically correct solutions to Eqs. (10.5.11) at the wall.

We advance our analysis by nondimensionalizing the dependent variables with u_τ as in Sections 9.2–9.4. Thus we let

$$\tilde{U} \equiv \frac{U_1}{u_\tau} \qquad \tilde{K} \equiv \frac{k}{u_\tau^2} \qquad \tilde{E} \equiv \frac{\varepsilon\, h}{u_\tau^3} \tag{10.5.14}$$

and obtain the dimensionless equations

$$\begin{aligned} &\left(\delta + c_\mu f_\mu \frac{\tilde{K}^2}{\tilde{E}}\right)\tilde{U}' = (1 - \eta) \\ &\left[\left(\delta + \frac{c_\mu f_\mu}{\sigma_k}\frac{\tilde{K}^2}{\tilde{E}}\right)\tilde{K}'\right]' + c_\mu\, f_\mu \frac{\tilde{K}^2}{\tilde{E}}(\tilde{U}')^2 - \tilde{E} = 0 \\ &\left[\left(\delta + c_\mu f_\mu \frac{\tilde{K}^2}{\tilde{E}}\right)\tilde{E}'\right]' + c_{\varepsilon 1}\, c_\mu\, f_\mu\, \tilde{K}(\tilde{U}')^2 - c_{\varepsilon 2}\, f_2 \frac{\tilde{E}^2}{\tilde{K}} = 0 \end{aligned} \tag{10.5.15}$$

Equations (10.5.15) are to be solved subject to the conditions $\tilde{U}(0) = \tilde{K}(0) = 0$ at the wall and to symmetry conditions at $\eta = 1$, namely, $\tilde{U}'(1) = \tilde{K}'(1) = \tilde{E}'(1) = 0$. If the three dependent variables are expanded about $\eta = 0$, we conclude that with the known behavior of f_μ and f_2 as $\eta \to 0$, $\tilde{K}''(0)$ and $\tilde{E}'(0)$ are arbitrary and can be selected so that *two* conditions at $\eta = 1$ are satisfied. However, from Eq. (10.5.15) we see that the symmetry condition on $\tilde{U}(\eta)$ is automatically satisfied, so these two parameters can be selected so that $\tilde{K}(\eta)$ and $\tilde{E}(\eta)$ are symmetric, i.e., so that $\tilde{K}'(1) = \tilde{E}'(1) = 0$. From the behavior of Eq. (10.5.16) as $\eta \to 0$, we find that—as expected from our earlier considerations [cf. Eq. (10.5.8)]—$\tilde{K}''(0) = \tilde{E}(0)$. Thus the value of $E(0)$ is obtained as part of the solution. If the predicted value is not in accord with either experiment or direct numerical simulations, some adjustment of the empirical constants in Eqs. (10.5.15) is called for.

Although our earlier remarks (cf. Section 10.3) regarding the numerical difficulties associate with the equations arising in the k–ε theory apply to Eqs. (10.5.15), we assume for our immediate purposes that numerical solutions calculated in some fashion are obtained for a range of δ. From such solutions we obtain the variation with Reynolds number, i.e., with δ, of $\tilde{U}(1) = U_1(1)/u_\tau$, the parameter determining the skin friction coefficient [cf. Eq. (9.4.2)]. Rather than pursue this approach, we take advantage, as before, of the small magnitude of the values of δ of applied interest by carrying out an asymptotic analysis. To simplify the presentation we restrict attention to the lowest orders in the expansion in terms of δ. Extension to include higher-order Reynolds-number effects can be carried out in a straightforward manner along the lines suggested in Sections 9.2–9.4.

10.6 AN ASYMPTOTIC ANALYSIS OF CHANNEL FLOW

The solutions for the outer flow, i.e., away from the wall, are obtained from Eqs. (10.5.15) with δ set to zero and $f_\mu = f_2 = 1$. However, to describe the inner layer as $\delta \to 0$, we must stretch the independent variable and rescale the mean viscous dissipation. Thus we let

$$\hat{U} = \tilde{U} \qquad \hat{K} = \tilde{K} \qquad \hat{E} = \frac{\varepsilon\,\nu}{u_\tau^4} = \tilde{E}\,\delta \tag{10.6.1}$$

The first two of these variables are introduced to assure identification with the wall layer, but are identical with their counterparts in the outer flow. However, we see that the dimensionless mean viscous dissipation must be scaled *down* by multiplication by the small parameter δ, the implication being that $\tilde{E}(\eta \to 0)$ is a large quantity; the physical basis for this requirement is that the turbulence scales which are proportional to $\tilde{K}^{3/2}/\tilde{E}$ *decrease* as the wall is approached and if, as we shall see, $\tilde{K}$ approaches a constant, then $\tilde{E}$ must increase. To resolve the wall layer we introduce our usual independent variable for that layer,

$$\hat{\eta} \equiv \frac{\eta}{\delta} \tag{10.6.2}$$

recalling that $\hat{\eta} = u_\tau x_2/\nu = y^+$.

When these definitions are introduced into Eqs. (10.5.15) and only the terms that are independent of δ are retained, there result

$$\left(1 + c_\mu f_\mu \frac{\hat{K}^2}{\hat{E}}\right)\hat{U}' = 1$$

$$\left[\left(1 + \frac{c_\mu f_\mu}{\sigma_k}\frac{\hat{K}^2}{\hat{E}}\right)\hat{K}'\right]' + c_\mu f_\mu \frac{\hat{K}^2}{\hat{E}}(\hat{U}')^2 - \hat{E} = 0 \qquad (10.6.3)$$

$$\left[\left(1 + \frac{c_\mu f_\mu}{\sigma_\varepsilon}\frac{\hat{K}^2}{\hat{E}}\right)\hat{E}'\right]' + c_{\varepsilon 1}c_\mu f_\mu \hat{K}(\hat{U}')^2 - c_{2\varepsilon} f_2 \frac{\hat{E}^2}{\hat{K}} = 0$$

where prime now denotes differentiation with respect to $\hat{\eta}$. Again we note the disappearance of the η term on the right side of the first of Eqs. (10.6.3), implying that, to lowest order in δ, the pressure gradient in channel flow does not influence the wall layer. In these equations the f-functions are given by Eqs. (10.5.13) with $N_{RT} = \hat{K}^2/\hat{E}$.

Boundary conditions are imposed on these equations at the wall, namely, $\hat{U}(0) = 0$ and $\hat{K}(\hat{\eta} \to 0) = \frac{1}{2}\hat{K}''(0)\,\hat{\eta}^2$, where $\hat{K}''(0) = \hat{E}(0)$ is arbitrary. Thus we select two parameters, $\hat{K}''(0)$ and $\hat{E}'(0)$, to meet the requirements for matching with the outer solutions, i.e., with Eqs. (10.5.15) with $\delta \to 0$ and $\eta \to 0$. As suggested earlier, in the discussion leading to Eq. (10.2.6), a consistent calculation results from the a-priori assumptions that

$$\hat{K}(\hat{\eta} \to \infty) = \frac{1}{c_\mu^{1/2}} \qquad \hat{E}(\hat{\eta} \to \infty) = \frac{1}{\kappa\hat{\eta}} \qquad (10.6.4)$$

where κ is the von Karman constant. With this behavior, the first of Eqs. (10.6.3) with $f_\mu(\hat{\eta} \to \infty) = 1$ gives

$$\hat{U}(\hat{\eta} \to \infty) = \frac{1}{\kappa}\ln\hat{\eta} + B \qquad (10.6.5)$$

where B is obtained as part of the solution. The second and third of Eqs. (10.6.3) are automatically satisfied provided the restraint of Eq. (10.2.6) is respected. We recognize Eq. (10.6.5) as the logarithmic law for the outer portion of the wall layer found earlier, but in this case the value of B can be adjusted only by changing the various empirical coefficients in Eqs. (10.6.3).

When the behavior given by Eq. (10.6.4) is considered in the second of Eqs. (10.6.3), we see that at the outer edge of the wall layer the production of turbulent kinetic energy is balanced by the mean viscous dissipation. We must thus expect that this same balance occurs at the inner edge of the outer layer, another example of the dominance of production and dissipation over diffusion and convection. Finally, with the behavior of Eq. (10.6.4) and with f_μ and f_2 approaching unity, each of the three terms in the third of Eqs. (10.6.3) is proportional to $\hat{\eta}^{-2}$, so a consistent solution is obtained.

Figure 10.6.1 shows the distributions of $\hat{K}(\hat{\eta})$ and $\hat{E}(\hat{\eta})$ found numerically with the recommended values of $c_{\varepsilon 1}$, $c_{\varepsilon 2}$, and c_μ but with $\sigma_\varepsilon = 1.11$ to reflect Eq. (10.2.7). The distribution of $\hat{U}(\hat{\eta})$ is essentially given by Fig. 9.3.1 but with $B = 3.7$ in the logarithmic portion, so this distribution need not be repeated here. Adjustments to the various empirical coefficients must be made if this value of B is to be brought in line with the generally accepted value of roughly 5, but we do not pursue this further.

From Fig. 10.6.1, $\hat{K}(\hat{\eta})$ increases quadratically from the wall, reaches a maximum at $\hat{\eta} \approx 20$, and then decreases so as to approach its asymptotic value of $c_\mu^{-1/2}$ from above. The corresponding variation of $\hat{E}(\hat{\eta})$ decreases slightly from a wall value of 0.144 before increasing to a maximum in the neighborhood of $\hat{\eta} \approx 9$ and then decreasing monotonically to zero. Both of these distributions are in good agreement with comparable calculations by others (cf. Myong and Kassagi, 1989).

Figure 10.6.2 shows the distribution of $c_\mu f_\mu \hat{K}^2/\hat{E}$, which is the ratio of ν_T/ν of the two exchange coefficients. For $\hat{\eta}$ less than approximately 4, this ratio is small compared to unity, implying a dominance of viscous transport; while for $\hat{\eta}$ greater than roughly 30, the reverse is true, and turbulent transport dominates. These limits are somewhat wider than, but roughly comparable to, those found earlier from the classical, mixing-length analysis (cf. Section 9.3).

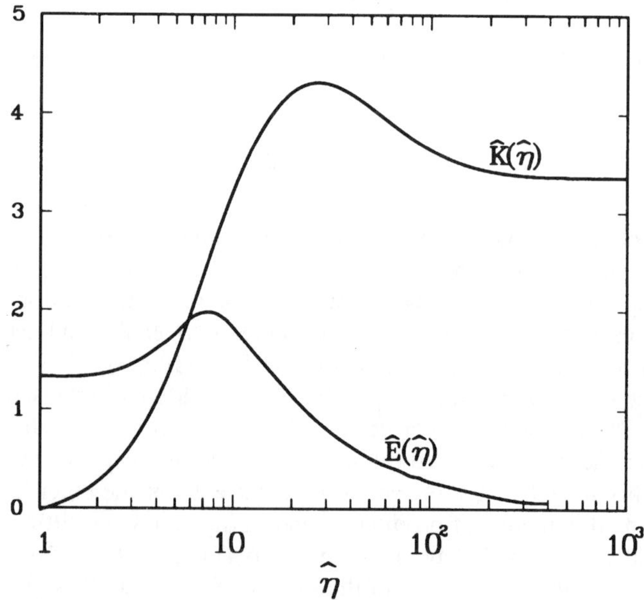

Figure 10.6.1 Turbulent kinetic energy in terms of $\hat{K}(\hat{\eta})$ and the mean viscous dissipation in terms of $\hat{E}(\eta)$ in the wall flow of a channel according to the k–ε theory.

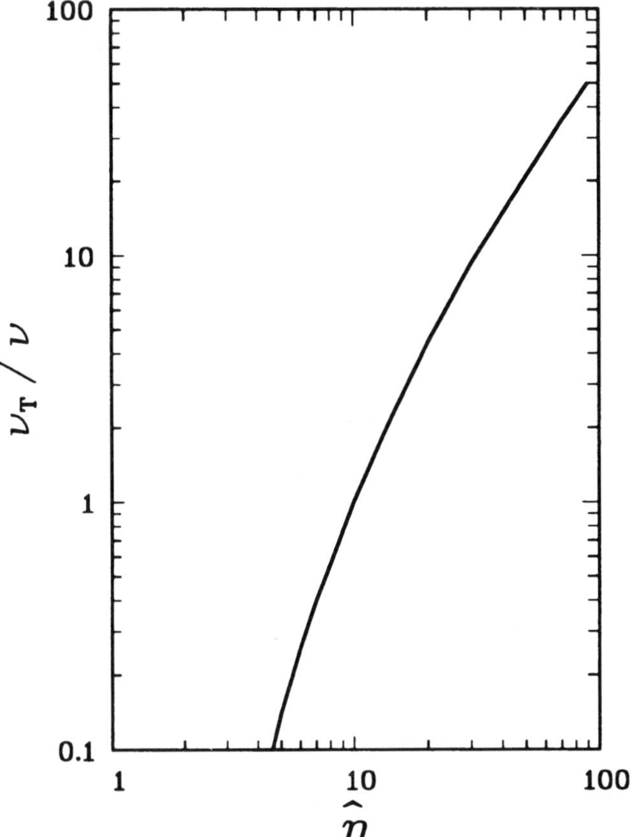

Figure 10.6.2 Ratio of exchange coefficients ν_T/ν in the wall layer of channel flow according to the k–ε theory.

Recall that from Eq. (10.1.5), the basic gradient transport relation, knowledge of the turbulent kinetic energy and the turbulent exchange coefficient permits *all* the Reynolds stresses to be determined. In channel flow this determination leads to the result that the turbulence is everywhere equipartitioned, that is, $\overline{u_i^2} = \frac{2}{3}k$ for $i = 1, 2, 3$. This expectation is indeed found to be satisfied. In channel flow, however, this is not only not in accord with experiment in the flow away from the walls, it is also not in accord with our earlier finding that as $x_2 \to 0$ the turbulence becomes two-dimensional with negligible values of $\overline{u_2^2}$. Again, if the physically correct behavior in this regard is required, a Reynolds stress theory as outlined later in this chapter must be employed.

Before concluding this discussion of the wall layer, several comments are appropriate. The description of the mean quantities within this layer does not suggest the complex fluid mechanical processes that are actually present there. Experiments starting with the hydrogen bubble photographs of Kline and coworkers (cf. Kline et al., 1967) and continuing with the ingeniously taken photographs in a smoke-filled turbulent boundary layer by Head and Bandyopadhyay (1978), and with the measurements and photographs of Brodkey, Falco, Willmarth, and others (cf. Smith and Abbott, 1978), establish that complex, three-dimensional structures exist in the wall layer, structures which periodically extend into the outer portions of the flow. Three-dimensionality within the wall layer implies fluctuations in the ω_1 vorticity component and an interaction with the dominant ω_3 vorticity (cf. Section 2.5 and Fig. 2.5.1). Clearly, such fluctuations lead to interchange in the wall layer of faster- and slower-moving fluid and to a significant influence on the mean shear stress $\rho\, \overline{u_1 u_2}$. The complexity of the flow in the wall layer is also indicated for relatively low Reynolds numbers in channel flow by the DNS of Monin and Kim (1982) and Kim et al. (1987). A further manifestation of complexity in the wall layer is the behavior of the shear stress at the wall; it is found from various experiments that the time dependence of this stress at a particular wall location $\mu \partial \tilde{u}_1 / \partial x_2(t; x_1, 0, x_3)$ involves extended periods during which the stress is relatively low and short periods during which the stress is relatively high. The mean values we calculate in this section simply reflect the average of these two levels and the periods in which those levels prevail. The short periods of high stress involve the interaction with the wall of relatively large fluid structures with high degrees of three-dimensional vorticity. In boundary layers these structures migrate to the outer edge and are associated with intermittency (cf. Section 5.7). The processes arising in the wall layer of turbulent channel and boundary layer flows are so complex and so important relative to our understanding of turbulence in the presence of walls that they continue to be the subject of study.

We now return to our analysis of channel flow. The solutions for the outer flow in terms of $\tilde{U}(\eta)$, $\tilde{K}(\eta)$, and $\tilde{E}(\eta)$ are obtained from Eqs. (10.5.15) with $\delta = 0$ and $f_\mu = f_2 = 1$. However, we see that an arbitrary constant can be added to $\tilde{U}(\eta)$ without altering a solution of these three equations. Thus we find numerically an intermediate solution for $\tilde{U}(\eta)$, denoted ${}^{(0)}\tilde{U}(\eta)$, such that ${}^{(0)}\tilde{U}(1) = 0$. Then $\tilde{U}(\eta) = \tilde{U}(1) + {}^{(0)}\tilde{U}(\eta)$. Note that ${}^{(0)}\tilde{U}(\eta)$ is the velocity defect of Eq. (9.2.8). The symmetry conditions $\tilde{K}'(1) = \tilde{E}'(1) = 0$ must be imposed, but $\tilde{U}'(1) = 0$ is automatically satisfied.

There remains to be discussed the matching of the solutions for the wall and outer flows, the procedure leading to the determination of $\tilde{U}(1)$, $\tilde{K}(1)$, and $\tilde{E}(1)$, and completing the determination of the outer solutions. As seen from the first of Eqs. (10.6.4), matching of the inner and outer turbulent kinetic energies clearly calls for

$$\tilde{K}(\eta \to 0) = \frac{1}{c_\mu^{1/2}} \tag{10.6.6}$$

while that for the viscous dissipation is more subtle. The starting point is the lowest-order matching requirement that

$$\lim_{\hat{\eta}\to\infty}\left[\frac{u_\tau^4}{\nu}\frac{1}{\kappa\hat{\eta}} - \frac{u_\tau^3}{h}\tilde{E}(\eta \to 0)\right] = 0$$

which leads to the requirement that

$$\tilde{E}(\eta \to 0) = \frac{1}{\kappa\eta} \tag{10.6.7}$$

It is this behavior which calls for the scaling of $\hat{E}(\hat{\eta})$ in Eq. (10.6.1).

With this behavior for $\tilde{K}(\eta)$ and $\tilde{E}(\eta)$ as $\eta \to 0$, the first of Eqs. (10.5.15) with $\delta = 0$ and $f_\mu = 1$ yields, for the intermediate result,

$$^{(0)}\tilde{U}(\eta \to 0) = \frac{1}{\kappa}\ln\eta - C \tag{10.6.8}$$

where C is determined as part of the solution. If we add the arbitrary values $\tilde{U}(1)$ to this and match the result to Eq. (10.6.5), we are left with a skin friction law relating $U_1(1)/u_\tau$ to the Reynolds number reflected in δ, that is, to Eq. (9.3.11), which we repeat for convenience:

$$B = \frac{U_1(1)}{u_\tau} + \frac{1}{\kappa}\ln\delta + C \tag{10.6.9}$$

The formulation of the outer flow is now complete.

In these outer solutions we could introduce a variation of σ_e from the value 1.11 called for by Eq. (10.2.7) as $\eta \to 0$ to the standard value of 1.3 elsewhere. However, with little loss in accuracy for our purposes, we keep σ_e fixed at its value in the wall layer.

Figure 10.6.3 gives the outer solutions for $\tilde{K}(\eta)$ and $\tilde{E}(\eta)$. The solution of $^{(0)}\tilde{U}(\eta)$ is essentially the same as that shown in Figs. 9.2.1 and 9.2.2 and is not repeated here. However, the value of C is slightly different, namely, -0.55. We see that the logarithmic portion of $\tilde{E}(\eta)$ is clearly evident for $\eta \leq 0.2$, while the approach of $\tilde{K}(\eta)$ to its wall value is much slower.

The combination of the results in Figs. 10.6.1 for the wall layer with those given in the figures for the outer flow in this section provides a complete description of the distributions of turbulent kinetic energy and mean viscous dissipation in turbulent channel flow. Qualitatively similar representations apply to other wall layers, e.g., to the turbulent boundary layers discussed in Section 10.8. In particular, the maximum turbulent kinetic energy and mean viscous dissipation occur in the wall layer. Thus the smallest Kolmogorov lengths exist in that layer. A significant consequence of this when DNS is applied to channel and other flows involving walls is that computational efficiency calls for a fine

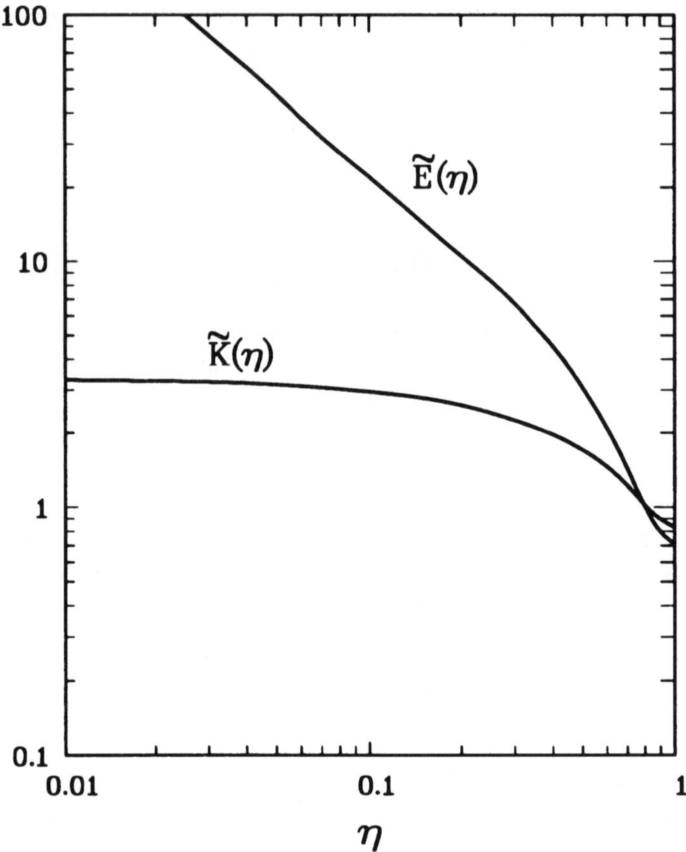

Figure 10.6.3 Turbulent kinetic energy and mean viscous dissipation in terms of $\tilde{K}(\eta)$ and $\tilde{E}(\eta)$, respectively, in the outer portion of channel flow according to the k–ε theory.

mesh near the wall and an expanded mesh away from the wall, so that the turbulence is equally well resolved throughout the flow.

10.7 HEAT TRANSFER IN CHANNEL FLOW

In Section 9.6 we discussed heat transfer in turbulent channel flow and noted in particular that if the statistical variables involving the temperature are symmetric about the plane of symmetry of the channel ($x_2 = h$), the usual notions of fully developed flow must be altered to account for streamwise evolution of the temperature field. Indeed, if we adopt the classical description for the mean turbulent flux of heat, we are forced to consider this symmetric case. Here, both for simplicity and for variety, we treat the asymmetric case of heat transfer in turbulent channel flow involving heat addition to the fluid at one wall and equal heat withdrawal from the opposite wall. The temperatures of the two walls are constant but only slightly unequal, so that fluid properties may still be assumed constant; thus we treat the temperature as a passive scalar. In this case the statistical properties of both the velocity and the temperature depend only on the normal coordinate x_2. The mean temperature distribution is shown schematically in Fig. 10.7.1; it resembles the corresponding distribution in turbulent Couette flow as treated in Section 9.4, although the velocity field is entirely different. We again have two thin, viscous sublayers with large gradients of mean temperature and a central region with considerably smaller gradients.

From Eqs. (7.2.1) and (9.6.1) we have, from the averaged energy equation,

$$\left[\left(\frac{\nu}{N_\sigma} + \frac{\nu_T}{\sigma_T}\right)\Theta'\right]' = 0 \tag{10.7.1}$$

where the prime denotes differentiation with respect to the dimensionless coordinate $\eta \equiv x_2/h$. The second and third of Eqs. (10.1.10) are again introduced to describe turbulent exchange coefficient in terms of k and ε.

The solution of Eq. (10.7.1) can be expressed as an integral of known quantities,

$$\Theta(\eta) = \Theta(0) + \Theta'_w \int_0^\eta \frac{d\eta'}{\delta + (c_\mu f_\mu N_\sigma/\sigma_T)(\tilde{K}^2/\tilde{E})} \tag{10.7.2}$$

where we introduce the nondimensionalization of Eqs. (10.5.14) and the inverse

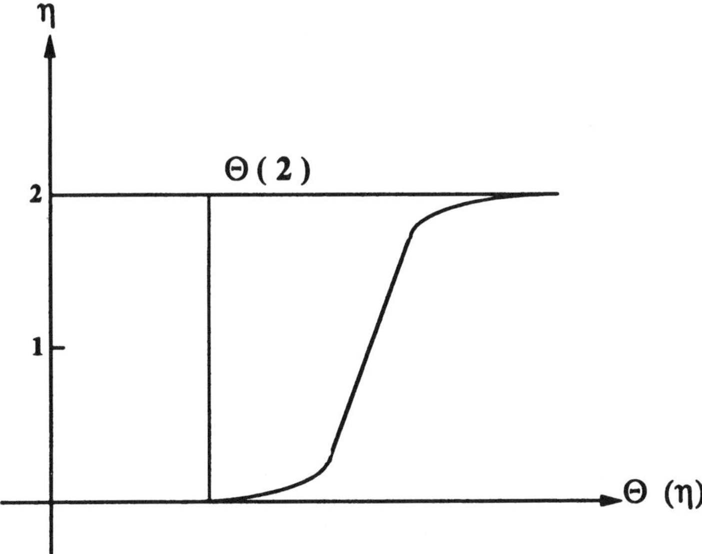

Figure 10.7.1 Schematic representation of the fully developed mean temperature distribution in turbulent channel flow with cooling and heating at the lower and upper walls, respectively.

Reynolds number parameter $\delta \equiv \nu/u_\tau h$ and where, here and *henceforth* in this section, $\Theta'_w \equiv d\Theta/d\hat{\eta}(0)$ is related to the heat transfer at the lower wall expressed in terms of $\hat{\eta}$, the variable for the wall layer.

When Eq. (10.7.2) is evaluated at $\eta = 2$, we obtain a relation involving the temperature difference $\Theta(2) - \Theta(0)$ and *determining* Θ'_w. Thus we find, as expected, that the heat transfer is proportional to the temperature difference across the channel. The symmetry of $\tilde{K}$ and $\tilde{E}$ with respect to the channel midplane and consideration of Fig. 10.7.1 imply that

$$\Theta(1) = \frac{1}{2}[\Theta(0) + \Theta(2)] \tag{10.7.3}$$

a relation we shall find useful later.

If we had earlier found solutions for the turbulent kinetic energy and the mean dissipation in terms of $\tilde{K}(\eta; \delta)$ and $\tilde{E}(\eta; \delta)$ for specific values of δ, the corresponding mean temperature distributions could now be found by quadrature from Eq. (10.7.2) in terms of $[\Theta(\eta; \delta)]$. However, in view of the asymptotic analysis involving $\delta \to 0$ actually carried out in Section 10.6, it is necessary and appropriate to consider separately the temperature distributions in the outer and wall flows. To do so for the latter flow, we first rewrite Eq. (10.7.2) in terms of the inner variables $\hat{\eta} = \eta/\delta$, $\hat{K} = \tilde{K} = k/u_\tau^2$, and $\hat{E} = \tilde{E}\,\delta = \varepsilon\nu/u_\tau^4$ and find

$$\Theta(\hat{\eta}) = \Theta(0) + \frac{d\Theta}{d\hat{\eta}}(0) \int_0^{\hat{\eta}} \frac{d\hat{\eta}'}{1 + (c_\mu f_\mu N_\sigma/\sigma_T)(\hat{K}^2/\hat{E})} \tag{10.7.4}$$

As $\hat{\eta} \to \infty$, we know that $f_\mu \to 1$ and, from Eqs. (10.6.6) and (10.6.7), that $\hat{K} \to 1/c_\mu^{1/2}$ and $\hat{E} \to 1/\kappa\hat{\eta}$, so Eq. (10.7.4) yields in this limit

$$\frac{\Theta(\hat{\eta}) - \Theta(0)}{\Theta'_w} = \frac{\sigma_T}{N_\sigma}\left(\frac{1}{\kappa}\ln\hat{\eta} + B_\theta\right) \tag{10.7.5}$$

where B_θ is a known constant and $\Theta'_w \equiv d\Theta/d\hat{\eta}(0)$ is related directly to the mean heat transfer at the wall. We thus find once again the logarithmic distribution of mean temperature in the outer portion of the wall layer, a distribution found earlier in Sections 9.4, 9.6, and 9.11.

Several comments are indicated. Equation (10.7.4) also applies close to the upper wall, provided $\hat{\eta}$ is redefined so as to determine distances close to $\eta = 2$. The parameter σ_T permits some adjustment to bring Eqs. (10.7.4) and (10.7.5) into agreement with experiment. Finally, because of its basis in an asymptotic analysis, Eq. (10.7.5) applies locally, i.e., for arbitrary values of x_1 in the wall layer of channel flows with variable wall temperatures and in turbulent boundary layers with heat transfer. Accordingly, we shall use this result in Section 10.9 to determine the boundary conditions at the inner edge of the outer flow applicable to equilibrium turbulent boundary layers with heat transfer.

The mean temperature in the outer flow is obtained to lowest order in δ from Eq. (10.7.1) by neglecting molecular transport, i.e., by setting $\delta = 0$. Thus we find

$$\Theta(1) - \Theta(\eta) = \frac{\sigma_T}{N_\sigma c_\mu} \Theta'_w \int_\eta^1 d\eta' \frac{\tilde{E}}{\tilde{K}^2} \tag{10.7.6}$$

As $\eta \to 0$, we know from Section 10.6 that $\tilde{K} \to 1/c_\mu^{1/2}$ and $\tilde{E} \to 1/\kappa\eta$, with the consequence that as the wall is approached Eq. (10.7.6) becomes

$$\Theta(1) - \Theta(\eta) \to \frac{\sigma_T}{c_\mu N_\sigma} \Theta'_w \left(-\frac{1}{\kappa} \ln \eta + C_\theta\right) \tag{10.7.7}$$

where C_θ is also known. Thus, if Eq. (10.7.7) is matched to Eq. (10.7.5), we obtain a heat transfer law of the form

$$\Theta(2) - \Theta(0) = \frac{2\,\sigma_T}{c_\mu N_\sigma} \Theta'_w \left(-\frac{1}{\kappa} \ln \delta + B_\theta + C_\theta\right) \tag{10.7.8}$$

where we eliminate $\Theta(1)$ by use of Eq. (10.7.3) so as to obtain the overall heat transfer between the two walls.

Figures 10.7.2 and 10.7.3 show the mean temperature distributions in the wall layer in terms of $[\Theta(\hat{\eta}) - \Theta(0)]/\Theta'_w$ and in the outer flow in terms of $(N_\sigma/\sigma_T)[\Theta(\eta) - \Theta(1)]/\Theta'_w$ obtained from Eqs. (10.7.4) and (10.7.6), respectively, using the distributions of the turbulent kinetic energy and viscous dissipation obtained from Section 10.6. In these calculations we take $N_\sigma = 0.7$ and $\sigma_T = 1$ and find $B_\theta = -4.20$. The value of C_θ in Eq. (10.7.6) is found to be -0.36.

10.8 THE EQUILIBRIUM TURBULENT BOUNDARY LAYER

We now reconsider the turbulent boundary layer with a distribution of external velocity $U_\infty(x_1)$ so that the similarity analysis of Section 9.10 applies but with a k–ε description of turbulent exchange. In this reconsideration we adopt the strategy of that section, focus on the outer flow, and subsequently reflect the influence of the wall layer by imposing the matching conditions of Eqs. (10.6.6) and (10.6.7). We thus assume, as in Section 9.10, that the Reynolds number is sufficiently high that the wall layer is the same for channel and boundary-layer flows, an assumption based on the insensitivity of the wall layer to streamwise pressure gradients under these conditions.

For this application of the k–ε theory, the notion of similarity must be extended to the turbulent kinetic energy and the mean viscous dissipation. To do so we require that

$$\tilde{K} \equiv \frac{k}{u_\tau^2} = \tilde{K}(\eta) \qquad \tilde{E} \equiv \frac{\varepsilon\,\Delta}{u_\tau^3} = \tilde{E}(\eta) \tag{10.8.1}$$

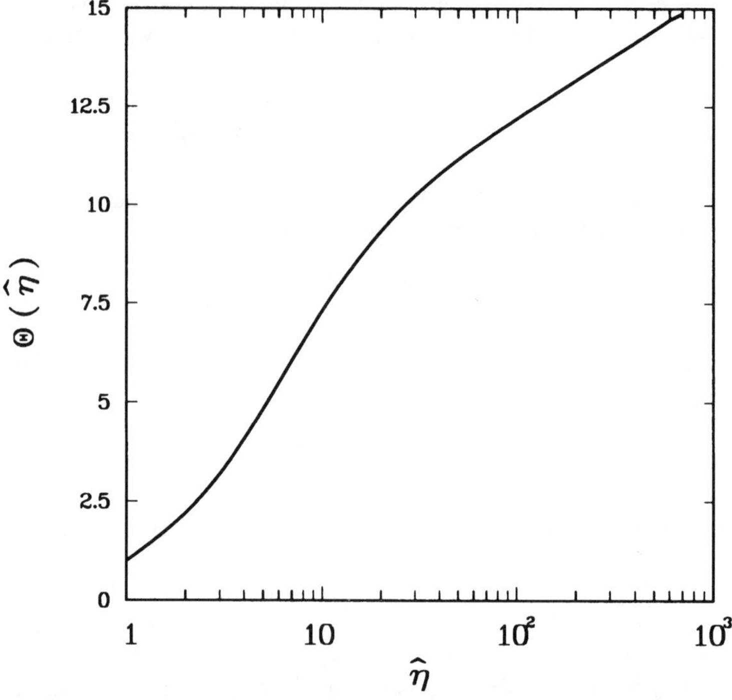

Figure 10.7.2 Mean temperature distribution in the wall layer of a turbulent channel flow according to the k–ε theory.

The implication of these equations is that the variations of k and ε with x_1 and x_2 and the variations of u_τ and Δ with x_1 are such that $\tilde{K}$ and $\tilde{E}$ vary only with the similarity variable $\eta = x_2/\Delta = \eta(x_1, x_2)$ [cf. Eq. (9.10.1)]. The Reynolds shear stress is thus given as

$$-\rho\, \overline{u_1 u_2} = c_\mu\, \rho\, u_\tau^2 \frac{\tilde{K}^2}{\tilde{E}} f'' = \rho\, u_\tau^2\, T\, f'' \tag{10.8.2}$$

where $T = T(\eta)$ is clearly but implicitly defined and where prime denotes differentiation with respect to η. The implication of Eq. (10.8.2) is that $\overline{u_1 u_2}/u_\tau^2$ varies only with η.

The considerations leading to Eqs. (9.9.10)–(9.9.13) again apply, but with modifications called for by the altered representation of the Reynolds shear stress. By following the steps leading to Eq. (9.9.14), it is found that

$$T\, f'' = -[(1 + 2\,\Pi)\,\eta\, f' + (1 - f)] \tag{10.8.3}$$

provided

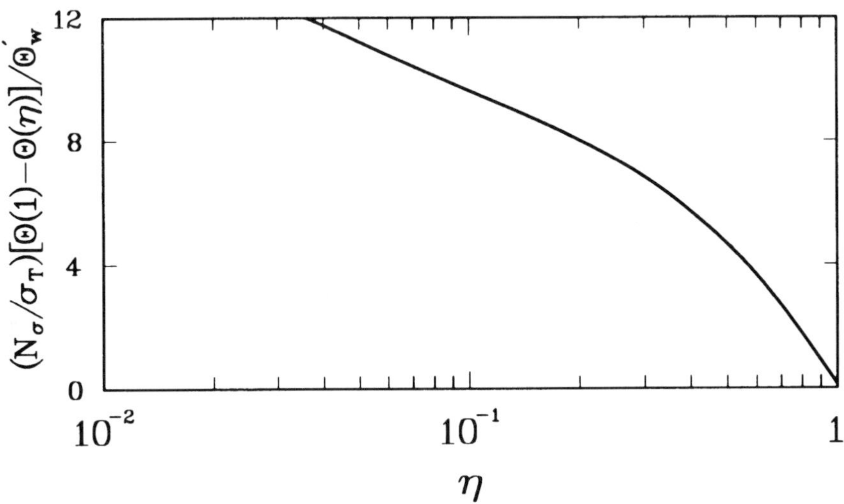

Figure 10.7.3 Mean temperature distribution in the outer region of a turbulent channel flow according to the k–ε theory.

$$\lim_{\eta \to \infty} T f'' = 0$$

a condition satisfied, for example, if $(\tilde{K}^2/\tilde{E})(\eta \to \infty)$ is constant and $f''(\eta \to \infty) = 0$. In Eq. (10.8.3) the parameter Π is defined by Eqs. (9.9.6) and determines the variation of the external velocity: If the pressure distribution is adverse, i.e., the pressure gradient is positive, then $\Pi > 0$; while if the pressure gradient is favorable and the pressure gradient is negative, then $\Pi < 0$. A further proviso connected with Eq. (10.8.3) relates to the behavior of the dependent variables as $\eta \to 0$; in particular, we recall that $f(0) = 0$ and that there are matching requirements developed in connection with our earlier analysis of turbulent channel flow [cf. Eqs. (9.9.12), (10.6.6), and (10.6.7)]. We thus have

$$\lim_{\eta \to 0} f' + \frac{1}{\kappa} \ln \eta + C = 0 \qquad \tilde{K}(\eta \to 0) = \frac{1}{C_\mu^{1/2}} \qquad \tilde{E}(\eta \to 0) = \frac{1}{\kappa\, \eta} \tag{10.8.4}$$

The discussion in Section 10.6 regarding the dominance of production and dissipation in determining the balance of turbulent kinetic energy at the inner edge of the outer flow pretains here.

Since $T(\eta)$ in Eq. (10.8.3) involves the turbulent kinetic energy and mean viscous dissipation, to proceed we next develop the similarity equations for $\tilde{K}(\eta)$ and $\tilde{E}(\eta)$, i.e., the forms of the second and third of Eqs. (10.1.10) that are appropriate for the present application. In doing so we use Eqs. (9.9.1) and (9.9.5) to express $U_1(x_1, x_2)$ and $U_2(x_1, x_2)$ in terms of $f(\eta)$ and invoke the inequalities leading to Eqs. (9.9.6) in order to simplify the convective terms. Thus in due course we find

$$2\,\Pi\,\tilde{K} - (1 + 2\,\Pi)\,\eta\,\tilde{K}' - T\,f''^2 = \left(\frac{T}{\sigma_k}\,\tilde{K}'\right)' - \tilde{E}$$
$$2\,\Pi\,\tilde{E} - (1 + 2\,\Pi)\,\eta\,\tilde{E}' - c_{\varepsilon 1}\,T\,\tilde{E}\,f''^2 = \left(\frac{T}{\sigma_\varepsilon}\,\tilde{E}'\right)' - c_{\varepsilon 2}\,\frac{\tilde{E}^2}{\tilde{K}} \qquad (10.8.5)$$

The first two terms and the third on the left side represent convection and production, respectively; the first term on the right side represents diffusion, while the last term describes dissipation. With the empirical constants assumed known, Eqs. (10.8.3), (10.8.5), and (10.8.6) determine $f(\eta)$, $\tilde{K}(\eta)$, and $\tilde{E}(\eta)$ with Π as the sole parameter.

The boundary conditions as $\eta \to 0$ are given by Eq. (10.8.4), while at the other limit we have

$$f(\eta \to \infty) = 1 \qquad \tilde{K}(\eta \to \infty) = 0 \qquad \tilde{E}(\eta \to \infty) = 0 \qquad (10.8.6)$$

The behavior of $\tilde{K}$ and $\tilde{E}$ in this limit must be such that the left side of Eq. (10.8.3) vanishes and Eqs. (10.8.5) are satisfied as $\eta \to \infty$. Our examination of this behavior closely follows the analysis of Section 10.3. Thus we expect a convective-diffusive balance to prevail in Eqs. (10.8.3) and (10.8.5) as $\eta \to \infty$. A consistent calculation to that end again results from the assumption that T approaches asymptotically a constant denoted T_∞. In the first of these equations we assume that

$$f \to 1 + A_f\,\eta^n \exp\left(-\frac{\alpha_f^2}{2}\,\eta^2\right) \qquad (10.8.7)$$

where A_f is arbitrary. Substitution and collection of powers of η into Eqs. (10.8.3) leads to

$$\alpha_f = \frac{1 + 2\,\Pi}{T_\infty} \qquad n = -\frac{2\,(1 + \Pi)}{1 + 2\,\Pi} \qquad (10.8.8)$$

The corresponding treatment of the other two equations is not as rigorous. However, if we assume an exponential decay so that

$$\tilde{K} \to A_K \exp\left(-\frac{\alpha_K}{2}\,\eta^2\right) \qquad \tilde{E} \to A_E \exp\left(-\frac{\alpha_E}{2}\,\eta^2\right) \qquad (10.8.9)$$

then the dominant terms in Eqs. (10.8.5) yield

$$\alpha_k = \frac{\sigma_K}{T_\infty}(1 + 2\,\Pi) \qquad \alpha_E = \frac{\sigma_e}{T_\infty}(1 + 2\,\Pi) \tag{10.8.10}$$

We see again the limitation $\Pi > -\frac{1}{2}$ found in Section 9.10. Now, the assumption regarding T_∞ again requires that $2\,\alpha_K - \alpha_E = 0$ and thus that $2\,\sigma_k = \sigma_\varepsilon$. If we retain the standard value of unity for σ_k, we require that $\sigma_\varepsilon \rightarrow 2$ as $\eta \rightarrow \infty$, compared with the standard value of $\sigma_\varepsilon = 1.3$. Since at the other limit of $\eta \rightarrow 0$, matching with the wall layer requires $\sigma_\varepsilon = 1.11$, we assume a distribution

$$\sigma_\varepsilon = 2 - 0.89 \exp\left(-4\frac{\eta}{\eta_e}\right) \tag{10.8.11}$$

where, as in Eq. (10.3.11), η_e is an effective edge and the multiplier in the argument of the exponential function is arbitrary. In the middle range of η spanning unity, σ_ε is roughly equal to its standard value.

Only the implementation of the boundary conditions on $f(\eta)$ given by Eqs. (10.8.4) calls for special attention; the following considerations apply. If the first of these equations is integrated from η to $\eta_0 << 1$ with the condition $f(0) = 0$ imposed, and if C is eliminated between the two equations, we obtain

$$f'(\eta_0) - \frac{f(\eta_0)}{\eta_0} + \frac{1}{\kappa} = 0 \tag{10.8.12}$$

Successively smaller values of η_0 are assumed until essentially no further change occurs in the solutions. The formulation for the numerical solution of equilibrium turbulent boundary layers according to the k–ε theory is now complete, with Π as the sole parameter.

Although, as suggested earlier, the nature of these equations calls for solution by finite-difference methods, in principle, numerical integration in two directions can be employed, a possibility that establishes the validity of the formulation. An integration in the direction of increasing η from $\eta_0 << 1$ with $f(\eta_0)$, $\tilde{K}(\eta_0)$, and $\tilde{E}(\eta_0)$ as unknowns, proceeds to an intermediate value of η denoted η_m. A second inward integration from $\eta_e >> 1$ with A_f, A_K, and A_E as unknowns likewise proceeds to η_m. The six unknowns are adjusted so that f, $\tilde{K}$, and $\tilde{E}$ and their first derivatives are continuous at η_m. After converged solutions are obtained, the appropriateness of the values of η_0 and η_e must be examined by reducing the former and increasing the latter to establish the insensitivity of the solutions to these changes.

Although we give the solutions for boundary layers with favorable, zero, and adverse pressure gradients, i.e., for $\Pi = -0.4$, 0, and 1, we consider first and in detail the results for $\Pi = 0$. In Fig. 10.8.1 the distributions of the turbulent exchange coefficient ν_T expressed in terms of the dimensionless $\nu_T/u_\tau\Delta$ as given

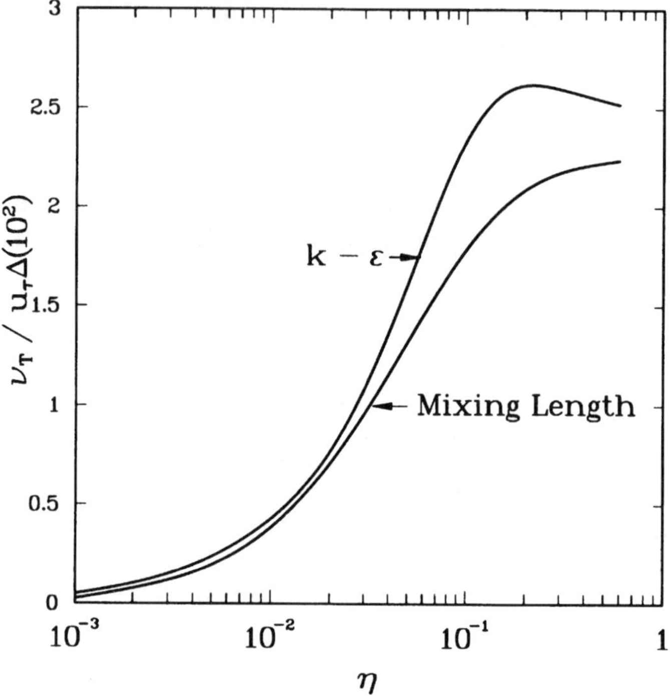

Figure 10.8.1 Distributions of the turbulent transport coefficient in terms of $\nu_T/u_\tau\Delta$ in equilibrium boundary layers with $\Pi = 0$ as given by the mixing-length and k–ε theories.

by this k–ε calculation and by the mixing-length calculation of Section 9.10 are presented. We see that, except at the outer edge of the boundary layer, the two distributions are in close agreement. The implication of this result is that the velocity distribution for $\Pi = 0$ as given in Fig. 9.10.1 is closely reproduced by the present theory and need not be repeated here. In considering all of the figures relative to $\Pi = 0$, it is useful to keep in mind that the edge of the boundary layer in this case is at $\eta \approx 0.5$.

Figure 10.8.2 shows the distributions of $K(\eta)$. The results for $\Pi = 0$ indicate a monotonic decrease from the vicinity of the wall to the external stream. When combined with the distribution in the wall layer shown in Fig. 10.6.1, we have a complete picture of the distribution of the turbulent kinetic energy in the turbulent boundary layer on a flat plate. Involved is an increase from zero at the wall to a maximum within the wall layer and then a continuous decrease to the external stream. The corresponding distribution of $\tilde{E}(\eta)$ is given in Fig. 10.8.3. The only noteworthy feature of this distribution is the rapid approach to zero at the outer edge, consistent with Eq. (10.8.10).

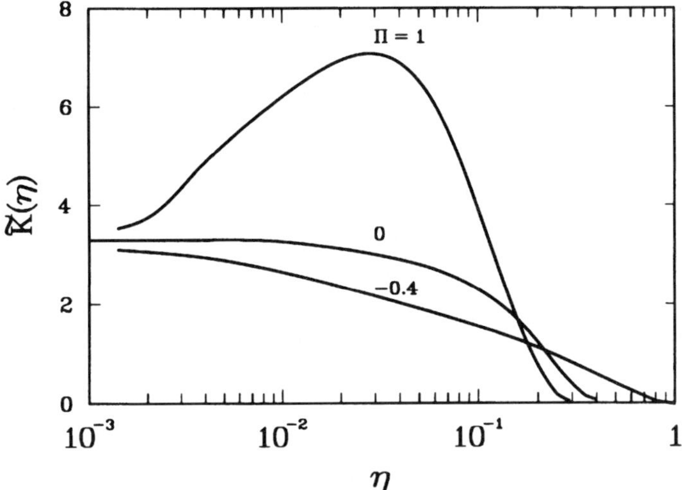

Figure 10.8.2 Distributions of turbulent kinetic energy in terms of $\tilde{K}(\eta)$ in equilibrium boundary layers.

We now turn to the solutions for turbulent boundary layers with streamwise pressure gradients, i.e., for $\Pi \neq 0$, and again consider Figs. 10.8.2 and 10.8.3. For a favorable pressure gradient ($\Pi < 0$), $K(\eta)$ decreases away from the wall somewhat more rapidly than for the flat-plate boundary layer, while for an adverse pressure gradient ($\Pi > 0$) there is a second maximum in the turbulent kinetic energy near the wall but outside of the wall layer, i.e., at $x_2/\delta \approx 0.1$, $\eta \approx 0.3$. The distributions of $\tilde{E}(\eta)$ with pressure gradient seen in Fig. 10.8.3 are uninterestingly similar to those for $\Pi = 0$, that is, a rapid and monotonic decrease with increasing η.

Comparison of all of these results suggests the complex effects influencing turbulent boundary layers, especially those involving streamwise pressure gradients. The view provided by this analysis of the outer portions of these boundary layers must be supplemented by the description of the wall layers provided by Section 10.6. The production–dissipation balance encountered repeatedly in our discussions prevails over large portions of such wall layers, but at the outer edge of the boundary layer these dominant contributors give way to a convective–diffusive balance. In turbulent boundary layers with adverse pressure gradients, turbulence production is diminished relative to dissipation, with the consequence that turbulent diffusion becomes more important and there is a second peak in the turbulent kinetic energy near the wall but outside the wall layer. Although these features of the turbulent boundary layer are developed here for equilibrium flows, they apply at least qualitatively to all boundary layers.

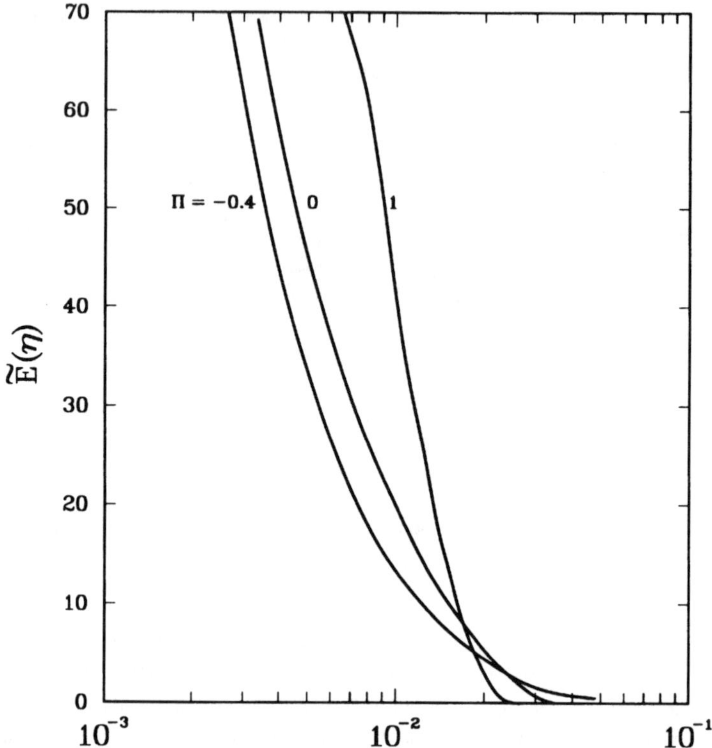

Figure 10.8.3 Distributions of the mean viscous dissipation in terms of $\tilde{E}(\eta)$ in equilibrium boundary layers.

10.9 HEAT TRANSFER IN THE EQUILIBRIUM TURBULENT BOUNDARY LAYER

In considering the extension of the analysis of the previous section to turbulent boundary layers involving small degrees of heat transfer, two distinct cases can be considered. One involves a boundary layer with a constant wall temperature for a sufficient upstream distance for the velocity and temperature distributions within the layer to correspond to similarity, i.e., to depend only on η. A second case arises if the wall temperature is changed abruptly at some streamwise station, with either that temperature or the heat transfer held constant downstream of that station. This latter case is reminiscent of heat transfer in channel flow as discussed in Sections 9.6 and 9.7 and is amenable to the eigenvalue-eigenfunction treatment suggested there with Π as a parameter.

It is sufficient for our purposes to treat the first case by the k–ε theory. The equation for the mean temperature with the normal heat flux $\rho\,\overline{u_2\theta}$ eliminated by a gradient assumption [cf. Eq. (7.2.1)] is

$$\frac{\partial}{\partial x_k} U_k \Theta = \frac{\partial}{\partial x_2}\left[\left(\frac{\nu}{N_\sigma} + \frac{\mu_T}{\sigma_T}\right)\frac{\partial \Theta}{\partial x_2}\right] \tag{10.9.1}$$

We again consider only the outer solutions, for which the transport of heat by thermal conductivity is negligible, but we impose a boundary condition as the wall is approached, i.e., at the inner edge of the outer flow, reflecting the distribution of mean temperature in the wall layer [cf. Eq. (10.7.5)]. Thus, if we assume that $\Theta = \Theta(\eta)$ and introduce $\mu_T = c_\mu \tilde{K}^2/\tilde{E}$ for the turbulent exchange coefficient, we obtain [cf. Eq. (9.12.2)]

$$-(1 + 2\,\Pi)\,\eta\,\Theta' = \frac{c_\mu}{\sigma_T}\left(\frac{\tilde{K}^2}{\tilde{E}}\,\Theta'\right)' \tag{10.9.2}$$

Since Eq. (10.9.2) must be solved numerically, it is convenient to change the dependent variable and define

$$\tilde{T}(\eta) = \frac{\Theta_\infty - \Theta(\eta)}{\Theta_\infty - \Theta_w} \tag{10.9.3}$$

so that $\tilde{T}$ approaches zero in the external stream. The equation for $\tilde{T}(\eta)$ is readily obtained from Eq. (10.9.2). As noted earlier, the boundary condition as $\eta \to 0$ is obtained from matching with the outer edge of the wall layer. Thus we have

$$\tilde{T}(\eta \to 0) = \frac{\Theta_w'}{\Theta_\infty - \Theta_w}\left(-\frac{\sigma_T}{\kappa N_\sigma}\ln\eta + C_\theta\right) \qquad \tilde{T}(\eta \to \infty) = 0 \tag{10.9.4}$$

where C_θ is a constant to be determined as part of the solution, and where Θ_w' is the derivative with respect to the independent variable for the wall layer $\hat{\eta} = \eta/\delta$ and is related to the mean heat transfer at the wall. The minus sign enters the right side of Eq. (10.9.4) because of the definition of the dimensionless temperature used here. Since $c_\mu \tilde{K}^2/\tilde{E}$ approaches a constant as $\eta \to \infty$, $\tilde{T}(\eta)$ approaches zero exponentially fast.

Figure 10.9.1 gives the solutions for the mean temperature distribution in terms of $\tilde{T}(\eta)(\Theta_\infty - \Theta_w)/\Theta_w'$ for the values of Π considered earlier. We see that for a favorable pressure gradient ($\Pi < 0$), the mean temperature expressed in this manner at a specified point in the boundary layer is higher than that for the flat-plate and adverse-pressure-gradient cases. The implication is that the mean heat transfer reflected in Θ_w' is increased if the pressure gradient is favorable.

10.10 REYNOLDS STRESS CLOSURE

To this point our closure of the describing equations has been based on gradient models, essentially Eqs. (7.1.3) and (7.2.1), with the various calculations differing in the form for the turbulent exchange coefficient μ_T. In Chapters 8 and 9 we used the classical forms for this coefficient, set forth in Chapter 7, while in

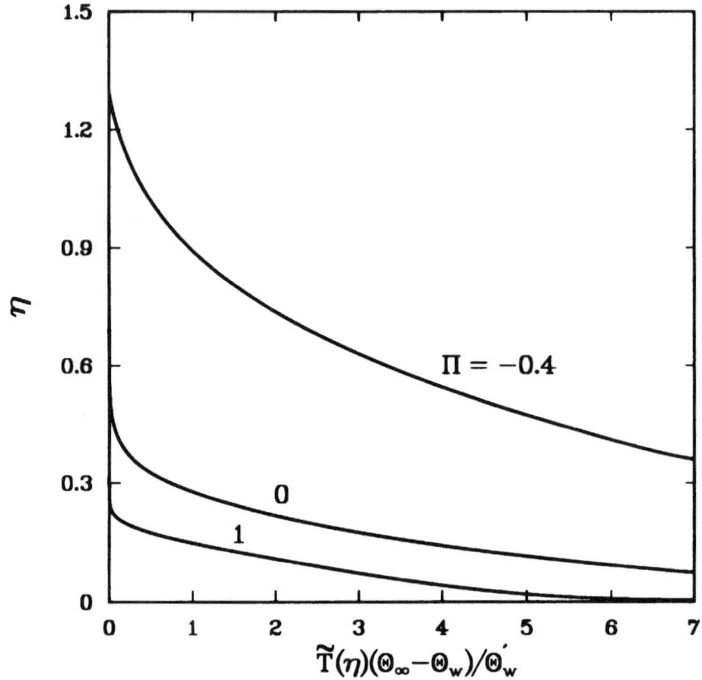

Figure 10.9.1 Mean temperature distribution in equilibrium turbulent boundary layers for various values of Π.

the previous sections of this chapter the turbulent kinetic energy and the mean viscous dissipation provided the requisite velocity and length scales. We now consider an alternative approach, which can be understood in a general way as follows. The closure problem for the mean momentum and mean temperature equations in a three-dimensional turbulent flow relates to the six Reynolds stresses and the three Reynolds fluxes, quantities that are replaced classically by gradient descriptions. Alternatively, we could supplement these three equations and the mean continuity equation with the six equations for the Reynolds stresses and the three equations for the Reynolds fluxes, i.e., with Eq. (4.4.1) for appropriate values for the indices i and j and with Eq. (4.6.1) for appropriate values for the index i. Such an approach represents a significant expansion of the scale of the formulation and of course can only be contemplated for general two- and three-dimensional flows within the context of large-scale computations.† In addition, to follow this strategy we must close these nine supplemental equations, e.g., eliminate the third-moment correlations of the form $\overline{u_i u_j u_m}$ and $\overline{u_i u_j \theta}$ *and* the various terms involving pressure fluctuations.

†Reynolds stress theory is applied to simple flows in Chapter 6.

The philosophical basis for this strategy resides in the expectation that suppression of modeling in the heirarchy of moment equations to the second level incorporates more of the mechanics of the turbulence and thus leads to a more accurate formulation. It can be argued that if models for the second-moment quantities in the first-moment equations lead to reasonable descriptions of first-moment quantities under a variety of conditions, as suggested by the discussions in Chapters 8 and 9 and to this point in Chapter 10, it might be expected that models for the third-moment quantities and for other effects which must be eliminated in the second-moment equations to achieve closure might lead to reasonable descriptions of second-moment quantities as well as to more accurate solutions of the mean momentum and mean temperature equations in flows with complex interactions, i.e., beyond the simple flows dealt with to this point. This expectation is widely held, with the consequence that Reynolds stress theory is believed to provide the framework for the ultimate moment description of turbulence, a framework warranting continuous efforts at improvement and refinement and benefiting from close coordination among theory, computation, and experiment. With respect to the present status of the theory, it is worth quoting Launder (1990): "From various comparative computations over the last four or five years, it has now been documented that, even with fairly rudimentary modeling of the processes, second-moment closure offers a far more reliable approach to predicting complex flows than any eddy-viscosity based model."

To appreciate the scope of the closure task we reproduce Eq. (4.4.1) in a compact form as follows:

$$C_{ij} = P_{ij} + F_{ij} - \varepsilon_{ij} - \left[\frac{\partial}{\partial x_k}\overline{u_i u_j u_k} + \frac{1}{\rho}\left(\frac{\partial}{\partial x_i}\overline{pu_j} + \frac{\partial}{\partial x_j}\overline{pu_i}\right)\right] + \overline{\frac{p}{\rho}\left(\frac{\partial u_i}{\partial x_j} + \frac{\partial u_j}{\partial x_i}\right)} \tag{10.10.1}$$

where

$$C_{ij} = \frac{\partial}{\partial x_k} U_k \overline{u_i u_j} \qquad P_{ij} = -\left(\overline{u_i u_k}\frac{\partial U_j}{\partial x_k} + \overline{u_j u_k}\frac{\partial U_i}{\partial x_k}\right)$$

$$F_{ij} = -g\,\beta\,(\delta_{i2}\,\overline{u_j\theta} + \delta_{j2}\,\overline{u_i\theta}) \qquad \varepsilon_{ij} = \frac{2}{3}\,\varepsilon\,\delta_{ij}$$

The P_{ij} terms describe the production of the correlation $\overline{u_i u_j}$ due to the mean rate of strain. The quantities C_{ij} and F_{ij} are compact representations of the convective and body force terms, respectively. Here we set the x_2 axis vertical, positive upward; use the Boussinesq approximation for the body force terms, β being the coefficient of volumetric expansion and equal to $\rho^{-1}\,\partial\rho/\partial\tilde{\theta}$, assumed a negative constant; and introduce the locally isotropic assumption for the mean viscous dissipation. Again we follow common practice and do not place a bar over the mean viscous dissipation ε.

Closure of this equation requires elimination of the last three sets of terms on the right side, those related to the effects of turbulent diffusion and of pressure–velocity and pressure–rate-of-strain correlations.[†] Note the role of both the pressure and the nonlinearity of convection in the closure problem. Because they involve gradients of mean quantities, the first two effects are grouped together so that

$$-\left[\frac{\partial}{\partial x_k}\overline{u_i u_j u_k} + \frac{1}{\rho}\left(\frac{\partial}{\partial x_i}\overline{p u_j} + \frac{\partial}{\partial x_j}\overline{p u_i}\right)\right] = c_s \frac{\partial}{\partial x_k}\left(\frac{k}{\varepsilon}\overline{u_k u_l}\frac{\partial}{\partial x_l}\overline{u_i u_j}\right) \qquad (10.10.2)$$

where c_s is an empirical coefficient. Here we see that a gradient model with all of its uncertainties is introduced to achieve closure of the *second-moment equations* and that the pressure–velocity correlation is modeled in terms of *local variables.*

Pressure–rate-of-strain effects are the subject of extensive study, of a considerable literature, and of continuing development (cf. Section 10.13). The general approach to modeling these effects is to utilize Eq. (2.4.1), to decompose the velocity components $\tilde{u}_i$ and the pressure into their mean and fluctuating components, and to integrate over the entire fluid domain, taking into account the presence of walls, to obtain a formal equation for the correlation of fluctuating pressure and the rate of strain at a given point in space at a given time. In this fashion we obtain

$$\frac{1}{\rho}\overline{p\frac{\partial u_i}{\partial x_j}}(\mathbf{x}, t) = \frac{1}{4\pi}\int\Bigg[\overline{\left(\frac{\partial^2 u_k u_l}{\partial x_k\,\partial x_l}\right)'\frac{\partial u_i}{\partial x_j}} + 2\left(\frac{\partial U_k}{\partial x_l}\right)'\overline{\left(\frac{\partial u_k}{\partial x_l}\right)'\frac{\partial u_i}{\partial x_j}}$$
$$+ \rho\, g\, \beta\, \overline{\left(\delta_{i2}\frac{\partial \theta}{\partial x_i} + \delta_{j2}\frac{\partial \theta}{\partial x_j}\right)'\frac{\partial u_i}{\partial x_j}}\Bigg]\frac{d\mathbf{x}'}{|\mathbf{x} - \mathbf{x}'|} \qquad (10.10.3)$$

where the ( )′ terms are functions of the integration variable $\mathbf{x}'$ and time, while $\partial u_i/\partial x_j$ depends on $\mathbf{x}$ and time. We can identify in Eq. (10.11.3) several influences of the pressure field: The first relates to turbulence–turbulence interactions; the second to the interaction of a mean rate of strain and turbulence, the pressure–rate-of-strain effect; and the third to the influence of buoyancy.

The turbulence–turbulence interactions in the "basic model" are represented by a linear return-to-isotropy term due to Rotta (1951), a term which reflects the role of pressure fluctuations in transferring energy associated with the fluctuations of the velocity in one coordinate direction into the other two directions, as seen in Section 6.3. Thus, for example, we have a term in the equation for

[†]In this discussion we have adopted for our purposes the "basic model" of Launder (1990) (cf. also Launder, 1989a).

the intensity $\overline{u_i^2}$ involving $(\overline{u_i^2} - \frac{2}{3}k)$. The modeling of the effects of the pressure–rate-of-strain and buoyancy in the "basic model" mimic this tendency of isotropization in terms of the operative production processes. Thus, for example, the former effect in the equation for $\overline{u_i^2}$ is taken to be proportional to the difference between the production of that intensity and the production of turbulent kinetic energy. There results from these considerations the model

$$\overline{\frac{p}{\rho}\left(\frac{\partial u_j}{\partial x_i} + \frac{\partial u_i}{\partial x_j}\right)} = -c_1 \frac{\varepsilon}{k}\left(\overline{u_i u_j} - \frac{2}{3}\delta_{ij}k\right) - c_2\left[\left(P_{ij} - \frac{2}{3}\delta_{ij}P\right) - \left(C_{ij} - \delta_{ij}\frac{2}{3}C\right) + \left(F_{ij} - \frac{2}{3}\delta_{ij}F\right)\right] \tag{10.10.4}$$

where c_1 and c_2 are empirical coefficients and

$$P = -\overline{u_l u_k}\frac{\partial U_l}{\partial x_k} = \frac{1}{2}P_{kk} \qquad C = \frac{1}{2}C_{kk}$$

$$F = -g\,\beta\,\overline{u_2\theta} = \frac{1}{2}F_{kk}$$

The influence of convection on the former effects is given by the C_{ij} terms, which are introduced to make the modeling independent of the coordinate system.[†] We see in Eq. (10.10.4) a description in terms of local variables, e.g., the correlations $\overline{u_i u_j}$ and the mean rates of strain $\partial U_i/\partial x_j$, of correlations involving the pressure given fundamentally by a *volume integral* as shown by Eq. (10.10.3). The adequacy of the Reynolds stress theory embodying such a description implies that the integrand must decrease rapidly with $|\mathbf{x} - \mathbf{x}'|$ to validate a local approximation to the volume integral.

When this modeling is substituted into Eq. (10.10.1), we obtain the closed equation

$$C_{ij} = P_{ij} + F_{ij} - \frac{2}{3}\varepsilon\,\delta_{ij} + c_s\frac{\partial}{\partial x_k}\left(\frac{k}{\varepsilon}\overline{u_k u_l}\frac{\partial\overline{u_i u_j}}{\partial x_l}\right) - c_1\frac{\varepsilon}{k}\left(\overline{u_i u_j} - \frac{2}{3}\delta_{ij}k\right) - c_2\left[\left(P_{ij} - \frac{2}{3}\delta_{ij}P\right) - \left(C_{ij} - \frac{2}{3}\delta_{ij}C\right) + \left(F_{ij} - \frac{2}{3}\delta_{ij}F\right)\right] \tag{10.10.5}$$

[†]As noted by Launder (1989a), the inclusion of the C_{ij} terms leads to significant improvement in ". . . the prediction of swirling flows while producing little change in simple shear flows." In all of our applications we neglect these terms.

If this equation is contracted by setting $i = j$ and is summed over i, we obtain another version of the turbulent kinetic energy equation:

$$\frac{\partial}{\partial x_k} U_k k = P + F - \varepsilon + c_s \frac{\partial}{\partial x_k}\left(\frac{k}{\varepsilon} \overline{u_k u_l} \frac{\partial k}{\partial x_l}\right) \tag{10.10.6}$$

Comparison with the second of Eqs. (10.1.9) establishes that this new equation possesses a more elaborate description of turbulent transport and includes the influence of body forces on the turbulent kinetic energy via F.

The mean dissipation needed in this formulation is given by an equation that resembles the third of Eqs. (10.1.9), differing only in a more elaborate description of turbulent transport and in inclusion of the influence of body forces; we have

$$\frac{\partial}{\partial x_k} U_k \varepsilon = c_\varepsilon \frac{\partial}{\partial x_k}\left(\frac{k}{\varepsilon} \overline{u_k u_l} \frac{\partial \varepsilon}{\partial x_l}\right) + c_{\varepsilon 1} \frac{\varepsilon}{k}(P + F) - c_{\varepsilon 2} \frac{\varepsilon^2}{k} \tag{10.10.7}$$

As suggested by the evolving nature of Reynolds stress closure, the various empirical coefficients in these equations do not have definitive values; indeed, as discussed in Section 10.13, some are given functional forms involving the invariants of the anisotropy tensor b_{ij} of Section 4.7 to make the formulation more accurate and more robust. However, the following may be considered indicative:

$$\begin{aligned} c_s &= 0.22 \qquad c_1 = 1.8 \qquad c_2 = 0.6 \\ c_\varepsilon &= 0.15 \qquad c_{\varepsilon 1} = 1.44 \qquad c_{\varepsilon 2} = 1.92 \end{aligned} \tag{10.10.8}$$

Our earlier comments concerning the need to modify these values in some circumstances applies in the present context.

In isothermal flows for which $F_{ij} \equiv 0$, Eqs. (10.10.5) and (10.10.7) supplement the continuity and momentum equations and determine the mean velocity components, $U_i(\mathbf{x})$; the six Reynolds stresses, $\rho\, \overline{u_i u_j}(\mathbf{x})$; the mean pressure, $P(\mathbf{x})$; and the mean viscous dissipation, $\varepsilon(\mathbf{x})$. When walls are present, a low-Reynolds-number version of these equations is needed (cf. So et al., 1991). In such applications a two-layer structure again prevails, one involving a wall layer with a characteristic length scale ν/u_τ and an outer layer devoid of viscous stresses and a length associated with the global scale L of the flow, e.g., a channel half-height or a boundary-layer thickness. Thus the asymptotic methods discussed earlier in connection with both classical methods of analysis and the k–ε theory can be applied to Reynolds stress formulations (cf., e.g., Champion and Libby, 1994).

10.11 REYNOLDS FLUX CLOSURE

There are second-moment equations for the Reynolds fluxes $\rho\, \overline{u_i \theta}(\mathbf{x})$ that are applicable to turbulent flows involving heat transfer and representing supplements to the equations of Section 10.11 for the Reynolds stresses. When buoyancy effects are dynamically significant, the two sets of equations are coupled. Thus we turn to Eqs. (4.6.1). Again rewriting these equations in a compact form, we obtain

$$C_{i\theta} = P_{i\theta} + F_{i\theta} - \left(\frac{\partial}{\partial x_k} \overline{u_i u_k \theta} + \frac{1}{\rho} \frac{\partial}{\partial x_i} \overline{p\theta} \right) + \frac{1}{\rho} \overline{p \frac{\partial \theta}{\partial x_i}} \tag{10.11.1}$$

where

$$C_{i\theta} \equiv \frac{\partial}{\partial x_k} U_k \overline{u_i \theta} \qquad P_{i\theta} \equiv -\left(\overline{u_i u_k} \frac{\partial \Theta}{\partial x_k} + \overline{u_k \theta} \frac{\partial U_i}{\partial x_k} \right) \qquad F_{i\theta} \equiv -g\,\beta\,\delta_{i2} \overline{\theta^2}$$

Here $C_{i\theta}$ describes the convection of the correlation $\overline{u_i\theta}$ and $P_{i\theta}$ the production of that correlation due to interaction of turbulence and mean gradients. The term $F_{i\theta}$ relates to the influence of buoyancy. Here we assume on the basis of local isotropy that the mean dissipation of the correlation $\overline{u_i\theta}$ is negligible. We see immediately that an additional second-moment equation, namely, an equation for the temperature intensity $\overline{\theta^2}$, is needed if buoyancy effects must be taken into account, and thus that Eq. (4.6.4) must be added to our system. On the other hand, if such effects are negligible so that we can set $\beta = 0$, the addition of the $\overline{\theta^2}$ equation is not mandatory but frequently is included and solved separately in order to permit comparison of prediction and experiment.

The closure of Eq. (10.11.1) calls for modeling of the last three sets of terms. The third-order moments and the pressure–temperature terms are grouped and treated in terms of a gradient model so that

$$-\left(\frac{\partial}{\partial x_k} \overline{u_i u_k \theta} + \frac{1}{\rho} \frac{\partial}{\partial x_i} \overline{p\theta} \right) = c_\theta \frac{\partial}{\partial x_k} \left(\frac{k}{\varepsilon} \overline{u_k u_l} \frac{\partial}{\partial x_l} \overline{u_i \theta} \right) \tag{10.11.2}$$

where c_θ is an empirical coefficient. The ongoing developments related to the effects of pressure fluctuations in Eq. (10.10.1) prevail here as well, in the last term of Eq. (10.11.1). However, in the "basic model," Launder (1990) takes

$$\frac{1}{\rho} \overline{p \frac{\partial \theta}{\partial x_i}} = -c_{\theta 1} \frac{\varepsilon}{k} \overline{u_i \theta} - c_{\theta 2} \left(\overline{u_k \theta} \frac{\partial U_i}{\partial x_k} - g\,\beta\,\delta_{i2} \overline{\theta^2} \right) \tag{10.11.3}$$

where $c_{\theta 1}$ and $c_{\theta 2}$ are empirical coefficients. We thus obtain the following closed equation for the Reynolds flux:

$$C_{i\theta} = P_{i\theta} + F_{i\theta} + c_\theta \frac{\partial}{\partial x_k} \left(\frac{k}{\varepsilon} \overline{u_k u_l} \frac{\partial}{\partial x_l} \overline{u_i \theta} \right) - c_{\theta 1} \frac{\varepsilon}{k} \overline{u_i \theta}$$
$$- c_{\theta 2} \left(\overline{u_k \theta} \frac{\partial U_i}{\partial x_k} - g\,\beta\,\delta_{i2} \overline{\theta^2} \right) \tag{10.11.4}$$

We also see, as noted in Section 10.10, that the correlation $\overline{u_i\theta}$ is influenced by convection, all temperature gradients, all mean rates of strain, etc., that is, all of the processes contained within Eq. (10.11.4). Contrast this with the simple gradient description of Eq. (7.2.1).

Equation (4.6.4) relates to the temperature intensity. Rewritten slightly, it becomes

$$\frac{\partial}{\partial x_k} U_k \overline{\theta^2} = -2\, \overline{u_k\theta} \frac{\partial \Theta}{\partial x_k} - \frac{\partial}{\partial x_k} \overline{u_k \theta^2} - 2\, \varepsilon_\theta \tag{10.11.5}$$

Here the third moment requires modeling; again, a gradient assumption is used, with the result that

$$-\frac{\partial}{\partial x_k} \overline{u_k \theta^2} = c_{s\theta} \frac{\partial}{\partial x_k} \left(\frac{k}{\varepsilon} \overline{u_k u_l} \frac{\partial}{\partial x_l} \overline{\theta^2} \right) \tag{10.11.6}$$

with $c_{s\theta}$ an empirical coefficient. Although a separate equation for the mean scalar dissipation ε_θ can be introduced and indeed is indicated in the treatment of flows in which the length scales characterizing the velocity and temperature fluctuations are different (cf. Jones and Musonge, 1988), a simpler alternative which is adequate for our purposes is to relate the two mean dissipations according to

$$2\, \varepsilon_\theta = c_{\varepsilon\theta} \frac{\varepsilon}{k} \overline{\theta^2} \tag{10.11.7}$$

where $c_{\varepsilon\theta}$ is another empirical constant. This model appears in Section 6.2, involving the temperature field downstream of a heated turbulence grid. The general validity of Eq. (10.11.7) is called into question by the uncertainty regarding the appropriate values for $c_{\varepsilon\theta}$, an uncertainty that is greater than that associated with the other empirical coefficients. Since Rodi (1980) quotes a range of values for this coefficient from 1 to 2 and we have no reason to select a particular value within that range, we assign $c_{\varepsilon\theta} = 1.5$.

Thus the closed equation for the temperature intensity is

$$\frac{\partial}{\partial x_k} U_k \overline{\theta^2} = -2\, \overline{u_k\theta} \frac{\partial \Theta}{\partial x_k} + c_{s\theta} \frac{\partial}{\partial x_k} \left(\frac{k}{\varepsilon} \overline{u_k u_l} \frac{\partial}{\partial x_l} \overline{\theta^2} \right) - c_{\varepsilon\theta} \frac{\varepsilon}{k} \overline{\theta^2} \tag{10.11.8}$$

Our earlier comment in Section 10.10 concerning uncertainties in the empirical coefficients apply in this section. However, indicative of current usage are the values recommended by Launder (1990) and Rodi (1987):

$$c_\theta = 0.15 \qquad c_{\theta 1} = 2.9 \qquad c_{\theta 2} = 0.4 \qquad c_{s\theta} = 0.13 \qquad c_{\varepsilon\theta} = 1.5 \tag{10.11.9}$$

With Eqs. (10.11.4) and (10.11.8) added to Eqs. (10.11.5) and (10.11.7) and to the mean continuity, momentum, and energy equations for three-dimensional flows, we deal with a system of 16 partial equations for the three mean velocities, $U_i(\mathbf{x})$; the mean pressure $P(\mathbf{x})$; the mean temperature $\Theta(\mathbf{x})$; the six Reynolds stresses, $\rho\, \overline{u_i u_j}$; the three Reynolds fluxes, $\rho\, \overline{u_i\theta}$; the temperature intensity $\overline{\theta^2}$;

and finally the mean viscous dissipation, $\varepsilon(\mathbf{x})$. With these equations specialized as appropriate for particular applications, we can systematically repeat all of our earlier calculations for free shear flows such as wakes and mixing layers, and for wall-bounded flows such as Couette, channel, and boundary-layer flows involving velocity and temperature variations and dynamic coupling between the two fields. With respect to the latter flows, our previous comments regarding the need for low-Reynolds-number corrections and the applicability of asymptotic methods pertain to this extended system as well. Examples of various applications of the Reynolds stress/flux theory are given by Launder (1990) and by Craft et al. (1994). In contrast to applications of the Reynolds stress theory to relatively simple flows, the numerical treatment of the full array of equations for three-dimensional flows is daunting. Indeed, if an *unsteady* three-dimensional flow is to be treated, e.g., the turbulence in the chamber of an internal combustion engine, a large eddy simulation formulation which limits modeling to the small unresolved scales might be more appropriate, since the scope of the computations required in the two approaches is comparable.

10.12 ALGEBRAIC STRESS MODELS AND TURBULENT EXCHANGE UNDER THE INFLUENCE OF BUOYANCY

A means for simplifying this imposing array of equations is provided by algebraic stress modeling (ASM), which is due to Rodi (1976). A further reason for discussing this simplification here is provided by our desire to examine in a more thorough way than was possible in Section 6.4 the influence of buoyancy on turbulent transport. Such an examination is facilitated by ASM. Because of its importance in geophysical flows, there is an extensive literature on this influence. In addition to the references cited in Section 6.4, the contributions of Rodi (1987) and of Craft et al. (1994) are noted.

The basic idea of ASM is to relate the terms characterizing convection and diffusion of the correlation $\overline{u_i u_j}$ in Eq. (10.10.4) in a simple fashion to the corresponding terms in the equation for the turbulent kinetic energy, Eq. (10.10.6). The physical basis for this approximation resides in the notion that production and dissipation are dominant in significant portions of turbulent shear flows, suggesting that the convective and diffusion terms can be approximated without serious consequences. Symbolically we can write

$$
\begin{aligned}
(\text{convection} - \text{diffusion})_{\overline{u_i u_j}} &\approx \frac{\overline{u_i u_j}}{k}(\text{convection} - \text{diffusion})_k \\
&= \frac{\overline{u_i u_j}}{k}(\text{production} - \text{dissipation})_k \\
&= (\text{production} - \text{dissipation})_{\overline{u_i u_j}}
\end{aligned}
$$

Replacement of the convection and diffusion terms in the $\overline{u_i u_j}$ equation with $\overline{u_i u_j}/k$ times the balance between production and dissipation of the *turbulent kinetic energy* results in the following *algebraic equation:*

$$\frac{\overline{u_i u_j}}{k}(P + F - \varepsilon) = P_{ij} + F_{ij} - \frac{2}{3}\delta_{ij}\,\varepsilon - c_1\frac{\varepsilon}{k}\left(\overline{u_i u_j} - \frac{2}{3}\delta_{ij}\,k\right)$$
$$- c_2\left[\left(P_{ij} - \frac{2}{3}\delta_{ij}\,P\right) + \left(F_{ij} - \frac{2}{3}\delta_{ij}\,F\right)\right] \qquad (10.12.1)$$

provided we drop the convective effects associated with the C_{ij} terms in Eq. (10.10.5). Here P, F, P_{ij}, and F_{ij} are defined in connection with Eqs. (10.10.1) and (10.10.4).

A similar treatment of the correlation $\overline{u_i\theta}$ given by Eq. (10.10.4) and of the intensity $\overline{\theta^2}$ given by Eq. (10.12.7) produces algebraic equations for the Reynolds fluxes and the intensity of the temperature fluctuations:

$$\frac{\overline{u_i\theta}}{k^{1/2}}(P + F - \varepsilon) = P_{i\theta} + F_{i\theta} - c_{\theta i}\frac{\varepsilon}{k}\,\overline{u_{i\theta}} - c_{\theta 2}\left(\overline{u_i\theta}\,\frac{\partial U_i}{\partial x_k} - g\,\beta\,\delta_{i2}\overline{\theta^2}\right)$$
$$\frac{\overline{\theta^2}}{k}(P + F - \varepsilon) = -2\,\overline{u_k\theta}\,\frac{\partial \Theta}{\partial x_k} - c_{\varepsilon\theta}\frac{\varepsilon}{k}\,\overline{\theta^2} \qquad (10.12.2)$$

The analysis of a three-dimensional turbulent flow with Eqs. (10.12.1)–(10.12.2) involves six partial differential equations, one from continuity, three from momentum transport and one each from the transport of turbulent kinetic energy and mean viscous dissipation, and ten *algebraic* equations for the six Reynolds stresses, the three Reynolds fluxes, and the intensity $\overline{\theta^2}$. Thus the original 16 partial differential equations are converted by this approximation into a mixed system of partial differential and algebraic equations. Numerical techniques are required that permit such a system to be dealt with, e.g., those based on finite differences for the partial derivatives so that the final equations are entirely algebraic, although nonlinear. In this case, either linearization followed by iteration or a direct solution via Newton-Rapheson and/or other numerical technique for dealing with nonlinear equations is needed.

The algebraic equations arising in ASM can be considered to be generalized relations connecting the stresses and fluxes to the rates of strain implicit in the production terms P_{ij}, to buoyancy effects reflected in the F_{ij} and $F_{i\theta}$ terms, and finally to the cross-effect involving rates of strain and the mean temperature gradient reflected in the $P_{i\theta}$ terms. Clearly these relations extend far beyond the classical gradient transport models of Eqs. (7.1.3) and (7.2.1). As noted in Section 7.3 for a two-dimensional flow in an inertial reference frame, Pope (1975) provides a generalized stress–rate-of-strain relation by inverting the equivalent of Eq. (10.12.1). Gatski and Speziale (1993) extend this idea and overcome the complex algebra involved by employing Mathematica to carry out the inversion for a three-dimensional flow in a noninertial frame, e.g., with rotation. There result complex, explicit relations giving the Reynolds stresses in terms of the rates of strain in contrast with the original implict relations; whether the numerical treatment of the combination of partial differential and algebraic equations is simplified by having these explicit relations available is unclear.

To illustrate the utilization of ASM, we revisit the simple shear flow with stratification dealt with initially in Section 6.4. There the influence of buoyancy was reflected in the flux Richardson number, a given parameter, and the resultant turbulent velocity field was discussed. Here we specify the mean rate of strain in a simple shear flow, that is, dU_1/dx_2 in the usual notation, and the coexisting mean temperature gradient $d\Theta/dx_2$, and *calculate* the flux Richardson number as well as the characteristics of the resultant velocity *and* temperature fields. As in Section 6.4, we restrict attention to flows in equilibrium, i.e., in balance among diffusion, production, and dissipation, but extension to homogeneous shear flows evolving with time from an arbitrary initial state is readily carried out.

It is useful to set forth the behavior we expect of these flows based on physical intuition and on our earlier analysis in Sections 6.3 and 6.4. When the temperature gradient is unstable, i.e., when $d\Theta/dx_2 < 0$, the mean shear and buoyancy *join* to produce turbulence against the destructive effect of mean viscous dissipation. Under conditions of stable stratification, $d\Theta/dx_2 > 0$, the competition is between production due to shear and destruction due to buoyancy *and* dissipation. In both cases we are concerned with the nature of the turbulence when these processes are in balance. To simplify the analysis we assume that homogeneity permits the neglect of the correlations $\overline{u_1u_3}$, $\overline{u_2u_3}$, $\overline{u_1\theta}$, and $\overline{u_3\theta}$.

With these assumptions, the convective terms $C_{ij} = C_{i\theta} \equiv 0$ and the only nonzero values of P_{ij}, $P_{i\theta}$, F_{ij}, and $F_{i\theta}$ of interest are

$$P_{11} = -2\,\overline{u_1u_2}\,\frac{dU_1}{dx_2} \qquad P = -\overline{u_1u_2}\,\frac{dU_1}{dx_2} \qquad P_{12} = -\overline{u_2^2}\,\frac{dU_1}{dx_2}$$

$$F_{22} = -2\,g\,\beta\,\overline{u_2\theta} \qquad F = -g\,\beta\,\overline{u_2\theta} \qquad F_{2\theta} = -g\,\beta^2\overline{\theta^2} \tag{10.12.3}$$

$$P_{2\theta} = -\,\overline{u_2^2}\,\frac{d\Theta}{dx_2}$$

We are now able to turn to Eq. (10.12.1) and to consider the combinations $i = j = 1$, $i = j = 2$, and $i = 1$, $j = 2$. We thus find balance equations in forms convenient for later developments as follows:

$$\frac{\overline{u_1^2}}{k}\left[\frac{\overline{u_1u_2}}{k}\left(\frac{k}{\varepsilon}\frac{dU_1}{dx_2}\right)+\left(\frac{g\,k^{1/2}}{\varepsilon}\right)\frac{\beta\overline{u_2\theta}}{k^{1/2}}+1\right]=2\,\frac{\overline{u_1u_2}}{k}\left(\frac{k}{\varepsilon}\frac{dU_1}{dx_2}\right)+\frac{2}{3}+c_1\left(\frac{\overline{u_1^2}}{k}-\frac{2}{3}\right)$$
$$-\,c_2\left[\frac{4}{3}\frac{\overline{u_1u_2}}{k}\left(\frac{k}{\varepsilon}\frac{dU_1}{dx_2}\right)-\frac{2}{3}\left(\frac{g\,k^{1/2}}{\varepsilon}\right)\frac{\beta\,\overline{u_2\theta}}{k^{1/2}}\right]$$
$$\frac{\overline{u_2^2}}{k}\left[\frac{\overline{u_1u_2}}{k}\left(\frac{k}{\varepsilon}\frac{dU_1}{dx_2}\right)+\left(\frac{g\,k^{1/2}}{\varepsilon}\right)\frac{\beta\,\overline{u_2\theta}}{k^{1/2}}+1\right]=2\left(\frac{gk^{1/2}}{\varepsilon}\right)\frac{\beta\,\overline{u_2\theta}}{k^{1/2}}+\frac{2}{3}+c_1\left(\frac{\overline{u_2^2}}{k}-\frac{2}{3}\right)$$
$$+\,c_2\left[\frac{2}{3}\frac{\overline{u_1u_2}}{k}\left(\frac{k}{\varepsilon}\frac{dU_1}{dx_2}\right)-\frac{4}{3}\left(\frac{g\,k^{1/2}}{\varepsilon}\right)\frac{\beta\overline{u_2\theta}}{k^{1/2}}\right] \qquad (10.12.4)$$
$$\frac{\overline{u_3^2}}{k}\left[\frac{\overline{u_1u_2}}{k}\left(\frac{k}{\varepsilon}\frac{dU_1}{dx_2}\right)+\left(\frac{g\,k^{1/2}}{\varepsilon}\right)\frac{\beta\,\overline{u_2\theta}}{k^{1/2}}+1\right]=\frac{2}{3}+c_1\left(\frac{\overline{u_3^2}}{k}-\frac{2}{3}\right)$$
$$+\,\frac{2}{3}c_2\left[\frac{\overline{u_1u_2}}{k}\left(\frac{k}{\varepsilon}\frac{dU_1}{dx_2}\right)+\left(\frac{g\,k^{1/2}}{\varepsilon}\right)\frac{\beta\,\overline{u_2\theta}}{k^{1/2}}\right]$$
$$\frac{\overline{u_1u_2}}{k}\left[\frac{\overline{u_1u_2}}{k}\left(\frac{k}{\varepsilon}\frac{dU_1}{dx_2}\right)+\left(\frac{g\,k^{1/2}}{\varepsilon}\right)\frac{\beta\,\overline{u_2\theta}}{k^{1/2}}+1\right]=(1-c_2)\frac{\overline{u_2^2}}{k}\left(\frac{k}{\varepsilon}\frac{dU_1}{dx_2}\right)+c_1\frac{\overline{u_1u_2}}{k}$$

In addition we consider Eq. (10.12.2) with $i = 2$ to find

$$\beta\frac{\overline{u_2\theta}}{k^{1/2}}\left[\frac{\overline{u_1u_2}}{k}\left(\frac{k}{\varepsilon}\frac{dU_1}{dx_2}\right)+\left(\frac{g\,k^{1/2}}{\varepsilon}\right)\frac{\beta\,\overline{u_2\theta}}{k^{1/2}}+1\right]=\frac{\overline{u_2^2}}{k}\left(\frac{\beta\,k^{3/2}}{\varepsilon}\frac{d\Theta}{dx_2}\right)$$
$$+\left(\frac{g\,k^{1/2}}{\varepsilon}\right)\beta^2\overline{\theta^2}+c_{\theta1}\frac{\beta\,\overline{u_2\theta}}{k^{1/2}}-c_{\theta2}\left(\frac{g\,k^{1/2}}{\varepsilon}\right)\beta^2\overline{\theta^2} \qquad (10.12.5)$$

and from the second of Eqs. (10.12.2) we have

$$\beta^2\overline{\theta^2}\left[\frac{\overline{u_1u_2}}{k}\left(\frac{k}{\varepsilon}\frac{dU_1}{dx_2}\right)+\left(\frac{g\,k^{1/2}}{\varepsilon}\right)\frac{\beta\,\overline{u_2\theta}}{k^{1/2}}+1\right]$$
$$=2\,\frac{\beta\,\overline{u_2\theta}}{k^{1/2}}\left(\frac{\beta\,k^{3/2}}{\varepsilon}\frac{d\Theta}{dx_2}\right)+c_{\varepsilon\theta}\,\beta^2\overline{\theta^2} \qquad (10.12.6)$$

Finally, we shall need an equation for the mean viscous dissipation. From Eq, (10.10.7) for a flow in local balance, we obtain

$$\frac{c_{\varepsilon1}}{c_{\varepsilon2}}\left[\frac{\overline{u_1u_2}}{k}\left(\frac{k}{\varepsilon}\frac{dU_1}{dx_2}\right)+\left(\frac{gk^{1/2}}{\varepsilon}\right)\frac{\beta\,\overline{u_2\theta}}{k^{1/2}}\right]+1=0 \qquad (10.12.7)$$

If we consider dU_1/dx_2, $d\Theta/dx_2$ the empiricial coefficients and the parameters such as g and β known, then Eqs. (10.12.4)–(10.12.7) represent seven equations for the four intensities, the correlations $\overline{u_1u_2}$ and $\overline{u_2\theta}$, and the mean viscous dissipation ε. The turbulent kinetic energy k is one-half the sum of the intensities of the velocity fluctuations. Now recall that in Section 6.4 our treatment of turbulent Couette flow, which also involves a local equilibrium between production and dissipation and called for the two coefficients $c_{\varepsilon 1}$ and $c_{\varepsilon 2}$ to be equal. Since we again deal with a flow in equilibrium, we must set these two coefficients equal.

To proceed, we nondimensionalize Eqs. (10.12.4)–(10.12.7) in terms of k, $S \equiv dU_1/dx_2$, and β. Except for the last parameter, which arises from consideration of the statistical behavior of the temperature, these are the quantities used for nondimensionalization in Section 6.4. Therefore, we introduce the following quantities:

$$I_{11} \equiv \frac{\overline{u_1^2}}{k} \qquad I_{22} \equiv \frac{\overline{u_2^2}}{k} \qquad I_{33} \equiv \frac{\overline{u_3^2}}{k} \qquad I_{12} \equiv \frac{\overline{u_1u_2}}{k} \qquad F \equiv \frac{\beta\,\overline{u_2\theta}}{k^{1/2}} \tag{10.12.8}$$

$$E \equiv \frac{\varepsilon}{k\,S} \qquad I_{\theta\theta} \equiv \beta^2\overline{\theta^2} \qquad G \equiv \frac{\beta\,k^{1/2}}{S}\frac{d\Theta}{dx_2} \qquad \hat{g} \equiv \frac{g}{k^{1/2}S}$$

In terms of these definitions, the flux Richardson number [cf. Eq. (6.4.2)] becomes

$$N_{Rf} = \frac{\hat{g}F}{-I_{12}} \tag{10.12.9}$$

We also note that the Brunt-Vaisala frequency, ω_B, a quantity of fundamental importance in stratified flows, namely, the frequency of oscillation of a statically stable object in a density gradient, is given by these definitions as

$$\omega_B^2 \equiv \frac{g}{\rho}\frac{\partial\rho}{\partial x_2} = \hat{g}\,G\,S^2 \tag{10.12.10}$$

When the definitions of Eqs. (10.12.8) are introduced into Eqs. (10.12.4)–(10.12.7) with the two empirical coefficients in the last equation set equal to one another, we obtain the following final equations:

$$0 = 2\,I_{12} + \frac{2}{3}E + c_1 E\left(I_{11} - \frac{2}{3}\right) - c_2\left(\frac{4}{3}I_{12} - \frac{2}{3}\hat{g}\,F\right)$$

$$0 = 2\,\hat{g}\,F + \frac{2}{3}E + c_1 E\left(I_{22} - \frac{2}{3}\right) + c_2\left(\frac{2}{3}I_{12} - \frac{4}{3}\hat{g}\,F\right)$$

$$0 = \frac{2}{3}E + c_1 E\left(I_{33} - \frac{2}{3}\right) + \frac{2}{3}c_2(I_{12} + \hat{g}\,F)$$

$$0 = (1 - c_2)\,I_{22} + c_1\,I_{12}\,E \tag{10.12.11}$$

$$0 = I_{22}G + \hat{g}I_{\theta\theta} + c_{\theta 1}E\,F - c_{\theta 2}\hat{g}I_{\theta\theta}$$

$$0 = 2\,F\,G + c_{\varepsilon\theta}E\,I_{\theta\theta}$$

$$I_{12} + \hat{g}\,F + E = 0$$

The last two parameters introduced by Eqs. (10.12.8), those for G and $\hat{g}$, define the flow treated by these equations. For a fixed rate of strain, i.e., a fixed value of S, buoyancy effects are negligible if $\hat{g}$ is negligibly small compared to unity. In this case the temperature is a passive scalar and the first four and last of Eqs. (10.12.11) determine the statistics of the velocity components via I_{11}, I_{22}, I_{33}, and I_{12}, as well as the mean viscous dissipation via E. Furthermore, in this case the fifth and sixth of these equations determine separately the temperature quantities F and $I_{\theta\theta}$, variables which in this case are proportional to G and G^2, respectively. In contrast, if $\hat{g}$ is not small, temperature is an active scalar and Eqs. (10.12.11) are all coupled, reflecting the interaction of the velocity and temperature fields.

The physical significance of G and $\hat{g}$ may be seen as follows. Since the quotient $k/g \equiv L_g$ can be considered a gravitational length which is small if the turbulence is weak, and $k^{1/2}/S \equiv L_S$, a length characterizing the mean rate of strain, which is also small if the turbulence is weak, we see that

$$\hat{g} = \frac{g/k}{S/k^{1/2}} = \frac{L_S}{L_g}$$

is the ratio of two lengths. The influence of buoyancy is small if the length characterizing the mean rate of strain is small compared with that characterizing gravity. Put differently, if g and S are fixed, buoyancy is ineffective if the turbulent kinetic energy is suitably large. Similarly, $[\beta(d\Theta/dx_2)]^{-1} \equiv L_\Theta$ is a length associated with the mean temperature gradient so that

$$G = \frac{\beta\,(d\Theta/dx_2)}{S/k^{1/2}} = \frac{L_S}{L_\Theta}$$

We thus see that the mean temperature gradient is ineffective if $L_\Theta >> L_S$. In this case we must expect the flux reflected in F and the temperature intensity reflected in $I_{\theta\theta}$ to be small in a sense to be exposed later. These considerations are consistent with and indeed reinforce our observations in Section 6.4 that buoyancy dominates turbulent flows involving small turbulent kinetic energy, small velocity gradients, and large temperature gradients.

Before proceeding, it is useful to review the physical situation described by Eqs. (10.12.11). We imagine that the parameters g and β are known and that a simple shear flow with a mean temperature gradient and with a turbulent kinetic energy k has been established so that the parameters G and $\hat{g}$ are known. These conditions are assumed to have prevailed for a sufficient time so that an equilibrium turbulent state exists. The central region of a turbulent Couette flow treated in Section 9.4 provides an example of the assumed state. Moreover, the atmospheric boundary layer can also be approximated by this analysis (cf. Lumley and Panofsky, 1964; Fernando, 1991). Given these conditions, we inquire of the nature of the turbulence in terms of I_{11}, I_{22}, etc., and of the associated flux Richardson number.

We now examine the solutions to Eqs. (10.12.11). If we make the realistic assumption that $\beta < 0$, stable distributions of mean temperature correspond to $G < 0$, $F > 0$, and unstable distributions to $G > 0$, $F < 0$. The corresponding values of the flux Richardson number are $N_{Rf} > 0$ and $N_{Rf} < 0$, respectively. We use the standard values of the empirical coefficients given by Eqs. (10.11.8)–(10.11.9).

If $\hat{g} = 0$, that is, if the temperature is a passive scalar, we readily find an analytic solution that closely resembles but supplements that in Section 6.4:

$$I_{11} = \frac{2}{3}\left[1 + 2\,\frac{(1 - c_2)}{c_1}\right] = 0.96 \qquad I_{22} = I_{33} = \frac{2}{3}\left(1 - \frac{1 - c_2}{c_1}\right) = 0.52$$

$$I_{12} = -E = \left(\frac{1 - c_2}{c_1}\, I_{22}\right)^{1/2} = -0.34 \qquad \frac{F}{G} = -\frac{I_{22}}{c_{\theta 1}E} = -0.53$$

$$\frac{I_{\theta\theta}}{G^2} = 2\,\frac{I_{22}}{c_{\theta 1}c_{\varepsilon\theta}E^2} = 2.1 \tag{10.12.12}$$

By adjusting the empirical coefficients slightly, we can require that $I_{11} = 1$, $I_{22} = I_{33} = \frac{1}{2}$, and $I_{12} = -E = -0.33$, the values found in Section 6.4 from a different perspective.

We now consider the more general cases of G and $\hat{g}$ arbitrary, for which numerical treatment is required. Examination of Eqs. (10.12.11) indicates that if $\hat{g}F/E$ is specified, the first four and seventh of these equations determine the fluid mechanical variables I_{11}, I_{22}, I_{33}, I_{12}, and E, and thus the flux Richardson number N_{Rf}. As a second step in the solution procedure, the fifth and sixth of these equations can be solved for F and $I_{\theta\theta}$ in the convenient form

$$\begin{aligned} \frac{F}{G} &= -\frac{I_{22} - 2\,(1 - c_{\theta 2})(\hat{g}F/E)/c_{\theta 2}}{c_{\theta 1}E} \\ \frac{I_{\theta\theta}}{G^2} &= 2\,\frac{I_{22} - 2\,(1 - c_{\theta 2})(\hat{g}F/E)/c_{\theta 2}}{c_{\theta 1}c_{\varepsilon\theta}E^2} \end{aligned} \tag{10.12.13}$$

If these equations are compared with Eq. (10.12.12), we see that they represent a generalization of the special forms of F/G and $I_{\theta\theta}/G^2$ and that they reduce properly when $\hat{g} = 0$. We also see the approach of F and $I_{\theta\theta}$ to zero as $G \rightarrow 0$.

A consequence of this strategy is that we can display the variation of all seven dependent variables in terms of the flux Richardson number with $\hat{g}F/E$ as the specified parameter. We do so in Figs. 10.12.1 and 10.12.2; in the former the characteristics of the velocity field are shown. The results in terms of the velocity variables resemble those obtained in Fig. 6.4.2. For unstable stratification, that is, $N_{Rf} < 0$, the contribution of the fluctuations of the vertical velocity component to the turbulent kinetic energy increases, while for stable stratification the turbulence tends to become two-dimensional in horizontal planes. The intensity in the cross-stream direction remains constant, while that in the vertical and streamwise directions exchanges importance. With unstable stratification,

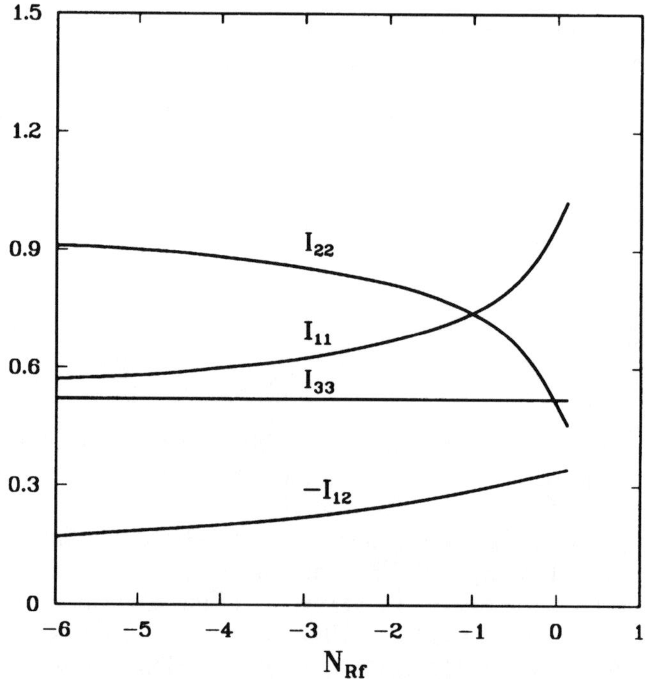

Figure 10.12.1 Velocity variables in a shear flow with buoyancy as predicted by ASM.

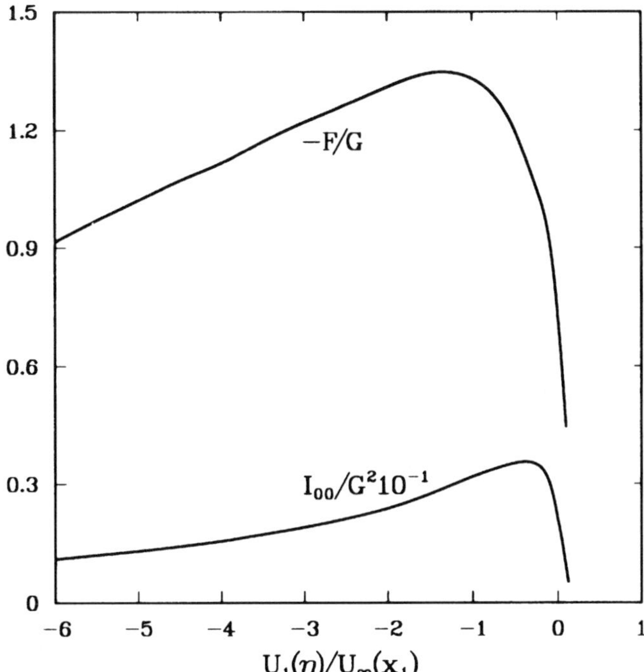

Figure 10.12.2 Temperature variables in a shear flow with buoyancy as predicted by ASM.

the Reynolds shear stress in terms of $-\ I_{12}$ decreases with increasing $-\ N_{Rf}$. The present calculation results in a limitation of $N_{Rf} = 0.13$, compared with the value 0.47 in Section 6.4. Here the intensity $I_{\theta\theta}$ goes to zero before I_{22}, as shown in Fig. 10.12.2, which relates to the temperature variables. If we want the same limiting value for the two calculations, some adjustment of the various empirical coefficients is called for. We see in Fig. 10.12.2 that there is a maximum in the flux and intensity of the temperature fluctuations as reflected in $-F/G$ and $I_{\theta\theta}/G^2$, respectively, in the neighborhood of $N_{Rf} = -1$.

Thus we have shown on the basis of ASM applied to a simple turbulent flow the significant influence of buoyancy on turbulent transport and the competition that influence faces from production due to mean shear and destruction due to mean viscous dissipation.

10.13 OTHER METHODS OF ANALYSIS

The k–ε, Reynolds stress, and closely related theories discussed in this chapter are the most widely employed advanced approaches to the analysis of turbulent, constant-density flows. In this closing section we call attention to related and alternative theories and provide references to their literature.

It will be recognized from our discussion of the closure methods in Chapter 7 and in this chapter that the development of second-moment methods calls for considerable physical insight, intuition, and support from experiment. Although refinements and improvements of the models which result from these processes continue to evolve, systematic approaches are attractive, either to support or to supplement insight, intuition, and experiment or, indeed, to illuminate the way forward when these processes are suspect. In particular, examination of the influence on turbulence of curvature, rotation, buoyancy, and low Reynolds numbers is problematical. From this perspective, the renomalization group (RNG) theory of turbulence currently attracts attention as a means for systematically analyzing these phenomena and for providing new turbulence models.

The original notion of this theory appears to be due to Forster et al. (1977), but the most significant advance and the basis for current developments is due to Yakhot and Orzag (1986). Critical assessments of the Yakhot-Orzag theory are given by Smith and Reynolds (1992) and by Lam (1992), while examples of recent applications of the theory are due to Speziale et al. (1991a) and to Avellaneda and Majda (1992).

The Yakhot-Orzag theory introduces a random force in the Navier-Stokes equations, a force which is homogeneous in space and time and isotropic in space and which respects certain statistical properties of turbulence, namely, the $-5/3$ decay law in the inertial subrange. A high-wave-number cutoff, Λ_0 is identified such that fluctuations with wave numbers greater than Λ_0 have little influence on the solution to the transport equations. On the other hand, as spectral slices less than Λ_0 are removed, corrections to the solution in the form of an eddy viscosity are introduced. After elimination of the small scales by these successive spectral slices, the random force is dropped.

The analysis to this point resembles the usual perturbation approach in terms of a small parameter associated with successive spectral slices. However, to achieve the $-5/3$ spectral decay in the inertial subrange, it is found that this parameter must be 4, a value which makes a perturbation perspective problematical. However, Lam provides alternative, physically based arguments supporting introduction of a forcing frequency and basis of the perturbation analysis.

The RNG theory yields theoretical estimates for the various empirical coefficients, e.g., in the k–ε theory. In the original Yakhot-Orzag paper the two coefficients in the dissipation equation, Eq. (10.1.13), are found to be 1.063 and 1.72 rather than the standard values of 1.44 and 1.92 [cf. Eqs. (10.2.8)]. Subsequent refinements of the theory provide new values of 1.42 and 1.68, closer to standard values. The theory also yields $c_\mu = 0.085$, close to the empirical standard value of 0.09 (cf. Yakhot et al., 1992).

Beyond the "basic model" of the Reynolds stress theory discussed in Section 10.11, more elaborate models for pressure–rate-of-strain effects are now under development (e.g., Speziale et al., 1991b; Launder, 1994). We discuss these developments briefly, to suggest the direction that turbulence modeling is currently taking. Equation (10.10.3) generalized to include the complementary rate of strain $\partial u_j / \partial x_i$ is formally written as

$$\Pi_{ij} \equiv \overline{p\left(\frac{\partial u_i}{\partial x_j} + \frac{\partial u_j}{\partial x_i}\right)} = \mathbf{A}_{ij} + \mathbf{M}_{ijkl} \frac{\partial U_k}{\partial x_l} \tag{10.13.1}$$

where $\mathbf{A}_{ij}$ is identified with turbulence–turbulence interaction and $\mathbf{M}_{ijmn}$ with the mean-rate-of-strain–turbulence interaction. The simplest dimensionally correct models for these contributions are found to be of the form

$$\mathbf{A}_{ij} = \varepsilon A_{ij}(\mathbf{b}) \qquad \mathbf{M}_{ijmn} = k\, \mathbf{M}_{ijmn}(\mathbf{b}) \tag{10.13.2}$$

where $\mathbf{b}$ identifies the anisotropy tensor b_{ij} of Eq. (4.7.9). When the mean rate of strain is decomposed into symmetric and antisymmetric components, when requirements of invariance under transformation of coordinates are imposed, and when experimental results for homogeneous shear flows are taken into account, there results the following elaboration of Eq. (10.11.4) according to the SSG theory (cf. Speziale et al., 1991):

$$\begin{aligned}\Pi_{ij} = &-(c_1\varepsilon + c_1^* P)\, b_{ij} + c_2\, \varepsilon \left(b_{ik}b_{kj} - \frac{1}{3}\delta_{ij} b_{kl} b_{kl}\right) \\ &+ (c_3 - c_3^* A_{\mathrm{II}}^{1/2})\, k\, S_{ij} + c_4\, k \left(b_{ik}S_{jk} + b_{jk}S_{ik} - \frac{2}{3}\delta_{ij} b_{kl} S_{kl}\right) \\ &+ c_5\, k\, (b_{ik}W_{ik} + b_{jk}W_{ik})\end{aligned} \tag{10.13.3}$$

where

$$S_{ij} \equiv \frac{1}{2}\left(\frac{\partial U_i}{\partial x_j} + \frac{\partial U_j}{\partial x_i}\right) \qquad W_{ij} \equiv \frac{1}{2}\left(\frac{\partial U_i}{\partial x_j} - \frac{\partial U_j}{\partial x_i}\right) + \varepsilon_{ijk}\Omega_k$$

are the mean rate of strain tensor and the rotation tensor, respectively, c_1–c_5 are empirical coefficients, and A_{II} is the second invariant of the tensor b_{ij} [cf. Eq. (4.7.12)]. Our early model for pressure–rate-of-strain effects in the absence of rotation is contained in Eq. (10.13.3) and is obtained by setting c_1^* and c_2–c_5 to zero, but we see that the influence of rotation on the pressure has now been incorporated in the modeling. The recommended values for the empirical coefficients in this model are

$$\begin{aligned}&c_1 = 3.4 \qquad c_1^* = 1.80 \qquad c_2 = 4.2 \qquad c_3 = \frac{4}{5} \\ &c_3^* = 1.30 \qquad c_4 = 1.25 \qquad c_5 = 0.40\end{aligned} \tag{10.13.4}$$

The other empirical coefficients, c_s, c_ε, $c_{\varepsilon 1}$, and $c_{\varepsilon 2}$, are assigned the values given by Eq. (10.11.9). The more general and elaborate modeling represented by Eq. (10.13.3) is found to "outperform" the simpler "basic model" in comparisons with DNS results for homogeneous shear flows with and without rotation.

Additional developments of Reynold stress theory are due to Shih and Lumley (1985), whose model respects the realizability constraint discussed in Section 4.7, and to Fu et al. (1987), whose model is *cubic in the anisotropy tensor*. Our purpose is simply to call attention to these ongoing efforts at improving Reynolds stress modeling so as to achieve a more accurate and more generally applicable formulation, i.e., one that is applicable to complex flows. For the relatively simple flows discussed in this chapter, the increased complexity associated with these improved models is probably unwarranted; but for flows involving, e.g., curvature, rotation, and streamline divergence, such models are expected to be more accurate.

In Section 3.3 we discussed probability density functions (pdf), their construction from experimental data, and their role in determining statistical quantities. In brief, if in a statistically stationary flow we know the multivariate pdf $\Phi(\tilde{\mathbf{u}}, \tilde{p}, \tilde{\theta}; \mathbf{x})$, that is, the pdf of the three velocity components, the pressure, and the temperature at a generic spatial location, then we could *calculate all statistical quantities* by quadrature, i.e., the mean values $U_i(\mathbf{x})$, $i = 1, 2, 3$, $P(\mathbf{x})$, and $\Theta(\mathbf{x})$; the Reynolds stresses $\rho\ \overline{u_i^2}$ and $\rho\ \overline{u_i u_j}$; the Reynolds fluxes $\rho\ \overline{u_i \theta}(\mathbf{x})$; and any other correlations of possible interest including those of higher order. Given the information it contains, the *calculation* of such a pdf would be an attractive alternative to the moment methods we have discussed. Indeed, a single equation for Φ can be formally developed; as indicated by its multivariate nature, it involves partial derivatives with respect to each velocity component $\tilde{u}_i$, the pressure $\tilde{p}$, etc., so tht its form and method of solution are quite different from the conservation equations of Chapter 2 and from the averaged conservation and transport equations in Chapters 4 and 8–10, all of which involve at most *three* independent variables. In fact, the solution for Φ is sought in physical space of coordinates x_i, $i = 1$–3, *and* in velocity space with coordinates u_i, $i = 1$–3, with the mean pressure eliminated in terms of an integral of the velocity [cf. Eq. (2.4.1)]. In addition, not surprisingly, the equation for the pdf is not closed but rather involves terms which must be modeled; again, not surprisingly, the terms involving pressure fluctuations must be approximated, but the viscous terms contribute to the closure problem in the direct pdf approach. Because of certain advantages possessed by these methods, they are usefully applied to the analysis of turbulent reacting flows (cf. O'Brien, 1980; Pope, 1985; Dopazo, 1994). Here we call attention to their applicability to constant-density flows.

To illustrate these methods, consider the equation for the pdf of velocity, i.e., for the limited problem associated with $\Phi(\tilde{\mathbf{U}}; \mathbf{x})$ (cf. Pope, 1985). We have the equation

$$\rho U_k \frac{\partial \Phi}{\partial x_k} - \frac{\partial P}{\partial x_k}\frac{\partial \Phi}{\partial U_k} = \frac{\partial}{\partial U_k}\left[\left\langle -\frac{\partial \tilde{\tau}_{lk}}{\partial x_l} + \frac{\partial p}{\partial x_k}\middle| \mathbf{U}\right\rangle \Phi\right] \tag{10.13.5}$$

The left side of this equation describes convective transport of Φ in physical space by the mean velocity components and in velocity space by the mean pressure gradient. The term on the right side represents the closure problem in the direct pdf approach; it describes transport in velocity space by the viscous stresses and fluctuations in the pressure gradient conditioned on the mean velocity $\mathbf{U}$. If the right side of Eq. (10.13.5) is expressed in terms of Φ and $\mathbf{U}$, and there are several models available to do so, then this equation is closed and determines $\Phi(\mathbf{U}; \mathbf{x})$. The mean velocities appearing there are obtained from Φ by quadrature, for example,

$$U_1(\mathbf{x}) = \int_{-\infty}^{\infty} dU_1' U_1' \int_{-\infty}^{\infty} dU_2' \int_{-\infty}^{\infty} dU_3' \, \Phi(\mathbf{U}'; \mathbf{x}) \tag{10.13.6}$$

with similar expressions for the other mean velocity components. Clearly, the multivariate nature of Eq. (10.13.5) and its partial differential–integral structure call for special numerical techniques distinct from those used to solve the partial differential equations discussed in Chapters 7–9 and earlier in this chapter. In brief, finite-difference methods will clearly not be applicable. For further details on closure approximations and the applicable numerical techniques associated with them, see Pope (1985) and the extensive list of references therein.

10.14 SUMMARY

This chapter has described the k–ε theory of turbulence, a theory widely applied at the present time and using $k^{1/2}$ and $k^{3/2}/\varepsilon$ for the velocity and length scales, respectively, which are needed in the turbulent exchange coefficient. The theory has the fundamental shortcomings of gradient transport but incorporates the *calculation* of length scales rather than their assumption as in classical closure. Using the k–ε theory, we repeat the analysis of most of the flows treated previously in Chapters 8 and 9, again employing asymptotic methods to deal with the viscous sublayer in those flows involving one or more walls. We then described the Reynolds stress theory, which suppresses the assumption of gradient transport to the next level in the hierarchy of moment equations; it thus avoids the fundamental difficulties cited earlier. While it is still under active development and improvement, this theory is believed to provide the basis for all future turbulent flow calculations of applied interest. An approximation based on the Reynolds stress theory permits inter alia an examination of the influence of buoyancy on turbulent transport. We closed with brief discussions of, and pertinent references to, some of the other methods of applied interest for the analysis of turbulent flows.

REFERENCES

Alfredsson, P. H., Johansson, A. V., Hanitonidis, J. H., and Ecklmann, H. (1988). The fluctuating wall-shear stress and the velocity field in the viscous sublayer. *Phys. Fluids* **31**:1026.

Anand, M. S., and Pope, S. B. (1985). Diffusion behind a line source in grid turbulence. In *Turbulent Shear Flows 4*. Springer-Verlag, Berlin, 46.

Anderson, P., LaRue, J. C., and Libby, P. A. (1979). Preferential entrainment in a two dimensional turbulent jet in a moving stream. *Phys. Fluids* **22**:1857.

Andersson, H. I., Bech, K. H., and Kristoffersen, R. (1992). On diffusion of turbulent energy in plane Couette flow. *Proc. R. Soc. London A* **438**:477.

Antonia, R. A., and Kim, J. (1994). A numerical study of local isotropy of turbulence. *Phys. Fluids* **6**:834.

Antonia, R. A., Spalart, P. R., and Mariani, P. (1994). Effect of suction on the near-wall anisotropy of a turbulent boundary layer. *Phys. Fluids* **6**:430.

Ashurst, W. T. (1979). Numerical simulation of turbulent mixing layers via vortex dynamics. In *Turbulent Shear Flows 1*. Springer-Verlag, Berlin, 402.

Aubry, N., Holmes, P., Lumley, J. L., and Stone, E. (1988). The dynamics of coherent structures in the wall region of a turbulent boundary layer. *J. Fluid Mech.* **192**:115.

Avellaneda, M., and Majda, A. J. (1992). Approximate and exact renormalization theories for a model for turbulent transport. *Phys. Fluids A* **4**:41.

Bardina, J., Ferziger, J. H., and Rogallo, R. S. (1985). Effect of rotation on isotropic turbulence; computation and modelling. *J. Fluid Mech.* **154**:321.

Batchelor, G. K. (1959). Small scale variation of convected quantities like temperature in turbulent fluid. Part 1. General discussion and case of small conductivity. *J. Fluid Mech.* **5**:113.

Batchelor, G. K. (1967). *The Theory of Homogeneous Turbulence*. Cambridge University Press, London.

Batchelor, G. K., Howells, I. D., and Townsend, A. A. (1959). Small scale variation of convected quantities like temperature in turbulent fluid. Part 2. The case of large conductivity. *J. Fluid Mech.* **5**:134.

Batchelor, G. K., and Proudman, I. (1956). The large-scale structure of homogeneous turbulence. *Trans. R. Soc. London A* **248**:369.

Bearman, P. W. (1972). Some measurements of the distortion of turbulence approaching a two-dimensional bluff body. *J. Fluid Mech.* **53**:451.

Bendat, J. S., and Piersol, A. G. (1980). *Engineering Applications of Correlation and Spectral Analysis.* John Wiley and Sons, New York.

Bilger, R. W., Antonia, R. A., and Sreenivasan, K. R. (1976). Determination of intermittency from the probability density function of a passive scalar. *Phys. Fluids* **19**:1471.

Bird, R. B., Stewart, W. E., and Lightfoot, E. N. (1960). *Transport Phenomena.* John Wiley, New York.

Blackwelder, R. F. (1978). The bursting process in turbulent boundary layers. In C. R. Smith and D. E. Abbott (eds.), *Workshop on Coherent Structure of Turbulent Boundary Layers.* Department of Mechanical Engineering, Lehigh University, Bethlehem, PA, 211.

Bradshaw, P. (1967a). Irrotational fluctuations near a turbulent boundary layer. *J. Fluid Mech.* **27**: 209.

Bradshaw, P. (1967b). The turbulence structure of equilibrium boundary layers. *J. Fluid Mech.* **29**: 625.

Bradshaw, P. (1969). The analogy between streamline curvature and buoyancy in turbulent shear flow. *J. Fluid Mech.* **36**:177.

Bradshaw, P. (1971). *An Introduction to Turbulence and Its Measurement.* Pergamon Press, Oxford.

Bradshaw, P. (1973). *Effects of Streamline Curvature on Turbulent Flow*, AGARDograph **169**. AGARD, Paris.

Bradshaw, P. (ed.). (1978). *Turbulence.* Springer-Verlag, Berlin.

Bradshaw, P. (1987). Coherent structure. In *Studying Turbulence Using Numerical Simulation Database*, Proc. 1987 Summer Program, Center for Turbulence Research, NASA Ames Research Center, Stanford University, Stanford, CA.

Bradshaw, P., Ferris, D. H., and Atwell, N. P. (1967). Calculation of boundary layer development using the turbulent energy equation. *J. Fluid Mech.* **28**:593.

Bradshaw, P., and Pontikos, N. S. (1985). Measurements in the turbulent boundary layer on an "infinite" swept wing. *J. Fluid Mech.* **159**:105.

Bray, K. N. C., and Libby, P. A. (1994). Recent developments in the BML model of premixed turbulent combustion. In P. A. Libby and F. A. Williams (eds.), *Turbulent Reacting Flows.* Academic Press, London, 115.

Brown, G. L., and Roshko, A. (1974). On density effects and large structure in turbulent mixing layers. *J. Fluid Mech.* **64**:775.

Browne, L. W. B., Antonia, R. A., and Shah, D. A. (1987). Turbulent energy dissipation in a wake. *J. Fluid Mech.* **179**:307.

Bruun, H. H. (1995). *Hot-Wire Anemometry—Principles and Signal Analysis.* Oxford University Press, London.

Cambon, C. (1994). Turbulent flows undergoing distortion and rotation. A report on EUROMECH 288. *Fluid Dynam. Res.* **13**:281.

Cantwell, B. J. (1981). Organized motion in turbulent flow. *Annu. Rev. Fluid Mech.* **13**:457.

Carnevale, G. F., and Vallis, G. K. (1990). Pseudo-advective relaxation to stable states of inviscid two-dimensional fluids. *J. Fluid Mech.* **213**:549.

Cazalbou, J. B., Spalart, P. R., and Bradshaw, P. (1994). On the behavior of two-equation models at the edge of a turbulent region. *Phys. Fluids* **6**:1793.

Cebeci, T., and Smith, A. M. O. (1974). *Analysis of Turbulent Boundary Layers.* Academic Press, New York.

Champagne, F. H. (1978). The fine-scale structure of the turbulent velocity field. *J. Fluid Mech.* **86**: 67.

Champagne, F. H., Harris, V. G., and Corrsin, S. (1970). Experiments on nearly homogeneous turbulent shear flow. *J. Fluid Mech.* **41**:81.

Champion, M., and Libby, P. A. (1991). Asymptotic analysis of stagnating turbulent flows. *AIAA J.* **29**:16.

Champion, M., and Libby, P. A. (1994). Reynolds stress description of opposed and impinging turbulent jets. Part II. Axisymmetric jets impinging on nearby walls. *Phys. Fluids* **6**:1805.

Chandrasekhar, S. (1956). Theory of turbulence. *Phys. Rev.* **102**:941.

Chasnov, J. R. (1993). Computation of the Loitsianski integral in decaying isotropic turbulence. *Phys. Fluids A* **11**:2579.

Choi, K.-S., and Lumley, J. L. (1984). Return to isotropy of homogeneous turbulence revisited. In T. Tatsumi (ed.), *Turbulence and Chaotic Phenomena in Fluids. Proc. Int. Symp.*, Kyoto, Japan, 5–10 September 1983, 267.

Clauser, F. H. (1956). The turbulent boundary layer. *Adv. Appl. Mech.* **4**, Academic Press, New York.

Clay, J. P. (1973). Turbulent mixing of temperature in water, air and mercury. Ph.D. thesis, University of California, San Diego.

Coantic, M. F. (1978). *An Introduction to Turbulence in Geophysics and Air-Sea Interactions*, AGARD-AG-232. AGARD, Paris.

Coles, D. (1956). The law of the wake in turbulent boundary layer. *J. Fluid Mech.* **1**:191.

Collier, F. S., Jr., and Schetz, J. A. (1984). Injection into a turbulent boundary layer through different porous surfaces. *AIAA J.* **22**:839.

Comte-Bellot, G. (1963). Turbulent flow between two parallel walls. Ph.D. thesis, University of Grenoble. Also available as A.R.C. 31 609 F.M. 4102, November 1969.

Comte-Bellot, G. (1976). Hot-wire anemometry. *Ann. Rev. Fluid Mech.* **8**:209.

Corrsin, S. (1943). Investigation of flow in an axially symmetrical heated jet of air. NACA WR W-94.

Corrsin, S. (1951). On the spectrum of isotropic temperature fluctuations in an isotropic turbulence. *J. Appl. Phys.* **11**:469.

Corrsin, S. (1958). On local isotropy in turbulent shear flow. NACA R&M 58B11.

Corrsin, S. (1961). Turbulent flow. *American Scientist*, September, 300 (from *The Johns Hopkins Magazine*).

Corrsin, S. (1974). Limitations of gradient transport models in random walks and in turbulence. *Adv. Geophys.* **18**:25

Corrsin, S., and Kistler, A. L. (1954). The free stream boundaries of turbulent flows. NACA TN 3133.

Craft, T. J., Ince, N. Z., and Launder, B. E. (1994). Recent developments in second-moment closure for buoyancy-affected flows. Fourth International Symposium on Stratified Flows, Grenoble. (To appear in *Dynamics of Atmospheres and Oceans.*)

Donaldson, C. duP. (1972). Calculation of turbulent shear flows for atmospheric and vortex motions. *AIAA J.* **10**:4.

Dopazo, C. (1994). Recent developments in pdf methods. In P. A. Libby and F. A. Williams (eds.), *Turbulent Reacting Flows.* Academic Press, London, 375.

Dryden, H. L. (1959). Transition from laminar to turbulent flow. In C. C. Lin (ed.), *Turbulent Flows and Heat Transfer.* Princeton University Press, Princeton, NJ.

Durbin, P. A., and Speziale, C. G. (1991). Local anisotropy in strained turbulence at high Reynolds numbers. *ASME J. Fluids Eng.* **113**:707.

Durst, F., Melling, A., and Whitelaw, J. H. (1976). *Principles and Practice of Laser-Doppler Anemometry.* Academic Press, New York.

Eckelmann, H. (1978). The structure of turbulence in the near wall region. In C. R. Smith and D. E. Abbott (eds.), *Workshop on Coherent Structure of Turbulent Boundary Layers.* Department of Mechanical Engineering and Mechanics, Lehigh University, Bethlehem, PA.

Eichelbrenner, E. A. (1973). Three-dimensional boundary layers. *Annu. Rev. Fluid Mech.* **5**:339.

Elghobashi, S., and Truesdell, G. C. (1992). Direst simulation of particle dispersion in a decaying isotropic turbulence. *J. Fluid Mech.* **242**:655.

El Telbany, M. M. M., and Reynolds, A. J. (1982). The structure of turbulent plane Couette flow. *ASME J. Fluid Eng.* **104**:367.

Falkner, V. M. (1943). The resistance of a smooth flat plate with turbulent boundary layer. *Aircraft Eng.* **15**:65.

Fannelop, T. K., and Krogstad, P. A. (1975). Three dimensional turbulent boundary layers in external flows: A report on Euromech 60. *J. Fluid Mech.* **71**:815.

Favre, A. (1965). Review on space-time correlations in turbulent fluids, *J. Appl. Mech.*, 241-257.

Fernando, H. J. S. (1991). Turbulent mixing in stratified fluids. *Annu. Rev. Fluid Mech.* **23**:455.

Forster, D., Nelson, D. R., and Stephen, M. J. (1977). Large-distance and long-time properties of a randomly stirred fluid. *Phys. Rev. A* **16**:732.

Fu, S., Launder, B. E., and Leschziner, M. A. (1987a). Modelling strongly swirling recirculating jet flow with Reynolds-stress transport closures. *Proc. 6th Turbulent Shear Flows Symp.*, Toulouse, Paper 17.6.

Fu, S., Launder, B. E., and Tselepidakis, D. P. (1987b). Accommodating the effects of high strain rates in modeling the pressure-strain correlation. Report TFD/87/5, Mech. Eng. Dept., UMIST, Manchester.

Galperin, B., and Orszag, S. A. (1993). *Large Eddy Simulation of Complex Engineering and Geophysical Flows.* Cambridge University Press, Cambridge.

Gatski, T. B., and Savill, A. M. (1989). An analysis of curvature effects for the control of wall bounded shear flows. AIAA-89-1014.

Gatski, T. B., and Speziale, C. G. (1993). On explicit algebraic stress models for complex turbulent flows. *J. Fluid Mech.* **254**:59.

Gence, J. N. (1983). Homogeneous turbulence. *Annu. Rev. Fluid Mech.* **15**:201.

Gence, J. N., and Mathieu, J. (1980). The return to isotropy of a homogeneous turbulence having been submitted to two successive plane strains. *J. Fluid Mech.* **101**:555.

George, W. K. (1992). The decay of homogeneous isotropic turbulence. *Phys. Fluids A* **4**:1492.

Gersten, K. (1985). The turbulent Couette flow from asymptotic theory point of view. In *Flow of Real Fluids—Lecture Notes in Physics, Vol. 235*, Springer-Verlag, Berlin.

Gerz, T., Schumann, U., and Elghobashi, S. E. (1989). Direct numerical simulation of stratified homogeneous turbulent shear flows. *J. Fluid Mech.* **200**:563.

Gibson, C. H., and Libby, P. A. (1972). On turbulent flows with fast chemical reactions. Part II. The distribution of reactants and products near a reacting surface. *Combustion Sci. Technol.* **6**:29.

Gibson, M. M., and Kanellopoulos, V. E. (1988). Turbulence measurements in a nearly homogeneous shear flow. In M. Hirata and N. Kasagi (eds.), *Transport Phenomena in Turbulent Flows.* Hemisphere, New York, 17.

Glauser, M. N., and Gatski, T. B. (1993). Near-wall reconstruction of high-order moments and length scales using the POD. In J. P. Bonnet and M. N. Glauser (eds.), *Eddy Structure Identification in Free Turbulent Shear Flows.* Kleuver, Dordrecht, The Netherlands.

Goldstein, S. (1965). *Modern Developments in Fluid Dynamics*, Vol. I. Dover, New York.

Grant, H. L., Stewart, R. W., and Moilliet, A. (1962). Turbulence spectra from a tidal channel. *J. Fluid Mech.* **12**:241.

Gurvich, A. S. (1980). Influence of the temporal evolution of turbulence inhomogeneities in frequency spectra. *Atmos. Ocean Phys.* **16**:23.

Hanjalic, K. (1994). Advanced turbulence closure models: A view of current status and future prospects. *Int. J. Heat Fluid Flow* **15**:178.

Hanjalic, K., and Launder, B. E. (1972a). Asymmetric flow in a plane channel. *J. Fluid Mech.* **51**: 301.

Hanjalic, K., and Launder, B. E. (1972b). A Reynolds stress model of turbulence and its application to thin shear layers. *J. Fluid Mech.* **52**:609.

Hanjalic, K., Launder, B. E., and Schiestel, R. (1980). Multiple time scale concepts in turbulent transport modeling. In L. J. S. Bradbury (ed.), *Turbulent Shear Flows—2*, Springer-Verlag, Berlin.

Hariri, A., Libby, P. A., and LaRue, J. C. (1982). Similarity solutions and experiment for turbulent wakes, *Phys. Fluids* **25**:1964.

Head, M. R., and Bandyopadhyay, P. (1978). Combined flow visualization and hot-wire measurements in turbulent boundary layers. In C. R. Smith and D. E. Abbott (eds.), *Workshop on Coherent Structure of Turbulent Boundary Layers*. Department of Mechanical Engineering, Lehigh University, Bethlehem, PA, 98.

Heisenberg, W. (1948). Zur statistichen Theoric der Turbulenz. *Z. Phys.* **124**:628.

Helland, K. N., and van Atta, C. W. (1977). Spectral energy transfer in high Reynolds number turbulence. *J. Fluid Mech.* **79**:337.

Hinze, J. O. (1975). *Turbulence*. McGraw-Hill, New York.

Ho, C.-M., and Huang, L.-S. (1982). Subharmonics and vortex merging in mixing layers. *J. Fluid Mech.* **119**:443.

Holmes, P. (1990). On Moffatt's paradox or can empirical projections approach turbulence? In J. L. Lumley (ed.), *Lecture Notes in Physics*. Springer-Verlag, Berlin.

Hossain, M. (1994). Reduction in the dimensionality of turbulence due to a strong rotation. *Phys. Fluids* **6**:1077.

Hunt, J. C. R., Leibovich, S., and Richards, K. J. (1988). Turbulent shear flow over low hills. *Quart. J. Roy. Meteor. Soc.* **114**:1435.

Hussain, A. K. M. F. (1986). Coherent structures and turbulence. *J. Fluid Mech.* **173**:303.

Hussain, A. K. M. F., Jeong, J., and Kim, J. (1987). Structure of turbulent shear flows. *Studying Turbulence Using Numerical Simulation Databases*. Proc. 1987 Summer Program, Center for Turbulence Research, NASA Ames Research Center, Stanford University, Stanford, CA.

Jeong, J., and Hussain, F. (1995). On the identification of a vortex. *J. Fluid Mech.* **285**:69.

Johnston, J. P. (1973). The suppression of shear layer turbulence in rotating systems. *J. Fluids Eng.* **95**:229.

Jones, W. P., and Musonge, P. (1988). Closure of the Reynolds stress and scalar flux equations. *Phys. Fluids* **31**:3589.

Keffer, J. F. (1965). The uniform distortion of a turbulent wake. *J. Fluid Mech.* **22**:135.

Keller, L. V., and Fridman, A. A. (1924). Differentialgleichung fur die turbulente bewegung einer Kompressiblen Flussigkeit. Proc. 1st Int. Congr. Applied Mechanics, Delft, 101.

Kim, J., Moin, P., and Moser, R. (1987). Turbulence statistics in fully developed channel flow at low Reynolds numbers. *J. Fluid Mech.* **177**:133.

Kline, S. J., Cantwell, B. J., and Lilley, G. M. (1982). Proceedings 1980–81 AFOSR–HTTM–Stanford Conference on Complex Turbulent Flows, Vols. II and III. Stanford University, Stanford, CA.

Kline, S. J., Reynolds, W. C., Schraub, F. A., and Runstadler, P. W. (1967). The structure of turbulent boundary layers. *J. Fluid Mech.* **30**:741.

Kolmogorov, A. N. (1941). The local structure of turbulence in incompressible viscous fluid at very large Reynolds numbers, *C. R. Acad. Sci. URSS* **30**:301.

Kolmogorov, A. N. (1962). A refinement of previous hypotheses concerning the local structure of turbulence in a viscous incompressible fluid at high Reynolds numbers. *J. Fluid Mech.* **13**:82.

Kovasznay, L. S. G. (1948). Spectrum of locally isotropic turbulence. *J. Aero. Sci.* **15**:745.

Kovasznay, L. S. G., Kiben, V., and Blackwelder, R. (1970). Large-scale motion in the intermittent region of a turbulent boundary layer. *J. Fluid Mech.* **41**:283.

Kraichnan, R. H. (1966). Dispersion of particle pairs in homogeneous turbulence. *Phys. Fluids* **9**: 1937.

Kraichnan, R. H., and Montgomery, D. (1980). Two-dimensional turbulence. *Rep. Prog. Phys.* **43**: 547.

Lam, S. H. (1992). On the RNG theory of turbulence. *Phys. Fluids A* **4**:1007.

LaRue, J. C., and Libby, P. A. (1974). Temperature fluctuations in the plane turbulent wake. *Phys. Fluids* **17**:1956.

LaRue, J. C., and Libby, P. A. (1976). Statistical properties of the interface in the turbulent wake of a heated cylinder. *Phys. Fluids* **19**:1864.

LaRue, J. C., and Libby, P. A. (1981). Thermal mixing layer downstream of a half-heated turbulence grid. *Phys. Fluids* **25**:597.

Launder, B. E. (1989a). Second-moment closure: Present. . .and future? *Int. J. Heat Fluid Flow* **10**: 282.

Launder, B. E. (1989b). The prediction of force field effects on turbulent shear flows via second-moment closure. In H. H. Fernholz and H. E. Fielder (eds.), *Advances in Turbulence 2*. Springer-Verlag, Berlin.

Launder, B. E. (1990). Phenomenological modelling: Present. . .and future. In J. L. Lumley (ed.), *Lecture Notes in Physics*. Springer-Verlag, Berlin.

Lee, M. J., and Reynolds, W. C. (1985). Numerical experiments on the structure of homogeneous turbulence, Department of Mechanical Engineering Report TF-24. Stanford University, Stanford, CA.

Leonard, A. (1974). Energy cascade in large-eddy simulations of turbulent fluid flows. *Adv. Geophys.* **18a**:237.

Libby, P. A. (1975). Diffusion of heat downstream of a turbulence grid. *Astronaut. Acta* **2**:867.

Libby, P. A. (1976). Prediction of the intermittent wake of a heated cylinder. *Phys. Fluids* **19**:494.

Libby, P. A., and Scragg, C. A. (1973). Diffusion of heat from a line source downstream of a turbulence grid. *AIAA J.* **11**:562.

Liepmann, H. W., and Laufer, J. (1947). Investigation on free turbulent mixing, NACA Tech. Note 1257.

Lin, C. C. (1953). On Taylor's hypothesis in wind tunnel turbulence. *Quart. Appl. Math.* **10**:295.

Linden, P. F., and Turner, J. S. (1975). Small-scale mixing in stably stratified fluids: A report on Euromech 51. *J. Fluid Mech.* **67**:1.

Loitsianski, L. G. (1939). Some basic laws for isotropic turbulent flow. *Trudy Tsentr. Aero.-Giedrodin. Inst.* **440**:31.

Lowery, P. S., Reynolds, W. C., and Mansour, N. N. (1987). Passive scalar entrainment and mixing in a forced, spatially-developing mixing layer, AIAA 87-0132.

Lugt, H. J. (1983). *Vortex Flow in Nature and Technology*. John Wiley, New York.

Lumley, J. L. (1965). Interpretation of time spectra measured in high-intensity shear flows. *Phys. Fluids* **8**:1056.

Lumley, J. L. (1967). The structure of inhomogeneous turbulent flows. In A. M. Yaglom and V. I. Tatarski (eds.), *Atmospheric Turbulence and Radio Wave Propogation*. Nauka, Moscow, 166.

Lumley, J. (1983). Turbulence modeling. *ASME J. Appl. Mech.* **50**:1097.

Lumley, J. L., and Panovsky, H. A. (1964). *The Structure of Atmospheric Turbulence*. Interscience, New York.

Mansour, N. N., Kim, J., and Moin, P. (1988). Reynolds-stress and dissipation rate budgets in a turbulent channel flow. *J. Fluid Mech.* **194**:15.

Mansour, N. N., Kim, J., and Moin, P. (1989). Near-wall k–ε turbulence modeling. *AIAA J.* **27**:1068.

Mariani, P., Spalart, P., and Kollmann, W. (1993). Direct simulation of a turbulent boundary layer with suction. In R. M. So, C. G. Speziale, and B. E. Launder (eds.), *Near-Wall Turbulent Flows*. Elsevier, Amsterdam.

Mellor, G. L., and Gibson, D. M. (1966). Equilibrium turbulent boundary layers. *J. Fluid Mech.* **24**: 225.

Metcalfe, R. W., Orszag, S. A., Brachet, M. E., and Riley, J. J. (1987). Secondary instability of a temporally growing mixing layer. *J. Fluid Mech.* **184**:207.

Mickley, H. S., and Davis, R. S. (1957). Momentum transfer for flow over a flat plate with blowing. NACA TN 4017.

Millionshtchikov, M. (1941). On the theory of homogeneous isotropic turbulence. *C. R. (Doklady) Acad. Sci. URSS* **32**:615.

Mohamed, M. S., and LaRue, J. C. (1990). The decay power law in grid-generated turbulence. *J. Fluid Mech.* **219**:195.

Moin, P., and Kim, J. (1982). Numerical investigation of turbulent channel flow. *J. Fluid Mech.* **118**: 341.

Moin, P., and Moser, R. D. (1989). Characteristic eddy decomposition of turbulence in a channel. *J. Fluid Mech.* **200**:471.

Monin, A. S., and Yaglom, A. M. (1971). *Statistical Fluid Mechanics: Mechanics of Turbulence.* MIT Press, Cambridge, MA.

Myong, H. K., and Kassagi, N. (1990). A new approach to the improvement of the k–ε turbulence model for wall-bounded shear flows. *JSME Int. J. Ser. II* **33**:93.

O'Brien, E. E. (1980). The probability density function (pdf) approach to reacting turbulent flows. In P. A. Libby and F. A. Williams (eds.), *Turbulent Reacting Flows.* Springer-Verlag, Berlin, 185.

Orzag, S. A., and Patterson, G. S. (1972). Numerical simulation of three-dimensional homogeneous isotropic turbulence. *Phys. Rev. Lett.* **28**:76.

Patel, V. C., Rodi, W., and Scheuerer, G. (1985). Turbulence models for near-wall and low Reynolds number flows: A review. *AIAA J.* **23**:1308.

Patel, V. C., and Scheuerer, G. (1982). Calculation of two-dimensional near and far wakes. *AIAA J.* **20**:900.

Paullay, A. J., Melnik, R. E., Rubel, A., Rudman, S., and Siclari, M. J. (1985). Similarity solutions for plane and radial jets using a k–ε turbulence model. *Trans. ASME, J. Fluids Eng.* **107**:79.

Pironneau, O., Rodi, W., Ryhming, I. L., Savill, A. M., and Trung, T. V. (1990). *Numerical Simulation of Unsteady Flows and Transition to Turbulence.* Cambridge University Press, Cambridge.

Pope, S. B. (1975). A more general effective eddy hypothesis. *J. Fluid Mech.* **72**:331.

Pope, S. B. (1985). PDF methods for turbulent reacting flows. *Prog. Energy Combustion Sci.* **11**:119.

Pope, S. B. (1988). Evolution of surfaces in turbulence. *Int. J. Eng. Sci.* **26**:445.

Prandtl, L. (1925). Bericht uber Untersuchungen zur ausgebildeten Turbulenz. *Z. Angew. Math. Mech.* **5**:136.

Prandtl, L. (1942). Bermerkungen zur Theorie der freien Turbulenz. *Z. Angew. Math. Mech.* **22**:241.

Praskovsky, A. A. (1992). Experimental verification of the Kolmogorov refined similarity hypothesis. *Phys. Fluids A* **4**:2589.

Press, W. H., Teukolsky, S. A., Vettering, W. T., and Flannery, B. P. (1992). *Numerical Recipes in Fortran.* Cambridge University Press, Cambridge.

Proudman, I., and Reid, W. H. (1954). On the decay of a normally distributed and homogeneous turbulent velocity field. *Phil. Trans. A* **247**:163.

Reichardt, H. (1959). Gesetzmassigkeiten der geradlinigen Turbulenten Couette-stromung. *Mitt. Max-Planck-Inst. fur Stromungsforschung* No. 22, Gottingen.

Reynolds, O. (1883). III. An experimental investigation of the circumstances which determine whether the motion of water shall be direct or sinuous, and of the law of resistance in parallel channels. *Phil. Trans. R. Soc. London* **174**:935.

Reynolds, O. (1894). IV. On the dynamical theory of incompressible viscous fluids and the determination of the criterion. *Phil. Trans. R. Soc. London A* **186**:123.

Reynolds, W. (1990). The potential and limitations of direct and large eddy simulations. In J. L. Lumley (ed.), *Whither Turbulence.* Springer-Verlag, Berlin.

Rhines, P. B. (1979). Geostrophic turbulence. *Annu. Rev. Fluid Mech.* **11**:401.

Richardson, L. F. (1922). *Weather Prediction by Numerical Process.* Cambridge University Press, Cambridge.

Robertson, J. M., and Johnson, H. F. (1970). Turbulence structure in plane Couette flow. *J. Eng. Mech. Div., Proc. ASCE, 1171.*

Robinson, S. K. (1991). Coherent motions in the turbulent boundary layer. *Annu. Rev. Fluid Mech.* **23**:601.

Rodi, W. A. (1975). A review of experimental data of uniform density free turbulent boundary layers. In B. E. Launder (ed.), *Studies in Convection.* Academic Press, London, 79.

Rodi, W. (1976). A new algebraic relation for calculating the Reynolds stresses. *Z. Angew. Math. Mech.* **56**:T219.

Rodi, W. (1980). *Turbulence Models and Their Application in Hydraulics—A State of the Art Review.* International Association for Hydraulic Research.

Rodi, W. (1987). Examples of calculation methods for flow and mixing in stratified fluids. *J. Geophys. Res.* **92**:5305.

Rogallo, R. S. (1981). Numerical experiments in homogeneous turbulence. NASA Tech. Memo. 81315.

Rogallo, R. S., and Moin, P. (1984). Numerical simulations of turbulent flows. *Annu. Rev. Fluid Mech.* **16**:99.

Rogers, M. M., and Moin, P. (1987). The structure of the vorticity field in homogeneous turbulent flow. *J. Fluid Mech.* **176**:33.

Rogers, M. M., Moin, P., and Reynolds, W. C. (1987). The structure of the vorticity in homogeneous turbulent flows. *J. Fluid Mech.* **176**:33.

Rohr, J. J., Itsweire, E. C., Helland, K. N., and van Atta, C. W. (1988). An investigation of the growth of turbulence in a uniform-mean-shear flow. *J. Fluid Mech.* **187**:1.

Rotta, J. (1951). Statistische Theorie nichthomogener Turbulenz. *Z. Phys.* **192**:547.

Rotta, J. (1979). A family of turbulence models for three dimensional boundary layers. In F. Durst et al. (eds.), *Turbulent Shear Flows 1.* Springer-Verlag, Berlin.

Ruelle, D., and Takens, R. (1971). On the nature of turbulence. *Commun. Math. Phys.* **20**:167.

Saddoughi, S. G. (1992). Local isotropy in high Reynolds number turbulent shear flows. *Center for Turbulence Research Annual Research Briefs*, Stanford University, 237.

Saddoughi, S. G. (1993). Local isotropy in distorted turbulent boundary layers at high Reynolds number. *Center for Turbulence Research Annual Research Briefs*, Stanford University, 347.

Saddoughi, S. G., and Veeravalli, S. V. (1994). Local isotropy in turbulent boundary layers at high Reynolds number. *J. Fluid Mech.* **268**:333.

Saffman, P. G. (1978). Problems and progress in the theory of turbulence. In H. Fiedler (ed.), *Structure and Mechanism of Turbulence II*, Lecture Notes in Physics, Vol. 7. Springer-Verlag, Berlin, 183.

Salmon, R. (1982). Geostrophic turbulence. *Topics in Ocean Physics. Enrico Fermi Corso LXXX.*

Savill, A. M. (1987). Recent developments in rapid distortion theory. *Annu. Rev. Fluid Mech.* **19**: 531.

Schetz, J. A. (1984). *Foundations of Boundary Layer Theory for Momentum, Heat and Mass Transfer.* Prentice-Hall, Englewood Cliffs, NJ.

Schlichting, H. (1955). *Boundary Layer Theory.* McGraw-Hill, London.

Schumann, U. (1977). Realizability of Reynolds stress turbulence models. *Phys. Fluids* **20**:721.

Shih, T.-H., and Lumley, J. L. (1985). Modeling of pressure correlation terms in Reynolds stress and scalar flux equations, Sibley School of Mechanics and Aero. Engineering Report No. FDA-85-3. Cornell University, Ithaca, NY.

Shih, T.-H., and Lumley, J. L. (1986). Influence of timescale ratio on scalar flux relaxation: Modelling Sirivat and Warhaft's homogeneous passive scalar fluctuations. *J. Fluid Mech.* **162**:211.

Siggia, E. D., and Patterson, G. S. (1978). Intermittency effects in a numerical simulation of stationary three-dimensional turbulence. *J. Fluid Mech.* **86**:567.

Silva-Freire, A. P. (1988). An asymptotic solution for transpired incompressible turbulent boundary layers. *Int. J. Heat Mass Transfer* **31**:1011.

Simpson, R. L. (1970). Characteristics of turbulent boundary layers at low Reynolds numbers with and without transpiration. *J. Fluid Mech.* **42**:769.

Simpson, R. L. (1989). Turbulent boundary layer separation. *Annu. Rev. Fluid Mech.* **21**:205.

Sirivat, A., and Warhaft, Z. (1982). The mixing of passive helium and temperature fluctuations in grid turbulence. *J. Fluid Mech.* **120**:475.

Sirovich, L. (1987). Turbulence and the dynamics of coherent structures. *Quart. Appl. Math.* **454**: 561.

Smith, C. R., and Abbott, D. E. (1978). *Workshop on Coherent Structure of Turbulent Boundary Layers*. Department of Mechanical Engineering, Lehigh University, Bethlehem, PA.

Smith, L. M., and Reynolds, W. (1992). On the Yakhot-Orzag renormalization group method for deriving turbulence statistics and models. *Phys. Fluids A* **4**:364.

Snyder, W. H., and Lumley, J. L. (1971). Some measurements of particle velocity autocorrelation functions in a turbulent flow. *J. Fluid Mech.* **48**:41.

So, R. M. C., Lai, Y. G., Zhang, H. S., and Hwang, B. C. (1991). Second-order near-wall turbulence closures: A review. *AIAA J.* **29**:1819.

So, R. M. C., and Yoo, G. J. (1987). Low Reynolds number modeling of turbulent flows with and without wall transpiration. *AIAA J.* **25**:1556.

Spalart, P. R. (1992). Direct simulation of a turbulent boundary layer up to $R_\theta = 1410$. *J. Fluid Mech.* **187**:61.

Speziale, C. G. (1989). Turbulence modeling in non-inertial frames of reference. *Theor. Comput. Fluid Dynam.* **1**:3.

Speziale, C. G., Abid, R., and Anderson, E. C. (1990a). A critical evaluation of two-equation models for near wall turbulence, AIAA 90-1481.

Speziale, C. G., Abid, R., and Durbin, P. A. (1993). New results on the realizability of Reynolds stress turbulence closures, ICASE Rep. 93-76.

Speziale, C. G., Sarkar, S., and Gatski, T. B. (1991b). Modeling the pressure-strain correlation of turbulence: An invariant dynamical systems approach. *J. Fluid Mech.* **227**:245.

Speziale, C. G., Gatski, T. B., and Fitzmaurice, N. (1991a). An analysis of RNG-based turbulence models for homogeneous shear flow. *Phys. Fluds A* **3**:2278.

Speziale, C. G., Gatski, T. B., and MacGiolla Mhuiris, N. (1990b). A critical comparison of turbulence models for homogeneous shear flows in a rotating frame. *Phys. Fluids A* **2**:1678.

Speziale, C. G., and MacGiolla Mhuiris, N. (1989). Scaling laws for homogeneous turbulent shear flows in a rotating frame. *Phys. Fluids A* **1**:294.

Stevenson, T. N. (1964). Turbulent boundary layers with transpiration. *AIAA J.* **2**:1500.

Strang, G. (1976). *Linear Algebra and Its Applications*. Academic Press, New York.

Tavoularis, S. (1985). Asymptotic laws for transversely homogeneous turbulent shear flows. *Phys. Fluids* **28**:999.

Tavoularis, S., and Karnik, U. (1989). Further experiments on turbulent stresses and scales in uniformly sheared turbulence. *J. Fluid Mech.* **204**:457.

Taylor, G. I. (1921). Diffusion by continuous movements. *Proc. London Math. Soc.* **20**:196.

Taylor, G. I. (1935a). Turbulence in a contracting stream. *Z. Angew. Math. Mech.* **15**:91.

Taylor, G. I. (1935b). Statistical theory of turbulence. *Proc. Roy. Soc. London A* **151**:421.

Taylor, G. I. (1938). The spectrum of turbulence. *Phil. Trans. Roy. Soc. London* **CLXIV-A**:476.

Tennekes, H., and Lumley, J. (1972). *A First Course in Turbulence*. MIT Press, Cambridge, MA.

Theodorsen, T. (1952). Mechanism of turbulence. *Proc. Midwestern Conf. Fluid Mechanics*, Ohio State University, OH.

Thoroddsen, S. T., and van Atta, C. W. (1992). Experimental evidence supporting Kolmogorov's refined similarity hypothesis. *Phys. Fluids A* **4**:2592.

Ting, L. (1959). On the mixing of two parallel streams, *J. Math. Phys.* **3**:153.

Townsend, A. A. (1954). The diffusion behind a line source in homogeneous turbulence. *Proc. Roy. Soc. London* **224A**:487.

Townsend, A. A. (1956). *The Structure of Turbulent Shear Flow*. Cambridge University Press, Cambridge.

Townsend, A. A. (1961). Equilibrium layers and wall turbulence. *J. Fluid Mech.* **11**:97.

Tselepidakis, D. P., Gatski, T. B., and Savill, A. M. (1992). A comparison of turbulence models for homogeneous shear flows with longitudinal curvature. In M. Y. Hussaini, A. Kumar, and C. L. Street (eds.), *Instability, Transition and Turbulence*. Springer-Verlag, Berlin, 544.

Tucker, H. J., and Reynolds, A. J. (1968). The distortion of turbulence by irrotational plane strain. *J. Fluid Mech.* **32**:657.

Uberoi, M. S. (1957). Equipartition of energy and local isotropy in turbulent flows. *J. Appl. Phys.* **28**:1165.

Uberoi, M. S. (1963). Energy transfer in isotropic turbulence. *Phys. Fluids* **6**:1048.

Uberoi, M. S., and Corrsin, S. (1954). Diffusion of heat from a line source in isotropic turbulence, NACA TR1142.

van Atta, C. W. (1974). Sampling techniques in turbulence. *Annu. Rev. Fluid Mech.* **6**:75.

van Atta, C. W. (1991). Local isotropy of the smallest scales of turbulent scalar and velocity fields. *Proc. R. Soc. London A* **434**:139.

van Atta, C. W., and Chen, W. Y. (1969). Measurements of spectral energy transfer in grid turbulence. *J. Fluid Mech.* **38**:743.

van Dyke, M. (1969). Higher-order boundary layer theory. *Annu. Rev. Fluid Mech.* **1**:265.

von Karman, T. (1930). Mechanische Ahnlichkeit und Turbulenz. *Nach. Ges. Wiss. Gottingen, Math.-Phys. Kl.* **58**:58.

von Karman, T. (1937). The fundamentals of the statistical theory of turbulence. *J. Aero. Sci.* **4**:131.

von Karman, T. (1948). Progress in the statistical theory of turbulence. *Proc. Natl. Acad. Sci. (USA)* **34**:530.

von Karman, T., and Howarth, L. (1938). On the statistical theory of isotropic turbulence. *Proc. R. Soc. London A* **164**:192.

Wang, K. C. (1974). Boundary layer over a blunt body at high incidence with an open-type of separation. *Proc. R. Soc. London A* **340**:33.

White, F. (1974). *Viscous Fluid Flow*. McGraw-Hill, New York.

Wigeland, R. A., and Nagib, H. M. (1978). Grid-generated turbulence with and without rotation about the streamwise direction, IIT Fluids and Heat Transfer Rep. R78-1.

Wilcox, D. C. (1993). *Turbulence Modeling for CFD*. DCW Industries, La Canada, CA.

Winter, M., Barber, T. J., Everson, R. M., and Sirovich, L. (1992). Eigenfunction analysis of turbulent mixing phenomena. *AIAA J.* **30**:1681.

Wygnanski, I. J., and Champagne, F. H. (1973). On transition in a pipe. Part I. The origin of puffs and slugs and the flow in a turbulent slug. *J. Fluid Mech.* **59**:281.

Wyngaard, J. C. (1981). Boundary layer modeling. In F. T. M. Nieuwstadt and H. van Dop (eds.), *Atmospheric Turbulence and Air Pollution Modelling*. R. Reidel, Dordrecht, The Netherlands.

Wyngaard, J. C., and Clifford, S. F. (1977). Taylor's hypothesis and high frequency turbulence spectra. *J. Atmos. Sci.* **34**:922.

Yakhot, V., and Orzag, S. A. (1986). Renormalization group analysis of turbulence. I. Basic theory. *J. Sci. Comput.* **1**:3.

Yakhot, V., Orzag, S. A., Thangam, S., Gatski, T. B., and Speziale, C. G. (1992). Development of turbulence models for shear flows by a double expansion technique. *Phys. Fluids A* **4**:1510.

INDEX

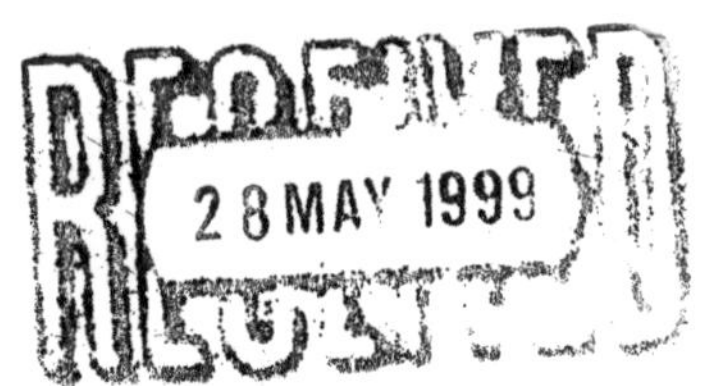

RECEIVED
28 MAY 1999